Stem Cells in Clinical Applications

Series Editor
Phuc Van Pham

More information about this series at http://www.springer.com/series/14002

Phuc Van Pham
Editor

Liver, Lung and Heart Regeneration

Editor
Phuc Van Pham
Laboratory of Stem Cell Research &
   Application, University of Science
Vietnam National University
Ho Chi Minh City, Vietnam

ISSN 2365-4198   ISSN 2365-4201   (electronic)
Stem Cells in Clinical Applications
ISBN 978-3-319-83563-1   ISBN 978-3-319-46693-4   (eBook)
DOI 10.1007/978-3-319-46693-4

© Springer International Publishing AG 2017
Softcover reprint of the hardcover 1st edition 2016
This work is subject to copyright. All rights are reserved by the Publisher, whether the whole or part of the material is concerned, specifically the rights of translation, reprinting, reuse of illustrations, recitation, broadcasting, reproduction on microfilms or in any other physical way, and transmission or information storage and retrieval, electronic adaptation, computer software, or by similar or dissimilar methodology now known or hereafter developed.
The use of general descriptive names, registered names, trademarks, service marks, etc. in this publication does not imply, even in the absence of a specific statement, that such names are exempt from the relevant protective laws and regulations and therefore free for general use.
The publisher, the authors and the editors are safe to assume that the advice and information in this book are believed to be true and accurate at the date of publication. Neither the publisher nor the authors or the editors give a warranty, express or implied, with respect to the material contained herein or for any errors or omissions that may have been made.

Printed on acid-free paper

This Springer imprint is published by Springer Nature
The registered company is Springer International Publishing AG
The registered company address is: Gewerbestrasse 11, 6330 Cham, Switzerland

Preface

The liver, lung and heart are major internal organs that control almost all biochemical physiology activities in the human body. The liver has a broad range of physiological functions such as detoxification, protein synthesis and biochemical production. The lung plays the respiratory function that extracts oxygen from the atmosphere and transfers it into the bloodstream to release carbon dioxide from the bloodstream into the atmosphere. And the heart is a pump that circulates blood through the blood vessels to all tissues. With these functions, the liver, lung and heart control some organ systems including the circulatory system, respiratory system, digestive system, urinary system, etc. Therefore, any diseases that relate to these organs can cause extreme health conditions.

There are two main groups of diseases related to these organs: infectious diseases and degenerative diseases (failure). Although there are different symptoms, they result from the significant loss of functional tissue cells. The regeneration of these functional tissue cells is considered a promising therapy to treat these conditions.

In recent years, stem cell therapies for liver, lung and heart regeneration are moved to the clinic with exciting results. Although there are some kinds of stem cells, including both pluripotent stem cells (induced pluripotent stem cells, embryonic stem cells) and adult stem cells used in the preclinical trials (in animals), a majority of clinical applications used mesenchymal stem cells—a kind of adult stem cells. Mesenchymal stem cells are used with two strategies including immune modulation and cell replacement. Indeed, some studies showed that mesenchymal stem cells could inhibit the inflammation in liver cirrhosis or are differentiated into the functional cells to replace the dead cells in ischemic heart disease.

This volume of "Stem Cells in Clinical Applications" book series with title *Liver, Lung and Heart Regeneration* aims to provide updated invaluable resource for advanced undergraduate students, graduate students, researchers and clinicians in stem cell applications for liver, lung and heart regeneration.

The book with 11 chapters covers almost the present applications of stem cells in liver, lung and heart regeneration. Chapters 1–4 discuss about liver regeneration. As the most popular liver disease, liver cirrhosis was treated by stem cell transplantation that is presented in Chaps. 2 and 3. And the mechanism of stem cell therapy for liver cirrhosis treatment also is suggested in Chap. 4. Chapters 5–7 discuss about applications of stem cells in lung regeneration. Both airway and lung regeneration are stated in Chaps. 5–7, respectively. And Chaps. 8–11 discuss about stem cell therapy and tissue engineering for heart regeneration.

We are indebted to our authors who graciously accepted their assignments and who have infused the text with their energetic contributions. We are incredibly thankful to the staff of Springer Science+Business Media who published this book.

Ho Chi Minh City, Vietnam   Phuc Van Pham

Contents

Part I Liver Regeneration

1 **Characteristics of Hepatic Progenitor Cells During Liver Development and Regeneration**.. 3
   Akihide Kamiya and Hiromi Chikada

2 **Cell Therapy in Chronic Liver Disease**.. 15
   Majid Alhomrani, Rebecca Lim, and William Sievert

3 **Clinical Applications of Stem Cells in Liver Cirrhosis**........................ 41
   Ahmer Irfan

4 **Mesenchymal Stem Cell Therapy for Liver Cirrhosis Treatment: Mechanisms and Bioeffects**... 51
   Nhung Hai Truong and Phuc Van Pham

Part II Lung Regeneration

5 **Mesenchymal Stem Cell Therapy for Airway Restoration Following Surgery**.. 69
   Francesco Petrella, Stefania Rizzo, Fabio Acocella, Stefano Brizzola, and Lorenzo Spaggiari

6 **Regenerative Potential of Mesenchymal Stem Cells: Therapeutic Applications in Lung Disorders**.................................... 77
   Kavita Sharma, Syed Yawer Husain, Pragnya Das, Mohammad Hussain, and Mansoor Ali Syed

7 **Recent Advances in Lung Regeneration**.. 119
   Kanwal Rehman and Muhammad Sajid Hamid Akash

Part III Heart Regeneration

8 Road to Heart Regeneration with Induced Pluripotent Stem Cells 137
Jun Fujita, Shugo Tohyama, Kazuaki Nakajima, Tomohisa Seki, Hideaki Kanazawa, and Keiichi Fukuda

9 Myocardial Tissue Engineering for Cardiac Repair 153
S. Pecha and Y. Yildirim

10 Stem Cell Therapy for Ischemic Heart Disease 165
Truc Le-Buu Pham, Ngoc Bich Vu, and Phuc Van Pham

11 Myocardial Tissue Engineering: A 5 Year—Update 197
Marie-Noelle Giraud and Inês Borrego

Index 211

Contributors

Fabio Acocella Department of Health, Animal Science and Public Health, University of Milan, Milan, Italy

Muhammad Sajid Hamid Akash Department of Pharmaceutical Chemistry, Government College University, Faisalabad, Pakistan

Majid Alhomrani Centre for Inflammatory Disease, Monash University, Melbourne, VIC, Australia

Hudson Institute of Medical Research, Melbourne, VIC, Australia

Medical College, Taif University, Taif, Saudi Arabia

Inês Borrego Department of Medicine, Cardiology, University of Fribourg, Fribourg, Switzerland

Stefano Brizzola Department of Health, Animal Science and Public Health, University of Milan, Milan, Italy

Hiromi Chikada Department of Molecular Life Sciences, Tokai University School of Medicine, Isehara, Kanagawa, Japan

Pragnya Das Drexel University School of Medicine, Philadelphia, PA, USA

Jun Fujita Department of Cardiology, Keio University School of Medicine, Tokyo, Japan

Keiichi Fukuda Department of Cardiology, Keio University School of Medicine, Tokyo, Japan

Marie-Noelle Giraud Department of Medicine, Cardiology, University of Fribourg, Fribourg, Switzerland

Syed Yawer Husain Stem Cell Research Lab, Rajasthan University of Health Sciences, Rajasthan, India

Mohammad Hussain Department of Biotechnology, Jamia Millia Islamia (Central University), New Delhi, India

Ahmer Irfan University of Edinburgh, Edinburgh, UK

Akihide Kamiya Department of Molecular Life Sciences, Tokai University School of Medicine, Isehara, Kanagawa, Japan

Hideaki Kanazawa Department of Cardiology, Keio University School of Medicine, Tokyo, Japan

Rebecca Lim Department of Obstetrics and Gynaecology, Monash University, Melbourne, VIC, Australia

Hudson Institute of Medical Research, Melbourne, VIC, Australia

Kazuaki Nakajima Department of Cardiology, Keio University School of Medicine, Tokyo, Japan

S. Pecha Department of Cardiovascular Surgery, University Heart Center Hamburg, Hamburg, Germany

DZHK (German Centre for Cardiovascular Research) Partner Site Hamburg/Kiel/Lübeck, Lübeck, Germany

Francesco Petrella Department of Thoracic Surgery, European Institute of Oncology, Milan, Italy

Truc Le-Buu Pham Laboratory of Stem Cell Research and Application, University of Science, Vietnam National University, Ho Chi Minh City, Vietnam

Phuc Van Pham Laboratory of Stem Cell Research and Application, University of Science, Vietnam National University, Ho Chi Minh City, Vietnam

Kanwal Rehman Institute of Pharmacy, Physiology and Pharmacology, University of Agriculture, Faisalabad, Pakistan

Stefania Rizzo Department of Radiology, European Institute of Oncology, Milan, Italy

Tomohisa Seki Department of Cardiology, Keio University School of Medicine, Tokyo, Japan

Kavita Sharma Department of Pathology, AIIMS, New Delhi, India

William Sievert Gastroenterology and Hepatology Unit, Monash Health, and Department of Medicine, Monash University, Melbourne, VIC, Australia

Centre for Inflammatory Disease, Monash University, Melbourne, VIC, Australia

Lorenzo Spaggiari Department of Thoracic Surgery, European Institute of Oncology, Milan, Italy

School of Medicine, University of Milan, Milan, Italy

Mansoor Ali Syed Stem Cell Research Lab, Rajasthan University of Health Sciences, Jaipur, Rajasthan, India

Department of Biotechnology, Jamia Millia Islamia (Central University), New Delhi, India

Shugo Tohyama Department of Cardiology, Keio University School of Medicine, Tokyo, Japan

Nhung Hai Truong Laboratory of Stem Cell Research and Application, University of Science, Vietnam National University, Ho Chi Minh City, Vietnam

Ngoc Bich Vu Laboratory of Stem Cell Research and Application, University of Science, Vietnam National University, Ho Chi Minh City, Vietnam

Y. Yildirim Department of Cardiovascular Surgery, University Heart Center Hamburg, Hamburg, Germany

Part I
Liver Regeneration

Chapter 1
Characteristics of Hepatic Progenitor Cells During Liver Development and Regeneration

Akihide Kamiya and Hiromi Chikada

1.1 Introduction

Liver is the central organ for the maintenance of homeostasis and consists of several types of cells, both parenchymal cells (hepatocytes) and non-parenchymal cells (i.e., sinusoidal endothelial cells, stellate cells, Kupffer cells, and cholangiocytes). Mature hepatocytes in adult liver perform many metabolic functions. In contrast, fetal liver barely expresses metabolic genes, instead it supports the expansion of hematopoietic cells such as erythrocytes (Kinoshita et al. 1999). Thus, a liver dramatically changes from being a hematopoietic organ to a metabolic organ during embryonic development. The embryonic liver bud, differentiating from the foregut endoderm, has many hepatic progenitor cells that are also known as the hepatoblasts. Hepatoblasts have a high proliferative ability and bipotency to differentiate into both hepatocytes and cholangiocytes (Lemaigre 2009). It was found that the characteristics of hepatic progenitor cells are different in the fetal and adult liver stages (Kamiya et al. 2009). Several cell surface molecules such as CD13, CD133, DLK1, and LIV2 are found as specific markers of the progenitor cells (Kamiya et al. 2009; Tanimizu et al. 2003; Watanabe et al. 2002; Rountree et al. 2007; Kakinuma et al. 2009). Recently, we found a new progenitor-marker gene during liver development. In addition, several new molecular mechanisms regulating the differentiation and proliferation of hepatic progenitor cells have been reported in recent studies. In this review, we discuss about the various characteristics of liver progenitor cells during embryonic development and liver regeneration.

A. Kamiya (✉) • H. Chikada
Department of Molecular Life Sciences, Tokai University School of Medicine,
143 Shimokasuya, Isehara, Kanagawa 259-1193, Japan
e-mail: kamiyaa@tokai-u.jp; chikadahiromi@tokai-u.jp

1.2 Regulation of Maturation and Proliferation of Liver Progenitor Cells During Embryonic Development

Fetal hepatic progenitor cells, or hepatoblasts, demonstrate limited metabolic activities but support the hematopoietic cells (Kinoshita et al. 1999). Oncostatin M (OSM), an interleukin 6 family cytokine, can promote hepatic maturation, as determined by the induction of metabolic enzymes, accumulation of glycogen and lipids, and detoxification of ammonia. In particular, OSM was expressed in $CD45^+$ hematopoietic cells in the mid-fetal livers, whereas the OSM receptor was mainly detected in hepatic cells (Kamiya et al. 1999). The interaction between extracellular matrix and integrin is important for terminal maturation of fetal hepatic cells (Kamiya et al. 2002). These results suggest that OSM produced from hematopoietic cells and extracellular matrices produced from non-parenchymal liver cells play pivotal roles in fetal liver development. However, these maturation factors, i.e., OSM and extracellular matrices could not induce high-level maturation of hepatic progenitor cells. Expression of several hepatic genes, in the progenitor-derived mature cells, stimulated by these factors was significantly lower than that in the adult hepatocytes (Kamiya et al. 1999; Si-Tayeb et al. 2010). It is thus suggested that unknown factors are required to induce high level of differentiation into mature hepatocytes.

Transcription factors are usually involved in maturation of stem/progenitor cells during embryonic development. Recently, we found that basic helix-loop-helix (bHLH) transcription factor Mist1 regulates the maturation of fetal hepatic progenitor cells in vitro (Chikada et al. 2015). Mist1 was identified as one of the bHLH transcription factors that binds to an E-box sequence and is known to play an important role in pancreatic acinar cell organization (Lemercier et al. 1997; Pin et al. 2001). At first, we analyzed expressional changes of Mist1 using our hepatic progenitor cell culture. Mist1 mRNA decreased in this culture without hepatic maturation factors, OSM and extracellular matrices. In contrast, this decrease was partially attenuated by the induction of hepatic maturation, suggesting that induction of Mist1 expression by OSM and extracellular matrices is involved in maturation of hepatic progenitor cells. Next, we analyzed the function of Mist1 for maturation of hepatic progenitor cells using the retrovirus-derived gene overexpression vector. Overexpression of Mist1 induced the expression of hepatic functional genes such as Cps1, Cytochrome P450 (Cyp) 3a11, Cyp2b9, and Cyp2b10. In contrast, the overexpression of Mist suppressed the expression of cholangiocytic markers such as Sox9, Sox17, cytokeratin (Ck) 19, and Grhl2. Next, we directly analyzed metabolism of a CYP3A-target substrate using mass spectrometry. The substrate of CYP3A, midazolam, was added into hepatic progenitor cell culture and the amounts of midazolam and the metabolite of midazolam, 1-hydroxymidazolam, were measured using liquid chromatography-tandem mass spectrometry. Overexpression of Mist1 in combination with OSM and extracellular matrices significantly increased the amount of the metabolite 1-hydroxymidazolam. These results suggested that hepatocyte-like cells with Mist1 overexpression have high expression and activity of drug metabolic enzymes in vitro.

During in vivo liver development, hepatic progenitor cells highly expand and differentiate into a large number of hepatocytes and cholangiocytes (Watanabe et al. 2002). These in vivo characteristics of hepatic progenitor cells have been elucidated by several transplantation experiments (Kakinuma et al. 2009; Oertel et al. 2008). However, high proliferative potential of hepatic progenitor cells is inhibited during in vitro culture. It is reported that genetic modification is required for long-term expansion of hepatic progenitor cells in the conventional culture system (Chiba et al. 2010). When purified hepatic progenitor cells are cultured at a low density, they can form several colonies on extracellular matrix-coated dishes. Fetal hepatic progenitor cells cultured on collagen-coated dishes mainly differentiated into CK19-positive cholangiocytic cells and the efficiency of colony formation is low. In contrast, when these cells were co-cultured with mesenchymal cells such as mouse embryonic fibroblasts (MEFs), most hepatic progenitor cells differentiated into albumin-positive hepatocyte-like cells and the efficiency of colony formation increased owing to the co-culturing with MEFs (Fig. 1.1). We also found that the addition of MEK inhibitor, PD0325901, can induce proliferation of hepatic progenitor cells in long-term culture. Primary hepatic progenitor cells derived from mid-fetal livers express several cell cycle related genes, such as p27cdkn1b, p57cdkn1c, p18cdkn2c, and p19cdkn2d. Expression of p21cdkn1a and p16/19cdkn2a is barely detected in primary hepatoblasts. However, expression of these cdk inhibitors is highly induced during the expansion in vitro, indicating that the up-regulation of these inhibitors is involved in the low proliferative ability of hepatic progenitor cells in long-term in vitro culture. We found that the up-regulation of these inhibitors is induced by MEK-ERK signaling pathway. Thus, the addition of PD0325901 can induce long-term proliferative ability through the suppression of these cdk

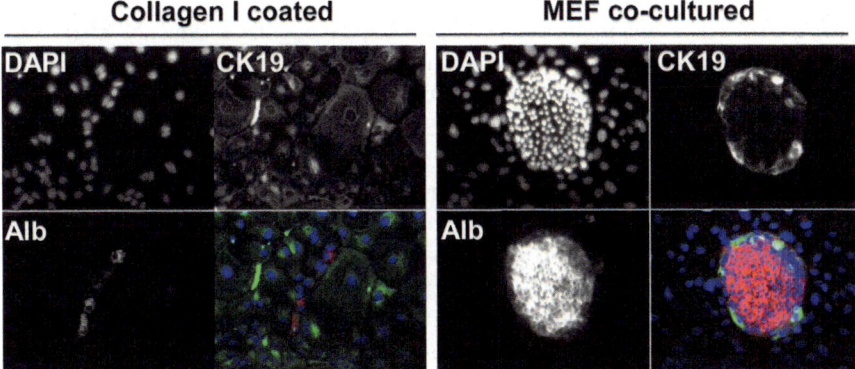

Fig. 1.1 Colony formation by hepatic progenitor cells co-cultured with mesenchymal cells. Expression of albumin (Alb, *red*) and CK19 (*green*) was detected. E13.5 CD45⁻Ter119⁻c-Kit⁻Dlk⁺CD133⁺ progenitor cells were cultured in the conventional H-CFU-C culture system (collagen coated) or MEF co-culture system (MEF co-cultured). After 6 days of culture, the cells were immunostained with anti-albumin and –CK19 antibodies. Nuclei were counterstained with DAPI (*blue*). (Reprinted with permission from ref. Kamiya et al. 2015 in STEM CELLS AND DEVELOPMENT, 2015, published by Mary Ann Liebert, Inc., New Rochelle, NY)

inhibitors (Kamiya et al. 2015). Primary hepatic progenitor cells can expand and differentiate into functional hepatocytes by transplantation into pre-conditioned recipient mouse livers. However, the transplantation and recovery of the injured liver usually requires many donor cells because liver is a large organ. A few hepatic progenitor cells have been expanded, using our co-culture system with PD0325901, and transplanted into the injured livers. Progenitor cells with PD0325901 maintain their proliferative ability after in vitro expansion and can proliferate in the recipient livers. Using this culture system, 100 mouse hepatic progenitor cells can proliferate into almost 3×10^5 cells in the long-term culture. It is thus suggested that this technique allows a small number of hepatic progenitor cells to be used for cell transplantation for future regenerative therapies in severe liver diseases. MEK-ERK pathway is known to be involved in the progression of G0-G1 cell cycle progression (Coutant et al. 2002; Fremin et al. 2012). In contrast, it is concluded that high activation of MEK-ERK pathway in fetal hepatic progenitor cells induced cell cycle arrest through the accumulation of p16/19cdkn2a and its downstream target p21cdkn1a.

1.3 Expression of the Brain Expressed X-Linked 2 (Bex2) Gene in Progenitor Cells During Liver Development

Several tissues such as hematopoietic cells, lung, heart, intestine, and liver contain somatic stem/progenitor cells having multipotent differentiation ability. Expression of specific markers is important for analyses of these stem/progenitor cells. Studies of chromosomal assignment of genes expressed during embryonic development revealed that stem cell-related genes are frequently located on the X chromosome (Forsberg et al. 2005; Ito et al. 2014). Part of these genes, Bex genes, are a family of genes that reside on the mammalian X chromosome and have sequence homology with transcription elongation factor A (SII)-like (TCEAL) genes (Jeon and Agarwal 1996; Thomas et al. 1998). Five members of the *BEX* family have been identified to date, including *BEX1*, *BEX2*, *BEX3*, *BEX4*, and *BEX5* (Alvarez et al. 2005). Expression of Bex family genes is detected in several tissues during embryonic development. In contrast, expression of these genes was detected at a high level only in the brain and testis of adult mice, but not in other tissues such as the endodermal organs and kidneys. Thus, it can be suggested that Bex family genes are involved in the embryonic organ differentiation. Recently, to analyze the functions and expression pattern of Bex2 in vivo, we generated Bex2 knock-in mice using a gene-targeting strategy, by replacing the entire open reading frame with enhanced green fluorescent protein (EGFP) (Ito et al. 2014). Male mice with the target allele (Bex2$^{EGFP/Y}$ mice) have no Bex2 expression and can grow into adulthood. In addition, we cannot find the differences between Bex2$^{EGFP/Y}$ male mice and wild-type male mice upon histological analysis of various tissues at the macroscopic level. These results suggest that Bex2 is not required for normal development, since expression patterns of Bex family genes in both fetal and adult tissues are similar. In particular, Bex1 and Bex2 have an extremely overlapping expression pattern in vivo. In addition, these genes exhibited ~90 % sequence similarity. Thus, the lack of phenotypes in Bex2-deficient mice might be compensated by the redundancy of Bex family genes such as Bex1.

Using this knock-in mouse, we can detect expression of Bex2 in vivo at a cellular level by expression of EGFP. As shown above, a colony-formation culture of fetal hepatic progenitor cells using mesenchymal cells as feeder cells was established (Kamiya et al. 2015; Okada et al. 2012). In this culture system, fetal hepatic progenitor cells expand and differentiate into either albumin-positive hepatocytic cells or CK19-positive cholangiocytic cells. EGFP-positive cells (Bex2-positive cells) but not negative cells can form large colonies containing both albumin$^+$ and CK19$^+$ cells. We compared the hepatic colony forming efficiency between Bex2-EGFP^{high} cells and conventional hepatic progenitor cells (Dlk1$^+$CD133$^+$ cells) using flow cytometry cell sorting method. It was found that Bex2-EGFP^{high} cells have higher hepatic colony forming efficiency than conventional hepatic progenitor cells (Fig. 1.2). Dlk1 and CD133 are known to be suitable markers for purification of hepatic progenitor cells in the fetal livers (Kamiya et al. 2009; Tanimizu et al. 2003). However, Dlk1 is also expressed in the mesenchymal cells (Ito et al. 2014; Tanaka

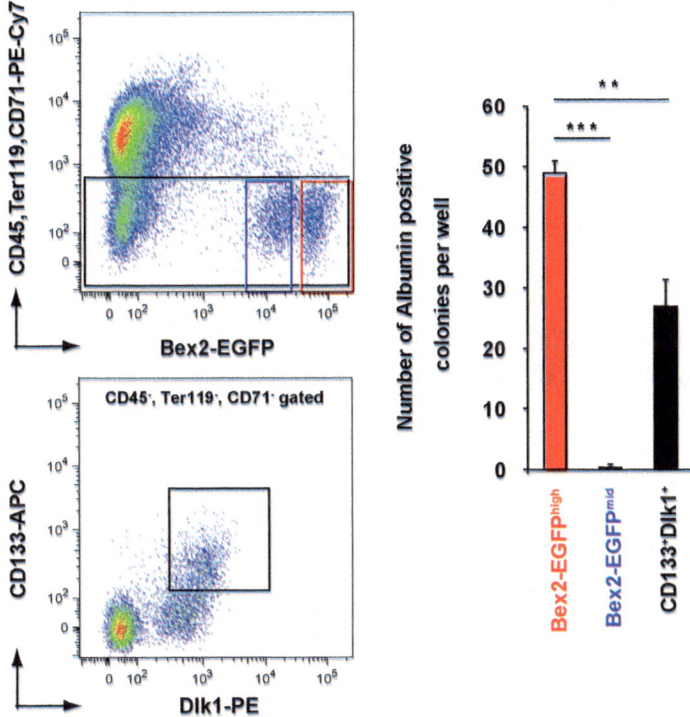

Fig. 1.2 Colony-formation ability of Bex2$^+$ hepatic progenitor cells. We compared the hepatic colony forming efficiency (elucidated by albumin expression) between Bex2^{high} fraction (gated in *red*), Bex2^{mid} fraction (gated in *blue*), and Dlk1$^+$CD133$^+$ fraction (gated in *black*) in the non-hematopoietic cell fraction. After purification using flow cytometry, cells were co-cultured with MEF at a low density (**$p<0.05$, ***$p<0.0001$). (Reproduced from ref Ito et al. 2014 with permission)

et al. 2009) and CD133 is expressed in a minor subset of mesenchymal cells, in addition to hepatoblasts. Therefore, Dlk1$^+$CD133$^+$ cell fraction derived from fetal livers harbors risks of contamination of mesenchymal cells at a certain frequency. Thus, the potential utility of Bex2 as a novel specific marker to purify fetal liver progenitor cells at higher frequency is elucidated. We also found that Bex2 was expressed in stem/progenitor cells derived from other tissues. Lgr5 and Pw1 are specific markers of stem/progenitor cells among various endodermal tissues (Barker et al. 2013; Besson et al. 2011). The base of the glands in the pyloric stomach contains Lgr5+ stem/progenitor cells (Barker et al. 2010). There are many EGFP-expressing Bex2-positive cells and these EGFP-positive cells highly express Lgr5 and Sox9, the markers of pyloric epithelial stem cells. We also analyzed expression of Bex2 in hematopoietic cells. The c-kit$^+$Sca-1$^+$Lineage marker$^-$ (KSL) hematopoietic cell fraction is known to be hematopoietic stem/progenitor cell fraction (Osawa et al. 1996). We found that these cells contain EGFP-positive cells at higher frequency than other hematopoietic cell fractions. These results suggest that Bex2 is a specific marker of stem/progenitor cells, not only for the hepatic lineages, but also for the pyloric stomach and hematopoietic tissues.

Liver can regenerate after hepatectomy or chemical-induced injury. Liver regeneration from hepatectomy and mild injury primarily depends on the proliferation of adult hepatocytes. However, severe and chronic liver injury suppresses proliferation of mature hepatocytes. In these conditions, stem and progenitor cells in the adult liver are the ones mainly involved in the regeneration steps. Recently, we found that chronic liver injury induced by 3,5-diethoxycarbonyl-1,4-dihydrocollidine significantly changed the expression of Bex2 (Ito, Kamiya et al., unpublished data), suggesting that Bex2-positive hepatic progenitor cells are involved in liver regeneration.

1.4 Differentiation and Proliferation of Hepatic Progenitor Cells Derived from Human Pluripotent Stem Cells

Stem cells have specific properties of self-renewal, multipotency (producing progeny belonging two or more lineages), and long-term tissue repopulation after transplantation. There are two types of stem cells: somatic and pluripotent stem cells. Somatic stem cells are differentiated into non-self-renewing progenitor cells with restricted differentiation potential during embryonic development. Thus, these cells contribute to organ formation and maintenance in embryonic and adult stages. In contrast, pluripotent embryonic stem (ES) cells are derived from the inner cell mass of the blastocyst. iPS cells, generated from somatic cells by simultaneous expression of specific transcription factors (Oct3/4, Klf4, Sox2, and c-Myc), are also pluripotent stem cells, similar to the ES cells. These cells can differentiate into several tissues derived from the three primary germ layers: ectoderm, mesoderm, and endoderm (Takahashi and Yamanaka 2006; Takahashi et al. 2007).

Mechanisms regulating the characteristics of hepatic progenitor cells are mainly studied by cells derived from rodent animals, such as mouse and rat. In contrast, regulation of proliferation and differentiation of human hepatic stem/progenitor cells remain largely unknown, due to the difficulty associated with analyzing cellular events using human samples. Thus, human iPS cells are likely to be more suitable for the analyses of characteristics of human hepatic progenitor cells. Sequential stimulation with several differentiation factors such as activin A, basic fibroblast growth factor, bone morphogenetic protein, and hepatocyte growth factor can induce pluripotent stem cells into hepatic-lineage cells (Gerbal-Chaloin et al. 2014). Our groups as well as others have reported about hepatic progenitor cells in the iPS cell-derived hepatocytic culture. For example, CD13 and CD133, which are known to be cell surface markers of stem/progenitor cells in mouse fetal livers (Kamiya et al. 2009), are useful for purification of human iPS cell-derived hepatic progenitor cells. Highly proliferative cells are found in the $CD13^+CD133^+$ fraction of human iPS cells stimulated by hepatocytic differentiation factors. Individual $CD13^+CD133^+$ cells formed large colonies containing more than 100 cells and expressed early hepatocytic marker genes (α-fetoprotein and hepatocyte nuclear factor 4α) (Yanagida et al. 2013). In addition, these cells can acquire mature hepatic and cholangiocytic characteristics in the suitable culture condition. Hepatic progenitor cells are self-assembled into aggregates called spheroids. This spheroid formation induced expression of several hepatic functional genes, such as cytochrome P450, in human iPS cell-derived hepatic progenitor cell culture. In addition, cholangiocyte differentiation derived from these progenitor cells was analyzed. Cholangiocytic cells can form epithelial cysts, demonstrating in vitro tubule formation in extracellular matrix gel supplemented with cytokines (Tanimizu et al. 2007). When hepatic progenitor cells derived from human iPS cells were cultured on feeder cells, part of these cells differentiated into CK7-positive cholangiocyte-like cells. These cells could form cysts with epithelial polarities. In addition, they expressed cholangiocytic marker CK7 and CK19, but did not express hepatocytic marker α-fetoprotein (Fig. 1.3). It is indicated that human iPS cell-derived hepatic progenitor cells have a bipotent differentiation ability, forming both hepatocyte- and cholangiocyte-like cells (Yanagida et al. 2013).

Next, we analyzed expression of cell surface molecules on human iPS cell-derived hepatic progenitor cells. Differentiation and proliferation of progenitor cells are sometimes regulated by soluble growth factors and cytokines through the binding to cell surface receptor proteins. Thus, we analyzed human $CD13^+CD133^+$ hepatic progenitor cells derived from human iPS cells using over 100 cell surface protein-specific antibodies by flow cytometry. According to the ratio of the positive cell population, cell surface molecules are classified into positive, split, or negative categories. Twenty cell surface antigens, which are categorized as "the positive marker," are highly expressed on human iPS cell-derived $CD13^+CD133^+$ hepatic progenitor cells. Part of these 20 positive antigens are the cell signaling receptors, which can bind to the cytokine or growth factor ligands. For example, CD221, CD340, and CD266 are highly expressed on hepatic progenitor cells. CD221 insulin-like growth factor (IGF) receptor 1 is a transmembrane tyrosine kinase receptor activated by IGF1 and IGF2. $CD13^+CD133^+$ hepatic progenitor cells stimulated with an anti-human CD221 blocking antibody suppressed their proliferative ability. Similarly, the addition of PQ401, a specific chemical inhibitor of CD221,

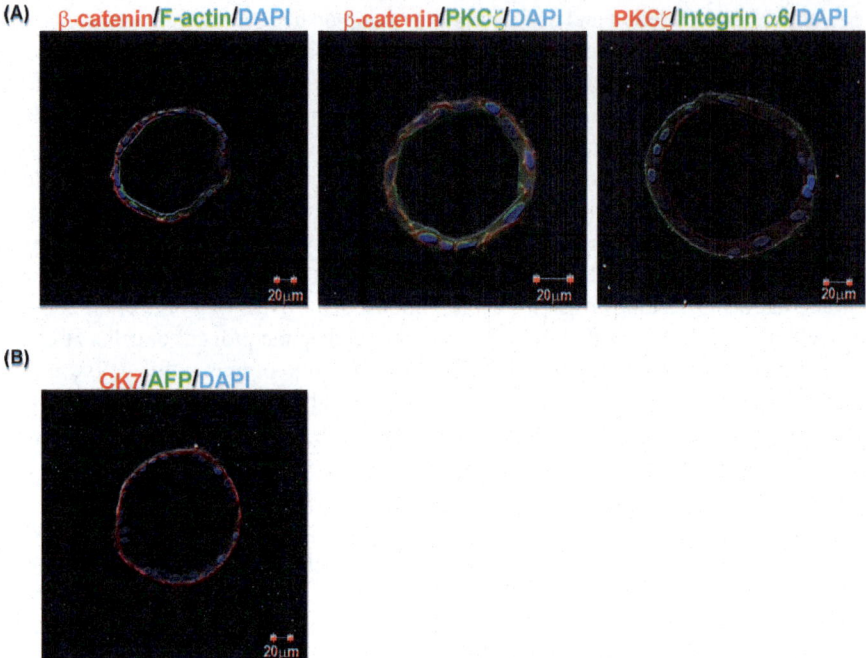

Fig. 1.3 Cholangiocytic differentiation from human iPS cell-derived hepatic progenitor cells. (**a**) Human iPS cell-derived colonies are dissociated and cultured in collagen-laminin mixed gel. After 10–12 days of culture, epithelial cysts are formed. Polarity of expression of β-catenin, F-actin, integrin α6, and PKCζ is detected in these cysts. (**b**) Expression of CK7 in human iPS cell-derived cysts. Cells were stained with antibodies against CK7 and α-feto protein. Nuclei were counterstained with DAPI. (Reproduced from ref. Yanagida et al. 2013 with permission)

also suppressed colony formation by these cells. CD340 is a member of the erbB transmembrane tyrosine kinase receptor family (Hynes and Lane 2005). We investigated the erbB family, the epidermal growth factor receptor (EGFR), erbB2 (CD340/HER2/neu), erbB3 (HER3), and erbB4 (HER4) on the $CD13^+CD133^+$ human iPS cell-derived hepatic progenitor cells. Expression of EGFR and ErbB2 is highly detected and ErtbB3 is moderately expressed on these cells. In addition, our colony culture system of hepatic progenitor cells required the addition of EGF or NRG1, the ligands of these receptors, suggesting that the cell signaling activated by the erbB family is critical for expansion of human hepatic progenitor cells. CD266 (Fn14) is a member of the tumor necrosis factor receptor superfamily and TWEAK is its ligand (Burkly et al. 2007). The addition of an anti-human Fn14 blocking antibody significantly suppressed the proliferative ability of human iPS cell-derived hepatic progenitor cells. These results suggest that IGF-1, IGF-2, EGF, NRG1, and TWEAK are involved in proliferation of human iPS cell-derived HPCs. As discussed earlier, our culture system is supported by mesenchymal cells. We found that paracrine factors such as IGF and TWEAK derived from supporting feeder cells directly regulate the proliferation of human iPS cell-derived hepatic progenitor cells through their specific receptors.

iPS cells derived from somatic cells have the characteristics regulated by its genetic background. Thus, cells generated from patient-derived iPS cells are expected to be useful for the analysis of genetic diseases. However, generating iPS cell clone derived from patients is associated with many problems such as the recruitment of patients, definition of guidelines to validate new clones, and the diversity of iPS cell characteristics from different genetic backgrounds. Novel methods to generate human cells having genetic mutations have been established by genome editing enzymes, which can induce double strand break at the specific site of genome DNA (Gaj et al. 2013). Human pluripotent stem cell-derived disease models carrying site-directed mutations reportedly have been established using these enzymes (Ding et al. 2013). Recently, we also established bile ductal disease model using cholangiocytic cells differentiated from mutated human iPS cells (Kamiya et al., unpublished data). These disease models are highly attractive to find new drugs for severe human diseases.

1.5 Summary and Future Perspective

In this review, we presented data showing that hepatic progenitor cells are regulated by several molecular mechanisms during liver development. Hepatic progenitor cells have high proliferative ability and can be transplanted into the injured liver tissues of the suitable mouse model. Therefore, these cells are good candidates to be used as regenerative therapies for severe liver diseases. However, function of hepatic cells maturated from progenitor cells in vitro is not high compared to that of the adult hepatocytes. In this study, we searched for the key factors regulating maturation of hepatic progenitor cells in vitro and found several candidate transcription factors. Therefore, the combination of knowledge of these factors and the methods of differentiation of human pluripotent stem cells into hepatic cells can contribute to the future transplantation therapy for liver injury.

Acknowledgements Our studies described in this review article were supported in part by Grants-in-Aid for Scientific Research from the Ministry of Education, Culture, Sports, Science, and Technology of Japan.

References

Alvarez E, Zhou W, Witta SE, Freed CR (2005) Characterization of the Bex gene family in humans, mice, and rats. Gene 357:18–28
Barker N, Huch M, Kujala P, van de Wetering M, Snippert HJ, van Es JH, Sato T, Stange DE, Begthel H, van den Born M, Danenberg E, van den Brink S, Korving J, Abo A, Peters PJ, Wright N, Poulsom R, Clevers H (2010) Lgr5(+ve) stem cells drive self-renewal in the stomach and build long-lived gastric units in vitro. Cell Stem Cell 6:25–36
Barker N, Tan S, Clevers H (2013) Lgr proteins in epithelial stem cell biology. Development 140:2484–2494

Besson V, Smeriglio P, Wegener A, Relaix F, Nait Oumesmar B, Sassoon DA, Marazzi G (2011) PW1 gene/paternally expressed gene 3 (PW1/Peg3) identifies multiple adult stem and progenitor cell populations. Proc Natl Acad Sci U S A 108:11470–11475

Burkly LC, Michaelson JS, Hahm K, Jakubowski A, Zheng TS (2007) TWEAKing tissue remodeling by a multifunctional cytokine: role of TWEAK/Fn14 pathway in health and disease. Cytokine 40:1–16

Chiba T, Seki A, Aoki R, Ichikawa H, Negishi M, Miyagi S, Oguro H, Saraya A, Kamiya A, Nakauchi H, Yokosuka O, Iwama A (2010) Bmi1 promotes hepatic stem cell expansion and tumorigenicity in both Ink4a/Arf-dependent and -independent manners in mice. Hepatology 52:1111–1123

Chikada H, Ito K, Yanagida A, Nakauchi H, Kamiya A (2015) The basic helix-loop-helix transcription factor, Mist1, induces maturation of mouse fetal hepatoblasts. Sci Rep 5:14989

Coutant A, Rescan C, Gilot D, Loyer P, Guguen-Guillouzo C, Baffet G (2002) PI3K-FRAP/mTOR pathway is critical for hepatocyte proliferation whereas MEK/ERK supports both proliferation and survival. Hepatology 36:1079–1088

Ding Q, Lee YK, Schaefer EA, Peters DT, Veres A, Kim K, Kuperwasser N, Motola DL, Meissner TB, Hendriks WT, Trevisan M, Gupta RM, Moisan A, E Banks, Friesen M, Schinzel RT, Xia F, Tang A, Xia Y, Figueroa E, Wann A, Ahfeldt T, Daheron L, Zhang F, Rubin LL, Peng LF, Chung RT, Musunuru K, Cowan CA (2013) A TALEN genome-editing system for generating human stem cell-based disease models. Cell Stem Cell 12:238–251

Forsberg EC, Prohaska SS, Katzman S, Heffner GC, Stuart JM, Weissman IL (2005) Differential expression of novel potential regulators in hematopoietic stem cells. PLoS Genet 1:e28

Fremin C, Ezan F, Guegan JP, Gailhouste L, Trotard M, Le Seyec J, Rageul J, Theret N, Langouet S, Baffet G (2012) The complexity of ERK1 and ERK2 MAPKs in multiple hepatocyte fate responses. J Cell Physiol 227:59–69

Gaj T, Gersbach CA, Barbas CF 3rd (2013) ZFN, TALEN, and CRISPR/Cas-based methods for genome engineering. Trends Biotechnol 31:397–405

Gerbal-Chaloin S, Funakoshi N, Caillaud A, Gondeau C, Champon B, Si-Tayeb K (2014) Human induced pluripotent stem cells in hepatology: beyond the proof of concept. Am J Pathol 184:332–347

Hynes NE, Lane HA (2005) ERBB receptors and cancer: the complexity of targeted inhibitors. Nat Rev Cancer 5:341–354

Ito K, Yanagida A, Okada K, Yamazaki Y, Nakauchi H, Kamiya A (2014) Mesenchymal progenitor cells in mouse foetal liver regulate differentiation and proliferation of hepatoblasts. Liver Int 34:1378–1390

Ito K, Yamazaki S, Yamamoto R, Tajima Y, Yanagida A, Kobayashi T, Kato-Itoh M, Kakuta S, Iwakura Y, Nakauchi H, Kamiya A (2014) Gene targeting study reveals unexpected expression of brain-expressed X-linked 2 in endocrine and tissue stem/progenitor cells in mice. J Biol Chem 289:29892–29911

Jeon C, Agarwal K (1996) Fidelity of RNA polymerase II transcription controlled by elongation factor TFIIS. Proc Natl Acad Sci U S A 93:13677–13682

Kakinuma S, Ohta H, Kamiya A, Yamazaki Y, Oikawa T, Okada K, Nakauchi H (2009) Analyses of cell surface molecules on hepatic stem/progenitor cells in mouse fetal liver. J Hepatol 51:127–138

Kamiya A, Kinoshita T, Ito Y, Matsui T, Morikawa Y, Senba E, Nakashima K, Taga T, Yoshida K, Kishimoto T, Miyajima A (1999) Fetal liver development requires a paracrine action of oncostatin M through the gp130 signal transducer. EMBO J 18:2127–2136

Kamiya A, Kojima N, Kinoshita T, Sakai Y, Miyaijma A (2002) Maturation of fetal hepatocytes in vitro by extracellular matrices and oncostatin M: induction of tryptophan oxygenase. Hepatology 35:1351–1359

Kamiya A, Kakinuma S, Yamazaki Y, Nakauchi H (2009) Enrichment and clonal culture of progenitor cells during mouse postnatal liver development in mice. Gastroenterology 137:1114–1126, 1126 e1111–1114

Kamiya A, Ito K, Yanagida A, Chikada H, Iwama A, Nakauchi H (2015) MEK-ERK activity regulates the proliferative activity of fetal hepatoblasts through accumulation of p16/19(cdkn2a). Stem Cells Dev 24:2525–2535

Kinoshita T, Sekiguchi T, Xu MJ, Ito Y, Kamiya A, Tsuji K, Nakahata T, Miyajima A (1999) Hepatic differentiation induced by oncostatin M attenuates fetal liver hematopoiesis. Proc Natl Acad Sci U S A 96:7265–7270

Lemaigre FP (2009) Mechanisms of liver development: concepts for understanding liver disorders and design of novel therapies. Gastroenterology 137:62–79

Lemercier C, To RQ, Swanson BJ, Lyons GE, Konieczny SF (1997) Mist1: a novel basic helix-loop-helix transcription factor exhibits a developmentally regulated expression pattern. Dev Biol 182:101–113

Oertel M, Menthena A, Chen YQ, Teisner B, Jensen CH, Shafritz DA (2008) Purification of fetal liver stem/progenitor cells containing all the repopulation potential for normal adult rat liver. Gastroenterology 134:823–832

Okada K, Kamiya A, Ito K, Yanagida A, Ito H, Kondou H, Nishina H, Nakauchi H (2012) Prospective isolation and characterization of bipotent progenitor cells in early mouse liver development. Stem Cells Dev 21:1124–1133

Osawa M, Nakamura K, Nishi N, Takahasi N, Tokuomoto Y, Inoue H, Nakauchi H (1996) In vivo self-renewal of c-Kit+Sca-1+Lin(low/-) hemopoietic stem cells. J Immunol 156:3207–3214

Pin CL, Rukstalis JM, Johnson C, Konieczny SF (2001) The bHLH transcription factor Mist1 is required to maintain exocrine pancreas cell organization and acinar cell identity. J Cell Biol 155:519–530

Rountree CB, Barsky L, Ge S, Zhu J, Senadheera S, Crooks GM (2007) A CD133-expressing murine liver oval cell population with bilineage potential. Stem Cells 25:2419–2429

Si-Tayeb K, Noto FK, Nagaoka M, Li J, Battle MA, Duris C, North PE, Dalton S, Duncan SA (2010) Highly efficient generation of human hepatocyte-like cells from induced pluripotent stem cells. Hepatology 51:297–305

Takahashi K, Yamanaka S (2006) Induction of pluripotent stem cells from mouse embryonic and adult fibroblast cultures by defined factors. Cell 126:663–676

Takahashi K, Tanabe K, Ohnuki M, Narita M, Ichisaka T, Tomoda K, Yamanaka S (2007) Induction of pluripotent stem cells from adult human fibroblasts by defined factors. Cell 131:861–872

Tanaka M, Okabe M, Suzuki K, Kamiya Y, Tsukahara Y, Saito S, Miyajima A (2009) Mouse hepatoblasts at distinct developmental stages are characterized by expression of EpCAM and DLK1: drastic change of EpCAM expression during liver development. Mech Dev 126:665–676

Tanimizu N, Nishikawa M, Saito H, Tsujimura T, Miyajima A (2003) Isolation of hepatoblasts based on the expression of Dlk/Pref-1. J Cell Sci 116:1775–1786

Tanimizu N, Miyajima A, Mostov KE (2007) Liver progenitor cells develop cholangiocyte-type epithelial polarity in three-dimensional culture. Mol Biol Cell 18:1472–1479

Thomas MJ, Platas AA, Hawley DK (1998) Transcriptional fidelity and proofreading by RNA polymerase II. Cell 93:627–637

Watanabe T, Nakagawa K, Ohata S, Kitagawa D, Nishitai G, Seo J, Tanemura S, Shimizu N, Kishimoto H, Wada T, Aoki J, Arai H, Iwatsubo T, Mochita M, Watanabe T, Satake M, Ito Y, Matsuyama T, Mak TW, Penninger JM, Nishina H, Katada T (2002) SEK1/MKK4-mediated SAPK/JNK signaling participates in embryonic hepatoblast proliferation via a pathway different from NF-kappaB-induced anti-apoptosis. Dev Biol 250:332–347

Yanagida A, Ito K, Chikada H, Nakauchi H, Kamiya A (2013) An in vitro expansion system for generation of human iPS cell-derived hepatic progenitor-like cells exhibiting a bipotent differentiation potential. PLoS One 8:e67541

Chapter 2
Cell Therapy in Chronic Liver Disease

Majid Alhomrani, Rebecca Lim, and William Sievert

2.1 Introduction

The liver is the second most complex organ after the brain. It controls critical processes within the body, including detoxification, metabolism, protein synthesis and digestion (Malarkey et al. 2005) and is responsible for cholesterol homeostasis and lipoprotein metabolism. The majority of serum proteins are produced and secreted by the liver including coagulation factors, albumin and hormones. The liver is also responsible for metabolism of endogenous compounds such as ammonia, and exogenous compounds such as drugs and toxic compounds. The catabolism of amino acids and glycogen also takes place in the liver.

The liver has a unique ability for regeneration and self-repair following reduction in volume, for example, by partial hepatectomy. The phenomenon of liver regeneration has been well documented, indeed the full restoration of liver mass within 20 days following 70 % partial hepatectomy was reported in rats over 80 years ago (Higgins 1931). Similarly, in human liver transplantation of the right lobe, the

---

M. Alhomrani
Centre for Inflammatory Disease, Monash University, Melbourne, VIC, Australia

Hudson Institute of Medical Research, Melbourne, VIC, Australia

Medical College, Taif University, Taif, Saudi Arabia

R. Lim
Department of Obstetrics and Gynaecology, Monash University, Melbourne, VIC, Australia

Hudson Institute of Medical Research, Melbourne, VIC, Australia

W. Sievert (✉)
Gastroenterology and Hepatology Unit, Monash Health, and Department of Medicine, Monash University, Melbourne, VIC, Australia

Centre for Inflammatory Disease, Monash University, Melbourne, VIC, Australia
e-mail: William.Sievert@monash.edu

restoration of normal liver mass in both donor and recipient has been achieved within 60 days post-transplantation (Marcos et al. 2000). Despite this capability for regeneration, chronic liver disease is associated with a decrease in hepatocyte replication and accumulation of extracellular matrix (ECM) proteins, a process referred to as fibrosis that can lead to destruction of the normal hepatic architecture (Bataller and Brenner 2005; Friedman 2003).

The proportion of the global population affected by chronic liver disease has increased significantly and the costs are huge. The Gastroenterological Society of Australia estimated that the annual cost for liver disease in 2012 was $50.7 billion in Australia for about 6.1 million patients (Zekry et al. 2013). A Scottish study showed that the mortality from liver cirrhosis has more than doubled in men and increased by half in women between 1990 and 2002 (Leon and McCambridge 2006). In the USA, estimates are that 360 in every 100,000 persons are diagnosed with liver cirrhosis and it is the most common cause of death in this group with approximately 30,000 deaths annually (Befeler and Di Bisceglie 2002; El-Serag and Mason 2000).

This chapter will review the current understanding of normal hepatic parenchymal and non-parenchymal cells involved in liver fibrosis, the most common causes of chronic liver injury and the associated cellular and inflammatory responses in the progression of injury. Finally, the current available treatments and potential applications of cell therapy will be outlined.

2.2 Liver Structure

The liver consists of a wide array of cell types. Hepatocytes are the main parenchymal cells comprising 80 % of the total cell population. Smaller populations of non-parenchymal cells include the liver sinusoidal endothelial cells (LSECs), resident hepatic macrophages (Kupffer cells), pericytes or hepatic stellate cells (HSCs) and liver progenitor cells (LPCs) (Malarkey et al. 2005; Wallace et al. 2008).

The hepatocyte is a polyhedral epithelial cell approximately 30–40 μm in diameter. About 5–10 % of its volume is the nucleus, 15 % is smooth and rough endoplasmic reticulum, 2 % is peroxisomes and 20 % of the cytoplasmic volume consists of mitochondria, as well as other cell organelles such as lysosomes, free ribosomes, cytoskeleton elements and Golgi complexes. This physical structure allows the hepatocyte to carry out many vital functions, including bile synthesis and secretion, glucose homoeostasis, metabolism of drugs and toxins, protein synthesis, proliferation and division (Crawford and Burt 2012).

The LSECs are a unique population of endothelial cells. They have small and large fenestrae as well as high endocytotic activity. LSECs act as a sieve between the blood and parenchymal cells, actively filtering fluid and solutes through receptor mediated endocytosis (Braet and Wisse 2002; Crawford and Burt 2012). They are surrounded by HSCs (liver pericytes) and interact closely with these cells (DeLeve 2013).

Hepatic macrophages, known as Kupffer cells, were first identified by von Kupffer in 1876 (Kupffer 1876). These specialised tissue macrophages account for

15 % of liver cells. Kupffer cells are the most abundant pool of resident tissue macrophages in the body; however, their origin remains a subject of debate. Some argue that a proportion of Kupffer cells arise from circulating bone marrow derived monocytes and have a short life span (Gale et al. 1978; Klein et al. 2007), while others argue that Kupffer cells represent a resident population of macrophages derived from foetal yolk sac precursors (Naito et al. 1997; Schulz et al. 2012). Kupffer cells are capable of proliferation, have significant phagocytic capacity and release numerous mediators that contribute not only to host defence but also to liver injury including transforming growth factor-β (TGF-β) and tumour necrosis factor (TNF) (Crawford and Burt 2012; McCuskey 2012).

HSCs, also initially described by Kupffer, comprise 5–8 % of the liver cells. They are spindle shaped cells, lying in the perisinusoidal space of Disse between the hepatocytes and sinusoidal endothelial cells. They were previously called Ito or fat storing cells as they play a pivotal role in the storage and metabolism of retinoids (80 % of the total body vitamin A is stored in HSCs) (Higashi and Senoo 2003). HSCs also play a valuable role in liver regeneration following injury by producing cytokines and growth factors such as fibroblast growth factor (FGF), hepatocyte growth factor (HGF), Wingless/Integrated (WNT), vascular endothelial growth factor (VEGF) and insulin-like growth factor (IGF). Furthermore, they regulate ECM remodeling by producing matrix components, as well as metalloproteinases (Crawford and Burt 2012; Friedman 2008a; Weiskirchen and Tacke 2014; Yin et al. 2013).

LPCs are bipotential progenitor cells that can differentiate into hepatocytes or cholangiocytes. In rodents, LPCs are referred to as oval cells due to their distinct oval shaped nuclei. They are thought to arise from the Canals of Hering and have the ability to proliferate and are associated with a process called the ductular reaction, which is associated with liver regeneration as well as with fibrosis (Newsome et al. 2004; Takiya et al. 2013; Vestentoft 2014). The expansion of LPCs is regulated by TNF-α and interferon-γ (INF-γ), TNF-like weak inducer of apoptosis (TWEAK) and lymphotoxin-β (LT-β). In vitro studies showed that transforming growth factor-β (TGF-β) has less effect on LPCs during chronic injury (Brooling et al. 2005; Knight et al. 2008; Nguyen et al. 2007; Viebahn et al. 2006). The role of LPCs in liver repair is described in greater detail in Sect. 2.4.

2.3 Aetiology and Pathology of Chronic Liver Disease

Chronic liver diseases can be attributed to a variety of etiologies. Both hepatitis B virus (HBV) and hepatitis C virus (HCV) can cause chronic liver disease. Additionally, steatohepatitis with or without alcohol, neonatal liver disease, parasitic infection (schistosomiasis) and metabolic disorders (Wilson disease) are risk factors for developing liver cirrhosis. While there has been significant improvement in the treatment and prevention of viral hepatitis, the prevalence of fatty liver disease remains on the rise. As such, this section will focus on alcoholic liver disease and non-alcoholic fatty liver disease (NAFLD).

2.3.1 Alcoholic Liver Disease

Ethanol is metabolised through two main pathways: alcohol dehydration by nicotinamide adenine dinucleotide (NAD$^+$) and microsomal ethanol oxidation system including cytochrome P450 and CYPE1. These oxidise ethanol into acetaldehyde, which is then converted into acetate. This results in the accumulation of reactive oxygen species (ROS) causing steatosis, oxidative stress and endoplasmic reticulum stress (Cederbaum 2012; Lieber 2000). These metabolites and toxins lead to hepatocyte inflammation and apoptosis (Luedde et al. 2014). High alcohol consumption also compounds steatosis through altered hepatic lipid metabolism by blocking both peroxisome proliferator activating receptor α (PPAR-α), 5' adenosine monophosphate-activated protein kinase (AMPK) pathways and activating sterol regulatory element-binding protein (SREP1c) (Louvet and Mathurin 2015).

2.3.2 Non-alcoholic Fatty Liver Disease

NAFLD refers to the presence of steatosis in the absence of alcohol consumption and is considered one of the leading causes of liver cirrhosis (Torres et al. 2012). Patients affected by NAFLD are at risk of developing hepatocellular carcinoma (HCC) (Michelotti et al. 2013). The incidence of NAFLD is high among the obese and type II diabetics and is also influenced by age and gender, increasing in older populations and in men (Than and Newsome 2015). NAFLD is generally categorised as either simple steatosis or NASH. Patients with simple steatosis are unlikely to develop liver fibrosis whereas approximately 10–20 % of those with NASH may develop advanced fibrotic liver disease including cirrhosis over time.

A two-hit hypothesis has been suggested for the development of NASH. The first hit refers to the presence of hepatic steatosis, which is characterised by the accumulation of triglycerides and free fatty acids and is largely related to insulin resistance. The second hit is characterised by hepatocyte damage and fibrosis arising from inflammation and ROS production (Than and Newsome 2015).

2.4 Pathology of Liver Fibrosis

Chronic liver disease is characterised by progressive hepatocyte injury, which results in a wound healing response with recruitment of inflammatory cells and the accumulation of ECM elaborated by HSC. The molecular composition of ECM includes glycoproteins such as hyaluronic acid, merosin, nidogen, laminin, fibronectin, and tenascin, collagen type I and III, and proteoglycans including biglycan, heparin, chondroitin sulphates, perlecan, decorin, dermatan and syndecan (Friedman 2003; Tsukada et al. 2006). This progressive ECM accumulation is seen morphologically as fibrotic bands (scar) with loss of normal hepatic parenchyma. The

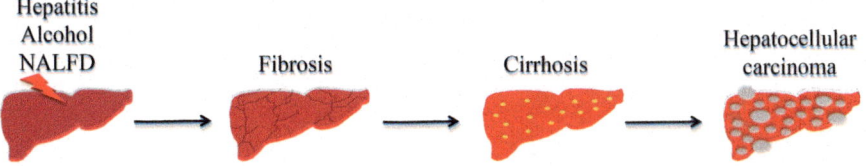

Fig. 2.1 Chronic liver disease

reversibility of fibrosis may be limited, presumably due to collagen cross-linkage (Hernandez-Gea and Friedman 2011) (Fig. 2.1). The persistence of this ECM then leads to end-stage liver fibrosis or cirrhosis.

The main features of cirrhosis include significant changes in hepatic architecture that lead to the development of portal hypertension (Pellicoro et al. 2014). Cirrhotic patients are at high risk of developing liver failure and HCC, which ranks third in cancer mortality worldwide (Forbes and Parola 2011). The pathophysiology of fibrosis and its resolution is a complex and dynamic process involving a heterogeneous population of cells including HSCs, macrophages and LPCs. Recent evidence suggests that innate lymphoid cells (ILCs) may also play a role in hepatic fibrogenesis. Each of these cell populations will be described in more detail in the following sections.

2.4.1 Hepatic Stellate Cells

The initiating step of hepatic fibrogenesis is the activation of quiescent HSCs and their transdifferentiation into myofibroblasts (MFs). HSCs are crucial fibrogenic cells and they are the most widely investigated cell population in experimental and clinical hepatic conditions (Friedman 2008a; Parola et al. 2008). HSC activation results in alterations in phenotypic and morphological hallmarks, including upregulation of the cytoskeletal protein α smooth muscle actin (α-SMA), loss of vitamin A storage, increase in rough endoplasmic reticulum content, enhanced cell migration, proliferation and DNA synthesis, secretion of ECM proteins as well as changes in proinflammatory cytokines and chemokines such as CCL2, CCL3 and CCL4. Change in gene expression includes an increase in collagen alpha 1 (I) mRNA stabilisation (Pellicoro et al. 2014).

The activation of HSCs can be defined in two major stages: the **initiation** stage arises due to paracrine stimulation as a result of hepatocyte damage, changes in the surrounding ECM and exposure to lipid droplets in patients with NAFLD (Friedman 2008a). It is characterised by expression of platelet-derived growth factor (PDGF) β receptor and the development of the fibrogenic phenotype. The second stage is **perpetuation**, which results from both paracrine and autocrine loops that are involved in different alteration of cell behaviour, such as fibrogenesis, proliferation, migration, contractility and matrix degradation (Forbes and Parola 2011; Friedman 2008a; Lee et al. 2015).

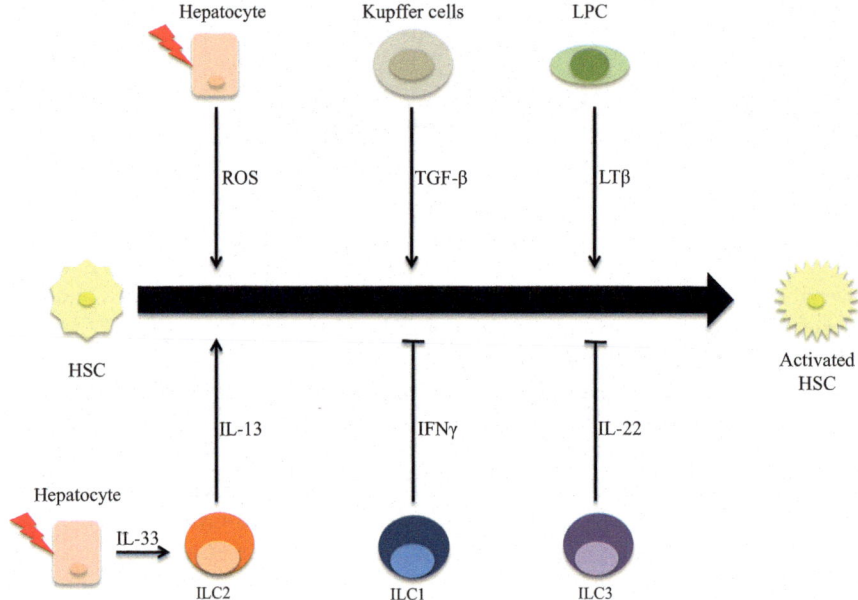

Fig. 2.2 Activation of hepatic stellate cells

During chronic liver disease HSCs express many mitogenic factors, particularly PDGF, as well as others such as FGF, thrombin and VEGF (Pinzani and Marra 2001). Both the number of HSCs and matrix production increase during fibrogenesis. These increases are mainly due to TGF-β1 released by HSCs and Kupffer cells (Novo and Parola 2008; Parola et al. 2008). The HSCs have the ability to migrate and accumulate at the sites of injury due to chemoattractants including PDGF-BB, monocyte chemo-attractant protein 1 (MCP-1), angiopoietin-1 and CXCR3 ligands (Friedman 2008a, b; Novo et al. 2011). The main contractile stimulus for HSCs is endothelin-1, which increases cellular contractility and contributes to the early stage of liver fibrosis. HSCs, Kupffer cells and endothelial cells produce nitric oxide, an antagonist to endothelin-1. However, the imbalance between the production of nitric oxide and endothelin-1 shifts to endothelin-1 during chronic liver disease (Tsukada et al. 2006). HSCs express high levels of matrix metalloproteinases (MMP), which degrade the matrix. These enzymes are inactivated by tissue inhibitors of metalloproteinases (Friedman 2008a; Visse and Nagase 2003) (Fig. 2.2).

2.4.2 Hepatic Macrophages

During the early stages of liver injury, Kupffer cells produce proinflammatory cytokines including IL-1β and TNFα, as well as chemokines including CCL2, CCL5 and CXCL16 (Luedde and Schwabe 2011). The cytokines induce apoptosis of

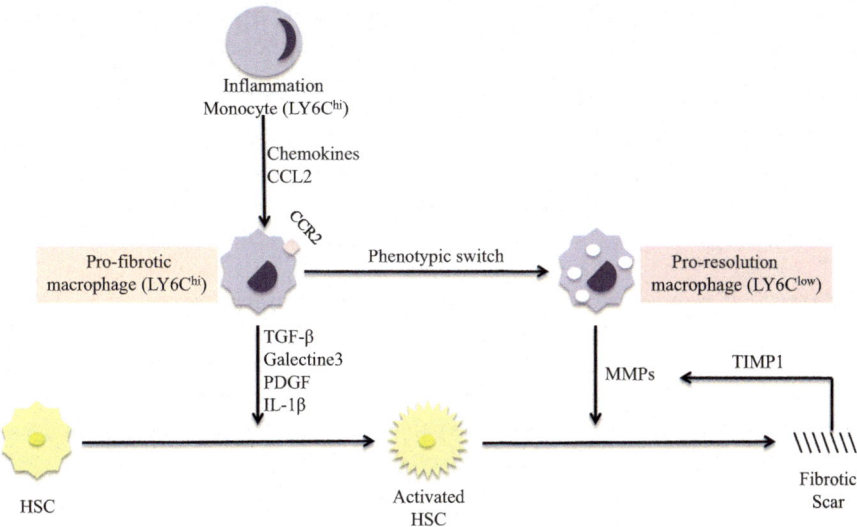

Fig. 2.3 Cascade of macrophages following liver injury

hepatocytes and attract NK cells (Luedde and Schwabe 2011). Through a paracrine mechanism, Kupffer cells have the ability to initiate HSC activation by producing TGF-β and PDGF and promote their survival through activation of nuclear factor (NF-kB) via IL-1β and TNF-α during liver fibrosis (Pradere et al. 2013). In addition, Kupffer cells contribute to the resolution of liver fibrosis through the production of MMPs (Tacke and Zimmermann 2014).

Further to this, a recent study has identified two populations of monocyte-derived macrophages based on their differential Ly6C expression in a mouse model of carbon tetrachloride (*CCL$_4$*) induced liver fibrosis (Ramachandran et al. 2012; Yang et al. 2014). One of these is the pro-fibrogenic Ly6C^{hi} CD11b$^+$F480$^+$ population, which reportedly produces proinflammatory cytokines such as IL-1β, TNFα, TGF-β and PDGF that are essential for proliferation, activation and survival of HSCs (Pellicoro et al. 2014). The other is Ly6C^{lo}CD11b$^+$F480$^+$, which has been reported as being an anti-fibrogenic subpopulation of macrophages that express MMP9, MMP12, MMP13 and the TNF-related apoptosis-inducing ligand (TRAIL), as well as anti-inflammatory cytokine such as CX$_3$CL1 (Pellicoro et al. 2014) (Fig. 2.3).

2.4.3 Liver Progenitor Cells

Liver regeneration occurs through two main mechanisms: the replication and division of mature hepatocytes during moderate liver injury and, in severe liver injury in which hepatocyte replication is impaired, the LPC compartment expands to regenerate the liver (Best et al. 2015). During chronic liver injury, macrophages and NK cells initiate LPC expansion by activating the NF-kB pathway through TNF-like weak inducer of apoptosis (TWEAK) signaling (Dwyer et al. 2014). Ductular reactions appear in the periportal regions and contain liver progenitor regions with increased TWEAK signaling and LPC expansion that correlates with the severity of fibrosis (Williams et al. 2014). LPCs have roles in chronic liver disease (Lanthier et al. 2013), fatty liver disease (Chiba et al. 2011; Richardson et al. 2007), chronic viral hepatitis (Clouston et al. 2005; Libbrecht et al. 2000; Wu et al. 2008) and alcohol-related liver disease (Lowes et al. 1999; Sancho-Bru et al. 2012). In chronic liver disease, the high proliferation rate of LPCs is linked to HCC development (Dumble et al. 2002; Fang et al. 2004). It is important to understand that there is no clear consensus on the mechanism of progenitor cell repair. Some argue that during liver injury in which hepatocyte replication is impaired, small progenitor cells present in the liver have the ability to generate significant number of mature hepatocyte and bile duct cells as well as expressing markers of stem cells (Huch et al. 2013; Lu et al. 2015). On the other hand, other researchers argue against the potency of LPCs as they are unable to generate hepatocytes (Grompe 2014; Schaub et al. 2014).

2.4.4 Innate Lymphoid Cells

The innate and adaptive immune systems play an important role in defence against infection. ILCs are non-B, non-T lymphocytes derived from Id2+ common helper-like ILCs progenitors characterised by lymphoid cell morphology and a lack of the recombination activating gene (Spits et al. 2013). Typically, ILCs have been divided into two subsets: the first is the cytotoxic natural killer (NK) cell subsets, which are responsible for protection against viruses and tumour cells (Kiessling et al. 1975). The second subset refers to the lymphoid tissue inducer (LTi) cells, which form the lymph nodes during embryogenesis (Mebius et al. 1997). Recently, ILCs have been subdivided into two classes: cytotoxic ILCs (including NK cells) and non-cytotoxic helper ILCs. The non-cytotoxic helper ILCs contain three groups: group 1 (ILC1s) expresses -IFNγ; group 2 (ILC2s) expresses IL-5, IL-9 and IL-13 and group 3 (ILC3s) expresses IL-22 and IL-17 (Artis and Spits 2015; McKenzie et al. 2014) (Fig. 2.4).

ILCs play an important role in defence mechanisms, adaptive immunity and wound healing (Eberl et al. 2015; McKenzie et al. 2014). Transcription factors and effector programs of the different ILC subsets are similar to the T-helper (Th) subsets. Interestingly, ILC populations secrete a wide array of cytokines that are similar to the cytokine profiles of T helper (Th) cells. For instance, some ILCs secrete IL-17 and IL-22 upon stimulation, similar to Th17, while other ILC subsets, when stimulated, secrete IL-5 and IL-13 similar to Th2. (Diefenbach et al. 2014). Recently, an IL-1 related cytokine IL-33 has been proposed to function as an alarmin or damage-associated molecular pattern, since it is released from necrotic cells and is thought to have a role in inflammatory and tissue injury response. IL-33 signaling is mediated via its receptor, ST2. Recently, ILC2 populations have been described to express ST2 and respond to IL-33 by producing IL-4, IL-5 and IL-13. In response

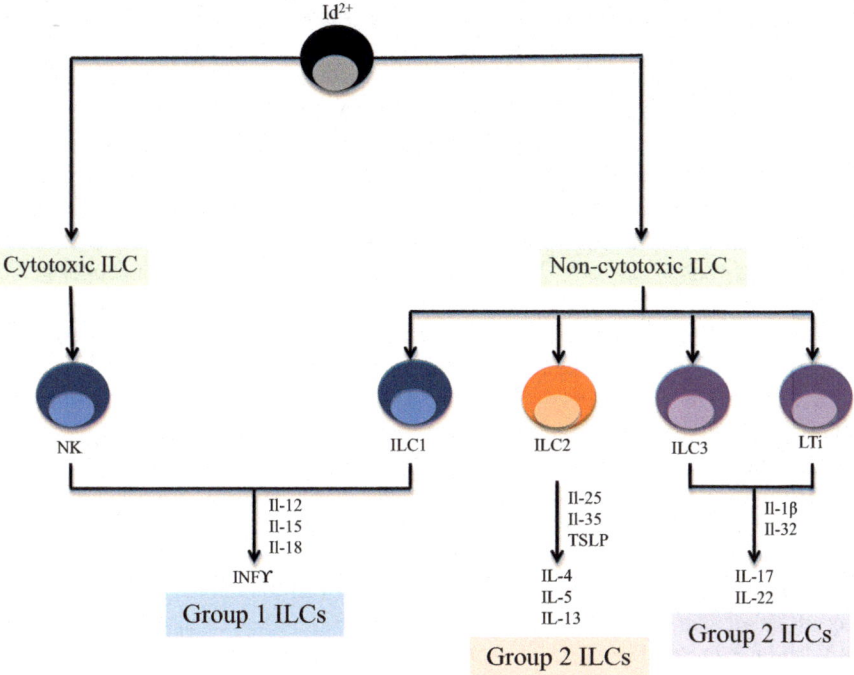

Fig. 2.4 The innate lymphoid cell

to chronic liver disease, IL-33 is released as a result of liver injury, which causes accumulation and activation of hepatic resident ILC2s. This will cause the ILC2s to produce IL-13, which then activates HSCs (Mchedlidze et al. 2013).

2.5 Treatment of Chronic Liver Fibrosis

The most effective way to treat chronic liver injury is to treat the primary cause. Long term treatment with entecavir for 57 HBV infected patients resulted in histological reduction in fibrosis or cirrhosis (Chang et al. 2010) and also reduced the incidence of HCC among HBV patients (Hosaka et al. 2013). Continuous treatment with the HBV antiviral drug, lamivudine, delayed advanced fibrosis and progression to cirrhosis in HBV patients and significantly decreased hepatic decompensation and risk of HCC development (Liaw et al. 2004). Long term treatment with another HBV antiviral drug, tenofovir, for up to 5 years has been shown to be safe, maintaining viral suppression and significantly decreasing fibrosis (Marcellin et al. 2013). The combination treatment of pegylated (PEG) INF-α-2b with ribavirin in HCV patients showed significant reduction in fibrosis progression (Poynard et al. 2002). HCV patients who achieved a sustained virological response (SVR) following treatment with the combination of INF-α-2b and ribavirin showed a decrease in the development of HCC (Hung et al. 2006). One third of HCV patients treated with Peg-INF combined with ribavirin achieved SVR which improved survival and reduced liver decompensation independently (Fernández-Rodríguez et al. 2010). However, HCV patients with cirrhosis who achieve SVR require continued monitoring for the development of HCC (Noureddin and Ghany 2010; Silva et al. 2014). In NAFLD clinical trials, pioglitazone decreased steatosis and ballooning necrosis, which indicates metabolic and histologic improvement (Belfort et al. 2006). Penicillamine is the medical treatment of Wilson disease, which is characterised by accumulation of copper in the liver (Medici et al. 2007). Prednisone and azathioprine are used in autoimmune liver disease (Manns et al. 2010). In addition, several liver diseases are preventable. For instance, HBV can be prevented by immunisation. Weight reduction and healthy life style can prevent the progression of NAFLD and cessation of alcohol consumption can improve alcohol-related liver disease (Su et al. 2014).

However, the progression of liver disease reduces the chances of both the reversal of the condition and the availability of effective treatment. The only effective treatment for end-stage liver disease is orthotopic liver transplantation (OLT). The limitations to OLT include the fact that it is an invasive procedure, requires the use of immunosuppression therapy and insufficient number of donors exist for the number of patients on the waiting list (Francoz et al. 2007). Therefore, there is an essential need for an anti-fibrotic treatment that is characterised by the ability to degrade ECM, reduce ECM deposition and decrease inflammation and promote liver regeneration.

2.5.1 Cell Therapy

Over the past years, many researchers have focused on adult cells and stem cells in an attempt to investigate their therapeutic and regenerative potential. Many cell types have been used in chronic liver disease treatment including hepatocytes, macrophages, embryonic stem cells (ESCs), induced pluripotent stem cells (iPSCs), mesenchymal stem cells (MSCs) and human amnion epithelial cells (hAECs).

2.5.1.1 Hepatocytes

Hepatocyte transplantation has been suggested as an alternative for OLT. Hepatocytes can be harvested from donor livers that are unsuitable for liver transplantation and infused through the portal vein or splenic artery of the recipient. Researchers have focused on using hepatocyte transplantation as a bridge for patients on organ waiting lists, as a support for acute failure, and as cell therapy for genetic and metabolic conditions such as Crigler–Najjar syndrome (Fisher and Strom 2006; Ito et al. 2009). Hepatocyte transplantation for Crigler–Najjar syndrome type I patients demonstrates the safety of using allogeneic hepatocytes with some clinical improvement

(Fox et al. 1998). Although hepatocyte transplantation is less invasive, expensive and has fewer complications compared to OLT, hepatocytes cannot be expanded in vitro, mainly due to limitations in quantity and quality of the cells.

2.5.1.2 Macrophages

During tissue injury, macrophages play an important role in both progression and resolution of the injury. Macrophages have the ability to produce MMPs which degrade the ECM and activate LPCs through paracrine TWEAK signaling (Bird et al. 2013). In an animal model of chronic liver disease induced by CCl_4, administration of bone marrow derived macrophages resulted in structural and functional improvements associated with increased expression of MMPs protein and upregulation of TWEAK. Serum albumin levels were elevated and fibrosis area reduced after 4 weeks of treatment (Thomas et al. 2011). Although macrophage therapy may be an effective therapy for chronic liver disease, macrophages are short-lived and their effect is transient.

2.5.1.3 Embryonic Stem Cells

ESCs are pluripotent cells derived from the inner cell mass that have the ability to self-renew and differentiate into all three embryonic germ layers, namely the ectoderm, mesoderm and endoderm (Johnson and McConnell 2004). Different protocols have been developed to differentiate human- and animal-derived ESCs into definitive endoderm, and then further differentiate these into hepatocyte-like cells (Basma et al. 2009; Duan et al. 2010; Han et al. 2011; Hay et al. 2008a, b; Kubo et al. 2010; Touboul et al. 2010). Human and animal ESCs engraft in the liver of injured mice and show hepatic differentiation (Cai et al. 2007; Moriya et al. 2007, 2008; Yamamoto et al. 2003). Despite the promise of using ESCs as a treatment for chronic liver disease, ESCs are unsuitable for cellular therapy due to limitations including ethical concerns regarding the use of human embryos, teratoma formation and immunogenicity. Indeed, the clinical use of ESCs has been prohibited since 2007 in the USA (Blum and Benvenisty 2007; Matthews and Rowland 2011).

2.5.1.4 Induced Pluripotent Stem Cells

As alternatives to ECSs, iPSCs are derived from autologous adult cells by reprogramming the mature somatic cells through the introduction of genes associated with pluripotency, such as the transcription factor Oct4. This avoids the ethical conundrums as well as the potential for immune rejection which accompany the use of ESCs (Takahashi et al. 2007). Human iPSCs have been effectively differentiated into hepatocyte-like cells (Song et al. 2009; Sullivan et al. 2010). Human iPSC-derived hepatocytes have been injected into animal models of chronic liver diseases induced by dimethylnitrosamine (DMN), where there was evidence of functional in vivo engraftment regardless of the source of iPSCs (Liu et al. 2011).

The iPSC-derived hepatocytes from Wilson disease patients also allow for the investigation of the disease pathology as well as the effects of therapy (Zhang et al. 2011b). Indeed, human iPSCs cultured with human MSCs and human umbilical vein endothelial cells (HUVECs) generated self-formed liver buds. The transplantation of these buds generate metabolic (including protein production) and vascular functions (by attachment to the host vessel) (Takebe et al. 2013). Thus, iPSCs have an obvious potential for the modeling of diseases and evaluation of new therapies in vitro. Therefore, iPSCs obtained from patients with chronic or genetic disease, which will differentiate into specific cell lineages, can be used to understand the disease as well as for drug screening (Hirschi et al. 2014). These cells are phenotypically stable but concern regarding the development of tumours in the long term complicates their use in clinical trials. More research is required to address the stability and tumour development prior to clinical trials using iPSCs.

2.5.1.5 Mesenchymal Stem Cells

General Features

MSCs are adult multipotent stem cells that have the ability to self-renew and form fibroblast-like colonies. These non-immunogenic cells were first described by Friedestein and were thought to be the structural component of bone marrow and the supporting elements for the haematopoietic stem cells (HSCs) (Friedenstein et al. 1968, 1973). The number of methods used for isolating and characterising these cells has increased dramatically which prompted the International Society of Cellular Therapy (ISCT) to propose a set of minimal criteria for identification to encourage some standardisation. These were a set of cell surface markers CD73, CD90 and CD105 and absence of CD14, CD11b, CD34, CD45, CD79 and HLA-DR. The criteria also included plastic adherence and differentiation potential to chondrocytes, adipocytes and osteocytes (Dominici et al. 2006).

MSCs are an important source for cell-based therapy. Despite their small representation as a percentage of all cells, they have the ability to expand in culture medium to a significant number in just a few weeks. MSCs have been studied and isolated mainly from bone marrow and other varieties of human tissues, including adipose tissue, umbilical cord blood, skeletal muscle, placental tissue, liver, spleen, amniotic fluid, pancreas and peripheral blood (De Coppi et al. 2007; Kruse et al. 2006; Lu et al. 2010; Meng et al. 2007; Noort et al. 2003; Scherjon et al. 2004).

Differentiation Potential

MSCs have the ability to differentiate into cell types including adipocytes, chondrocytes, osteocytes (Jiang et al. 2002), endothelial cells (Oswald et al. 2004) and cardiomyocytes (Makino et al. 1999). Indeed, MSCs can differentiate into hepatocyte-like cells with functional abilities in vitro such as urea and albumin production (Hong et al. 2005; Schwartz et al. 2002; Taléns-Visconti et al. 2006) and show the ability to differentiate into hepatocytes-like cells in rodents (Sato et al. 2005; Shu et al. 2004; Theise et al. 2000), sheep (Chamberlain et al. 2007) and humans (Alison et al. 2000). Also, BM-MSC can be cultured with nanofibre scaffolds that would enhance and support their differentiation into hepatocyte-like cells (Hashemi et al. 2009; Kazemnejad et al. 2009).

Role of Action

Homing is the process by which cells migrate to the tissue in which they will exert functional effects. Under normal conditions, MSCs injected intravenously preferentially home to bone marrow (Bensidhoum et al. 2004; Wynn et al. 2004). However, under inflammatory conditions, a cell responds to tissue or cellular injury. These inflammatory conditions produce a continuous source of inflammatory mediators, including chemokines, growth factors and cytokines (Spaeth et al. 2008).

Chapel et al. reported homing of labelled MSCs to different tissues with localisation correlating to the geometry and severity of injury in models of multiple organ failures (Chapel et al. 2003). MSCs have the ability to modulate the immune system and act as immunomodulatory cells that support and regulate T cell proliferation, suppress the formation of cytotoxic T cells and B cells, as well as interact with dendritic cells (DCs) to regulate their secretory and migratory properties (Abumaree et al. 2012; Aggarwal and Pittenger 2005; Corcione et al. 2006).

MSCs are able to exert an anti-inflammatory effect through reducing the production of TNF-α, INF, IL-12 and IL-4 and producing IL-10 (Aggarwal and Pittenger 2005; Jiang et al. 2005; Zhang et al. 2004). Indeed, administration of MSC conditioned media in animal models showed results similar to administration of MSC indicating the presence of cytokines and other factors in the conditioned media (Khubutiya et al. 2014).

Another exciting area of MSC biology involves exosomes, which are microvesicles 40–100 nm in size that carry a cargo of nucleic acid, protein and lipid (Théry et al. 2002). They play a role in cellular communication, tissue homeostasis and tissue regeneration (Lai et al. 2015). In a CCl_4 animal model of liver disease, an intrasplenic injection of 0.4 μg of exosomes improved liver enzyme levels (alanine transaminase (ALT) and aspartate transaminase (AST), maintained liver homeostasis and regenerated hepatocytes (Tan et al. 2014). The role of exosomes and their potential therapeutic effect requires further investigation.

Preclinical and Clinical Trials for MSCs

There have been several preclinical animal studies reporting on the use of MSCs in liver injury. In rat models of liver fibrosis, the animals received intraperitoneal injection of CCl_4 for 8 weeks to induce liver injury. Then animals were treated with human umbilical cord blood derived MSCs (HMSCs) by injecting 1×10^6

HMSCs intravenously. HMSCs reduced liver fibrosis, TGF-β_1, α-SMA, collagen I and differentiated into hepatocyte-like cells (Jung et al. 2009). In another study, rats were subcutaneously injected with CCl_4 for 4 weeks followed by portal vein injection of 1×10^6 human bone marrow MSCs (hBMMSCs). hBMMSCs induced reduction in liver fibrosis and improved liver function through differentiating into hepatocyte-like cells and producing MMP (Chang et al. 2009). In mouse models of liver fibrosis induced by CCl_4 for 6 weeks, treatment with 5×10^5 hBMMSCs reduced TNFα, TGF-β, αSMA and fibrosis and increased MMP-9 expression (Tanimoto et al. 2013).

Several human clinical trials treating liver fibrosis using human MSCs have been published (Table 2.1). Mohamadnejad et al. reported that 4 patients between 34 and 56 years of age with liver cirrhosis were treated with a single dose of autologous BM-MSCs. The results showed no complications or side-effects and improvements in both the model for end-stage liver disease (MELD) score and in the quality of life by 36-item Short Form Health Survey (SF-36) measurement (Mohamadnejad et al. 2007). Kharaziha et al. treated 8 cirrhotic patients using a single dose of BM-MSCs. The treatment showed improvement in MELD score and both bilirubin and albumin level (Kharaziha et al. 2009). Amin et al. evaluated the results of BM-MSCs injection of 20 cirrhotic patients. The results indicated a significant increase in albumin level and a significant reduction in total bilirubin (TBIL), AST, ALT, prothrombin time (PT) and international normalised ratio (INR) (Amin et al. 2013). Furthermore, in another randomised controlled trial, Peng et al. noted the improvement in 53 patients diagnosed with liver cirrhosis who received a single dose of BM-MSCs. These improvements involved albumin, TBIL, ALT levels and MELD score compared with control groups (Peng et al. 2011). Amer et al. injected a single dose of differentiated BM-MSCs through intrahepatic or intrasplenic routes in 20 patients. The treated groups showed a significant improvement compared with controls regarding albumin level, MELD and Child scores and fatigue score. Also, they found no significant difference in the results between routes of administration (Amer et al. 2011). Interestingly, El-Ansary et al. compared differentiated and undifferentiated BM-MSCs, which were injected into cirrhotic patients based on their weight. The follow-up indicated improvements in serum albumin, bilirubin and MELD score without differences between differentiated and undifferentiated BM-MSCs (El-Ansary et al. 2012). Zhang et al. noted similar improvements in cirrhosis patients after the injection of umbilical cord MSCs (UC-MSCs). After treatment the patients showed improved albumin and total bilirubin levels, as well as MELD score (Zhang et al. 2012). An interesting study by Mohamadnejad et al. evaluated the use of BM-MSCs in cirrhosis patients compared with placebo. The results concluded that the cirrhosis patients who received BM-MSCs had no beneficial effect compared with patients who received placebo (Mohamadnejad et al. 2013).

Indeed, treating chronic liver injury with MSCs is encouraging. However, standardised administration routes, dose and frequency are lacking. More research is needed to address these issues and to further investigate the mechanism of action of MSCs.

Table 2.1 Some clinical trials of stem cell therapy for chronic liver diseases

Source of MSC	Sample size	Disorder	Methods	Results	Reference
BM	4 Patients	3 Cryptogenic and 1 AHI	Single dose of mean 31.73×10^6 via peripheral vein	Safe Improvement in MELD	Mohamadnejad et al. (2007)
BM	8 Patients	4 Chronic HBV 1 Chronic HCV 1 Alcoholic cirrhosis 2 Cryptogenic	Single dose of $30-50 \times 10^6$ portal vein	Safe Improve MELD and liver function	Kharaziha et al. (2009)
BM	53 Patients 105 Control	Chronic HBV	Single dose of $3.4-3.8 \times 10^8$ via hepatic artery	Safe Decrease TBIL, MELD Increase ALB	Peng et al. (2011)
BM (differentiated into hepatocyte-like cells)	20 Patients 20 Control	Chronic HCV	Single dose of 20×10^6 10 intrahepatic 10 intrasplenic	Safe Increase ALB Less ascites/edema No different between intrahepatic and intrasplenic	Amer et al. (2011)
UC	35 Patients 15 Control	Chronic HBV	5×10^5/kg, three times Peripheral vein	Safe Increase ALB Decrease TBIL, MELD Reduce ascites	Zhang et al. (2012)
BM and differentiation into hepatic like cells	15 Patients 10 Control	Chronic HCV	1×10^6/Kg Peripheral vein	Safe Increase PT, ALB Decrease TBIL, MELD No different between BM and hepatocyte-like cell	El-Ansary et al. (2012)
BM	15 Patients 12 Placebo	Cirrhosis	$1.2-2.9 \times 10^8$ Peripheral vein	No different between groups	Mohamadnejad et al. (2013)
BM	20	Chronic HCV	10×10^6 Intrasplenic	Decrease TBIL, ALT, AST, PT, INR Increase ALB	Amin et al. (2013)

2.5.1.6 Human Amnion Epithelial Cells

The placenta plays a crucial role in fetal development including delivery of nutrients, gas exchange and waste product disposal; however, the placenta is discarded after delivery. Epithelial cells derived from the placental amniotic membrane (hAEC) express pluripotent markers including SSEA-3, TRA1-60 and SSEA-4 (Miki et al. 2005, 2007). The average yield isolated from amniotic membrane is 120 $\times$ 10^6 (Murphy et al. 2010). They are less likely to be rejected after transplantation due to the absence of HLA-DR expression, and do not form tumours after transplantation into immunocompromised mice (Akle et al. 1981; Wolbank et al. 2007). Regarding their use in liver diseases, hAEC have the ability to differentiate into hepatocyte-like cells (Marongiu et al. 2011; Miki et al. 2002, 2009; Sakuragawa et al. 2000; Takashima et al. 2004). In chronic liver fibrosis mouse models induced by injection of CCL_4, hAECs have been shown to reduce the fibrosis area in treated animals (Manuelpillai et al. 2010; Zhang et al. 2011a) and also activate M2 macrophages (Manuelpillai et al. 2012). In vitro culture of hAECs conditioned media with HSCs showed reduction in HSCs activity and proliferation (Hodge et al. 2014). Therefore, hAECs hold a promising future for the treatment of human liver diseases, particularly because of their safety, availability and absence of ethical concerns regarding their use. However, their mechanism of action requires further investigation.

2.6 Conclusion

There is a need for alternative treatments for end-stage liver injury as the only effective treatment currently available is liver transplantation. Cellular therapy including stem cells is a potential alternative to whole organ transplantation that is currently being pursued on several fronts. The outcomes from early phase clinical trials using MSC indicate that this approach is well tolerated and safe. Understanding the molecular mechanisms of MSCs in this context as well as that of other stem cells such as hAECs will inform the design of efficacy trials and uncover biological settings in which cell therapies can best be exploited.

References

Abumaree M, Al Jumah M, Pace RA, Kalionis B (2012) Immunosuppressive properties of mesenchymal stem cells. Stem Cell Rev Rep 8(2):375–392

Aggarwal S, Pittenger MF (2005) Human mesenchymal stem cells modulate allogeneic immune cell responses. Blood 105(4):1815–1822

Akle C, Welsh K, Adinolfi M, Leibowitz S, McColl I (1981) Immunogenicity of human amniotic epithelial cells after transplantation into volunteers. Lancet 318(8254):1003–1005

Alison MR, Poulsom R, Jeffery R, Dhillon AP, Quaglia A, Jacob J, Novelli M, Prentice G, Williamson J, Wright NA (2000) Cell differentiation: hepatocytes from non-hepatic adult stem cells. Nature 406(6793):257

Amer M-EM, El-Sayed SZ, El-Kheir WA, Gabr H, Gomaa AA, El-Noomani N, Hegazy M (2011) Clinical and laboratory evaluation of patients with end-stage liver cell failure injected with bone marrow-derived hepatocyte-like cells. Eur J Gastroenterol Hepatol 23(10):936–941

Amin MA, Sabry D, Rashed LA, Aref WM, Ghobary MA, Farhan MS, Fouad HA, Youssef YAA et al (2013) Short-term evaluation of autologous transplantation of bone marrow – derived mesenchymal stem cells in patients with cirrhosis: Egyptian study. Clin Transplant 27(4): 607–612

Artis D, Spits H (2015) The biology of innate lymphoid cells. Nature 517(7534):293–301

Basma H, Soto–Gutiérrez A, Yannam GR, Liu L, Ito R, Yamamoto T, Ellis E, Carson SD, Sato S, Chen Y (2009) Differentiation and transplantation of human embryonic stem cell–derived hepatocytes. Gastroenterology 136(3):990–999. e4

Bataller R, Brenner DA (2005) Liver fibrosis. J Clin Invest 115(2):209

Befeler AS, Di Bisceglie AM (2002) Hepatocellular carcinoma: diagnosis and treatment. Gastroenterology 122(6):1609–1619

Belfort R, Harrison SA, Brown K, Darland C, Finch J, Hardies J, Balas B, Gastaldelli A, Tio F, Pulcini J (2006) A placebo-controlled trial of pioglitazone in subjects with nonalcoholic steatohepatitis. N Engl J Med 355(22):2297–2307

Bensidhoum M, Chapel A, Francois S, Demarquay C, Mazurier C, Fouillard L, Bouchet S, Bertho JM, Gourmelon P, Aigueperse J (2004) Homing of in vitro expanded Stro-1-or Stro-1+ human mesenchymal stem cells into the NOD/SCID mouse and their role in supporting human CD34 cell engraftment. Blood 103(9):3313–3319

Best J, Manka P, Syn W-K, Dollé L, van Grunsven LA, Canbay A (2015) Role of liver progenitors in liver regeneration. Hepatobiliary Surg Nutr 4(1):48

Bird TG, Lu W-Y, Boulter L, Gordon-Keylock S, Ridgway RA, Williams MJ, Taube J, Thomas JA, Wojtacha D, Gambardella A (2013) Bone marrow injection stimulates hepatic ductular reactions in the absence of injury via macrophage-mediated TWEAK signaling. Proc Natl Acad Sci 110(16):6542–6547

Blum B, Benvenisty N (2007) Clonal analysis of human embryonic stem cell differentiation into teratomas. Stem Cells 25(8):1924–1930

Braet F, Wisse E (2002) Structural and functional aspects of liver sinusoidal endothelial cell fenestrae: a review. Comp Hepatol 1(1):1

Brooling JT, Campbell JS, Mitchell C, Yeoh GC, Fausto N (2005) Differential regulation of rodent hepatocyte and oval cell proliferation by interferon γ. Hepatology 41(4):906–915

Cai J, Zhao Y, Liu Y, Ye F, Song Z, Qin H, Meng S, Chen Y, Zhou R, Song X (2007) Directed differentiation of human embryonic stem cells into functional hepatic cells. Hepatology 45(5):1229–1239

Cederbaum AI (2012) Alcohol metabolism. Clin Liver Dis 16(4):667–685

Chamberlain J, Yamagami T, Colletti E, Theise ND, Desai J, Frias A, Pixley J, Zanjani ED, Porada CD, Almeida - Porada G (2007) Efficient generation of human hepatocytes by the intrahepatic delivery of clonal human mesenchymal stem cells in fetal sheep. Hepatology 46(6): 1935–1945

Chang Y-J, Liu J-W, Lin P-C, Sun L-Y, Peng C-W, Luo G-H, Chen T-M, Lee R-P, Lin S-Z, Harn H-J (2009) Mesenchymal stem cells facilitate recovery from chemically induced liver damage and decrease liver fibrosis. Life Sci 85(13):517–525

Chang TT, Liaw YF, Wu SS, Schiff E, Han KH, Lai CL, Safadi R, Lee SS, Halota W, Goodman Z (2010) Long - term entecavir therapy results in the reversal of fibrosis/cirrhosis and continued histological improvement in patients with chronic hepatitis B. Hepatology 52(3):886–893

Chapel A, Bertho JM, Bensidhoum M, Fouillard L, Young RG, Frick J, Demarquay C, Cuvelier F, Mathieu E, Trompier F (2003) Mesenchymal stem cells home to injured tissues when co -

infused with hematopoietic cells to treat a radiation - induced multi - organ failure syndrome. J Gene Med 5(12):1028–1038

Chiba M, Sasaki M, Kitamura S, Ikeda H, Sato Y, Nakanuma Y (2011) Participation of bile ductular cells in the pathological progression of non-alcoholic fatty liver disease. J Clin Pathol 64(7):564–570

Clouston AD, Powell EE, Walsh MJ, Richardson MM, Demetris AJ, Jonsson JR (2005) Fibrosis correlates with a ductular reaction in hepatitis C: roles of impaired replication, progenitor cells and steatosis. Hepatology 41(4):809–818

Corcione A, Benvenuto F, Ferretti E, Giunti D, Cappiello V, Cazzanti F, Risso M, Gualandi F, Mancardi GL, Pistoia V (2006) Human mesenchymal stem cells modulate B-cell functions. Blood 107(1):367–372

Crawford J, Burt A (2012) Anatomy, pathophysiology and basic mechanisms of disease. In: Burt A, Portmann B, Ferrell (eds) Pathology of the liver. Churchill Livingstone, Edinburgh, pp 1–78

De Coppi P, Bartsch G, Siddiqui MM, Xu T, Santos CC, Perin L, Mostoslavsky G, Serre AC, Snyder EY, Yoo JJ (2007) Isolation of amniotic stem cell lines with potential for therapy. Nat Biotechnol 25(1):100–106

DeLeve LD (2013) Liver sinusoidal endothelial cells and liver regeneration. J Clin Invest 123(5):1861

Diefenbach A, Colonna M, Koyasu S (2014) Development, differentiation, and diversity of innate lymphoid cells. Immunity 41(3):354–365

Dominici M, Le Blanc K, Mueller I, Slaper-Cortenbach I, Marini F, Krause D, Deans R, Keating A, Prockop D, Horwitz E (2006) Minimal criteria for defining multipotent mesenchymal stromal cells. The International Society for Cellular Therapy position statement. Cytotherapy 8(4):315–317

Duan Y, Ma X, Zou W, Wang C, Bahbahan IS, Ahuja TP, Tolstikov V, Zern MA (2010) Differentiation and characterization of metabolically functioning hepatocytes from human embryonic stem cells. Stem Cells 28(4):674–686

Dumble ML, Croager EJ, Yeoh GC, Quail EA (2002) Generation and characterization of p53 null transformed hepatic progenitor cells: oval cells give rise to hepatocellular carcinoma. Carcinogenesis 23(3):435–445

Dwyer BJ, Olynyk JK, Ramm GA, Tirnitz-Parker JE (2014) TWEAK and LTβ signaling during chronic liver disease. Front Immunol 5:39

Eberl G, Di Santo JP, Vivier E (2015) The brave new world of innate lymphoid cells. Nat Immunol 16(1):1–5

El-Ansary M, Abdel-Aziz I, Mogawer S, Abdel-Hamid S, Hammam O, Teaema S, Wahdan M (2012) Phase II trial: undifferentiated versus differentiated autologous mesenchymal stem cells transplantation in Egyptian patients with HCV induced liver cirrhosis. Stem Cell Rev Rep 8(3):972–981

El-Serag HB, Mason AC (2000) Risk factors for the rising rates of primary liver cancer in the United States. Arch Intern Med 160(21):3227–3230

Fang C-H, Gong J-Q, Zhang W (2004) Function of oval cells in hepatocellular carcinoma in rats. World J Gastroentol 10(17):2482–2487

Fernández-Rodríguez CM, Alonso S, Martinez SM, Forns X, Sanchez-Tapias JM, Rincón D, Rodriguez-Caravaca G, Bárcena R, Serra MA, Romero-Gómez M (2010) Peginterferon plus ribavirin and sustained virological response in HCV-related cirrhosis: outcomes and factors predicting response. Am J Gastroenterol 105(10):2164–2172

Fisher RA, Strom SC (2006) Human hepatocyte transplantation: worldwide results. Transplantation 82(4):441–449

Forbes SJ, Parola M (2011) Liver fibrogenic cells. Best Pract Res Clin Gastroenterol 25(2):207–217

Fox IJ, Chowdhury JR, Kaufman SS, Goertzen TC, Chowdhury NR, Warkentin PI, Dorko K, Sauter BV, Strom SC (1998) Treatment of the Crigler–Najjar syndrome type I with hepatocyte transplantation. N Engl J Med 338(20):1422–1427

Francoz C, Belghiti J, Durand F (2007) Indications of liver transplantation in patients with complications of cirrhosis. Best Pract Res Clin Gastroenterol 21(1):175–190

Friedenstein AJ, Petrakova KV, Kurolesova AI, Frolova GP (1968) Heterotopic transplants of bone marrow. Transplantation 6(2):230–247

Friedenstein A, Deriglasova U, Kulagina N, Panasuk A, Rudakowa S, Luria E, Ruadkow I (1973) Precursors for fibroblasts in different populations of hematopoietic cells as detected by the in vitro colony assay method. Exp Hematol 2(2):83–92

Friedman SL (2003) Liver fibrosis–from bench to bedside. J Hepatol 38:38–53

Friedman SL (2008a) Hepatic stellate cells: protean, multifunctional, and enigmatic cells of the liver. Physiol Rev 88(1):125–172

Friedman SL (2008b) Mechanisms of hepatic fibrogenesis. Gastroenterology 134(6):1655–1669

Gale RP, Sparkes RS, Golde DW (1978) Bone marrow origin of hepatic macrophages (Kupffer cells) in humans. Science 201(4359):937–938

Grompe M (2014) Liver stem cells, where art thou? Cell Stem Cell 15(3):257–258

Han S, Dziedzic N, Gadue P, Keller GM, Gouon - Evans V (2011) An endothelial cell niche induces hepatic specification through dual repression of Wnt and Notch signaling. Stem Cells 29(2):217–228

Hashemi SM, Soleimani M, Zargarian SS, Haddadi-Asl V, Ahmadbeigi N, Soudi S, Gheisari Y, Hajarizadeh A, Mohammadi Y (2009) In vitro differentiation of human cord blood-derived unrestricted somatic stem cells into hepatocyte-like cells on poly (ε-caprolactone) nanofiber scaffolds. Cells Tissues Organs 190(3):135–149

Hay DC, Fletcher J, Payne C, Terrace JD, Gallagher RC, Snoeys J, Black JR, Wojtacha D, Samuel K, Hannoun Z (2008) Highly efficient differentiation of hESCs to functional hepatic endoderm requires ActivinA and Wnt3a signaling. Proc Natl Acad Sci 105(34):12301–12306

Hay DC, Zhao D, Fletcher J, Hewitt ZA, McLean D, Urruticoechea - Uriguen A, Black JR, Elcombe C, Ross JA, Wolf R (2008) Efficient differentiation of hepatocytes from human embryonic stem cells exhibiting markers recapitulating liver development in vivo. Stem cells 26(4):894–902

Hernandez-Gea V, Friedman SL (2011) Pathogenesis of liver fibrosis. Annu Rev Pathol 6:425–456

Higashi N, Senoo H (2003) Distribution of vitamin A - storing lipid droplets in hepatic stellate cells in liver lobules—A comparative study. Anat Rec A Discov Mol Cell Evol Biol 271(1):240–248

Higgins G (1931) Experimental pathology of the liver: I. Restoration of the liver of the white rat following surgical removal. Arch Pathol 12:186–202

Hirschi KK, Li S, Roy K (2014) Induced pluripotent stem cells for regenerative medicine. Annu Rev Biomed Eng 16:277

Hodge A, Lourensz D, Vaghjiani V, Nguyen H, Tchongue J, Wang B, Murthi P, Sievert W, Manuelpillai U (2014) Soluble factors derived from human amniotic epithelial cells suppress collagen production in human hepatic stellate cells. Cytotherapy 16(8):1132–1144

Hong SH, Gang EJ, Jeong JA, Ahn C, Hwang SH, Yang IH, Park HK, Han H, Kim H (2005) In vitro differentiation of human umbilical cord blood-derived mesenchymal stem cells into hepatocyte-like cells. Biochem Biophys Res Commun 330(4):1153–1161

Hosaka T, Suzuki F, Kobayashi M, Seko Y, Kawamura Y, Sezaki H, Akuta N, Suzuki Y, Saitoh S, Arase Y (2013) Long - term entecavir treatment reduces hepatocellular carcinoma incidence in patients with hepatitis B virus infection. Hepatology 58(1):98–107

Huch M, Dorrell C, Boj SF, van Es JH, Li VS, van de Wetering M, Sato T, Hamer K, Sasaki N, Finegold MJ (2013) In vitro expansion of single Lgr5+ liver stem cells induced by Wnt-driven regeneration. Nature 494(7436):247–250

Hung CH, Lee CM, Lu SN, Wang JH, Hu TH, Tung HD, Chen CH, Chen WJ, Changchien CS (2006) Long - term effect of interferon alpha - 2b plus ribavirin therapy on incidence of hepatocellular carcinoma in patients with hepatitis C virus - related cirrhosis. J Viral Hepat 13(6):409–414

Ito M, Nagata H, Miyakawa S, Fox IJ (2009) Review of hepatocyte transplantation. J Hepatobiliary Pancreat Surg 16(2):97–100

Jiang Y, Jahagirdar BN, Reinhardt RL, Schwartz RE, Keene CD, Ortiz-Gonzalez XR, Reyes M, Lenvik T, Lund T, Blackstad M (2002) Pluripotency of mesenchymal stem cells derived from adult marrow. Nature 418(6893):41–49

Jiang X-X, Zhang Y, Liu B, Zhang S-X, Wu Y, Yu X-D, Mao N (2005) Human mesenchymal stem cells inhibit differentiation and function of monocyte-derived dendritic cells. Blood 105(10):4120–4126

Johnson MH, McConnell JM (2004) Lineage allocation and cell polarity during mouse embryogenesis. Semin Cell Dev Biol 15(5):583–597

Jung KH, Shin HP, Lee S, Lim YJ, Hwang SH, Han H, Park HK, Chung JH, Yim SV (2009) Effect of human umbilical cord blood - derived mesenchymal stem cells in a cirrhotic rat model. Liver Int 29(6):898–909

Kazemnejad S, Allameh A, Soleimani M, Gharehbaghian A, Mohammadi Y, Amirizadeh N, Jazayery M (2009) Biochemical and molecular characterization of hepatocyte - like cells derived from human bone marrow mesenchymal stem cells on a novel three - dimensional biocompatible nanofibrous scaffold. J Gastroenterol Hepatol 24(2):278–287

Kharaziha P, Hellström PM, Noorinayer B, Farzaneh F, Aghajani K, Jafari F, Telkabadi M, Atashi A, Honardoost M, Zali MR (2009) Improvement of liver function in liver cirrhosis patients after autologous mesenchymal stem cell injection: a phase I–II clinical trial. Eur J Gastroenterol Hepatol 21(10):1199–1205

Khubutiya MS, Vagabov AV, Temnov AA, Sklifas AN (2014) Paracrine mechanisms of proliferative, anti-apoptotic and anti-inflammatory effects of mesenchymal stromal cells in models of acute organ injury. Cytotherapy 16(5):579–585

Kiessling R, Klein E, Wigzell H (1975) "Natural" killer cells in the mouse. I. Cytotoxic cells with specificity for mouse Moloney leukemia cells. Specificity and distribution according to genotype. Eur J Immunol 5(2):112–117

Klein I, Cornejo JC, Polakos NK, John B, Wuensch SA, Topham DJ, Pierce RH, Crispe IN (2007) Kupffer cell heterogeneity: functional properties of bone marrow–derived and sessile hepatic macrophages. Blood 110(12):4077–4085

Knight B, Tirnitz–Parker JE, Olynyk JK (2008) C-kit inhibition by imatinib mesylate attenuates progenitor cell expansion and inhibits liver tumor formation in mice. Gastroenterology 135(3):969–979. e1

Kruse C, Kajahn J, Petschnik AE, Maaß A, Klink E, Rapoport DH, Wedel T (2006) Adult pancreatic stem/progenitor cells spontaneously differentiate in vitro into multiple cell lineages and form teratoma-like structures. Ann Anat 188(6):503–517

Kubo A, Kim YH, Irion S, Kasuda S, Takeuchi M, Ohashi K, Iwano M, Dohi Y, Saito Y, Snodgrass R (2010) The homeobox gene Hex regulates hepatocyte differentiation from embryonic stem cell–derived endoderm. Hepatology 51(2):633–641

Kupffer C (1876) Ueber Sternzellen der Leber. Archiv für Mikroskopische Anatomie 12(1):353–358

Lai RC, Yeo RWY, Lim SK (2015) Mesenchymal stem cell exosomes. Semin Cell Dev Biol. 40:82–88

Lanthier N, Rubbia-Brandt L, Spahr L (2013) Liver progenitor cells and therapeutic potential of stem cells in human chronic liver diseases. Acta Gastroenterol Belg 76(1):3–9

Lee YA, Wallace MC, Friedman SL (2015) Pathobiology of liver fibrosis: a translational success story. Gut 64:830–841

Leon DA, McCambridge J (2006) Liver cirrhosis mortality rates in Britain from 1950 to 2002: an analysis of routine data. Lancet 367(9504):52–56

Liaw Y-F, Sung JJ, Chow WC, Farrell G, Lee C-Z, Yuen H, Tanwandee T, Tao Q-M, Shue K, Keene ON (2004) Lamivudine for patients with chronic hepatitis B and advanced liver disease. N Engl J Med 351(15):1521–1531

Libbrecht L, Desmet V, Van Damme B, Roskams T (2000) Deep intralobular extension of human hepatic 'progenitor cells' correlates with parenchymal inflammation in chronic viral hepatitis: can 'progenitor cells' migrate? J Pathol 192(3):373–378

Lieber CS (2000) Alcohol: its metabolism and interaction with nutrients. Annu Rev Nutr 20(1):395–430

Liu H, Kim Y, Sharkis S, Marchionni L, Jang Y-Y (2011) In vivo liver regeneration potential of human induced pluripotent stem cells from diverse origins. Sci Transl Med 3(82):82ra39–82ra39

Louvet A, Mathurin P (2015) Alcoholic liver disease: mechanisms of injury and targeted treatment. Nat Rev Gastroenterol Hepatol 12:231–242

Lowes KN, Brennan BA, Yeoh GC, Olynyk JK (1999) Oval cell numbers in human chronic liver diseases are directly related to disease severity. Am J Pathol 154(2):537–541

Lu SH, Yang AH, Wei CF, Chiang HS, Chancellor MB (2010) Multi - potent differentiation of human purified muscle - derived cells: potential for tissue regeneration. BJU Int 105(8):1174–1180

Lu W-Y, Bird TG, Boulter L, Tsuchiya A, Cole AM, Hay T, Guest RV, Wojtacha D, Man TY, Mackinnon A (2015) Hepatic progenitor cells of biliary origin with liver repopulation capacity. Nat Cell Biol 17:971–983

Luedde T, Schwabe RF (2011) NF-kB in the liver—linking injury, fibrosis and hepatocellular carcinoma. Nat Rev Gastroenterol Hepatol 8(2):108–118

Luedde T, Kaplowitz N, Schwabe RF (2014) Cell death and cell death responses in liver disease: mechanisms and clinical relevance. Gastroenterology 147(4):765–783.e4

Makino S, Fukuda K, Miyoshi S, Konishi F, Kodama H, Pan J, Sano M, Takahashi T, Hori S, Abe H (1999) Cardiomyocytes can be generated from marrow stromal cells in vitro. J Clin Invest 103(5):697

Malarkey DE, Johnson K, Ryan L, Boorman G, Maronpot RR (2005) New insights into functional aspects of liver morphology. Toxicol Pathol 33(1):27–34

Manns MP, Czaja AJ, Gorham JD, Krawitt EL, Mieli - Vergani G, Vergani D, Vierling JM (2010) Diagnosis and management of autoimmune hepatitis. Hepatology 51(6):2193–2213

Manuelpillai U, Tchongue J, Lourensz D, Vaghjiani V, Samuel CS, Liu A, Williams ED, Sievert W (2010) Transplantation of human amnion epithelial cells reduces hepatic fibrosis in immunocompetent CCl4-treated mice. Cell Transplant 19(9):1157–1168

Manuelpillai U, Lourensz D, Vaghjiani V, Tchongue J, Lacey D, Tee J-Y, Murthi P, Chan J, Hodge A, Sievert W (2012) Human amniotic epithelial cell transplantation induces markers of alternative macrophage activation and reduces established hepatic fibrosis. PLoS One 7(6):e38631

Marcellin P, Gane E, Buti M, Afdhal N, Sievert W, Jacobson IM, Washington MK, Germanidis G, Flaherty JF, Schall RA (2013) Regression of cirrhosis during treatment with tenofovir disoproxil fumarate for chronic hepatitis B: a 5-year open-label follow-up study. Lancet 381(9865):468–475

Marcos A, Fisher RA, Ham JM, Shiffman ML, Sanyal AJ, Luketic VA, Sterling RK, Fulcher AS, Posner MP (2000) Liver reggeneration and function in donor and recipient after right lobe adult to adult living donor liver transplantation12. Transplantation 69(7):1375–1379

Marongiu F, Gramignoli R, Dorko K, Miki T, Ranade AR, Paola Serra M, Doratiotto S, Sini M, Sharma S, Mitamura K (2011) Hepatic differentiation of amniotic epithelial cells. Hepatology 53(5):1719–1729

Matthews KR, Rowland ML (2011) Stem cell policy in the Obama age: UK and US perspectives. Regen Med 6(1):125–132

McCuskey R (2012) Anatomy of the liver. In: Sanyal TDBPMJ (ed) Zakim and Boyer's hepatology, 6th edn. W.B. Saunders, Saint Louis, pp 3–19

Mchedlidze T, Waldner M, Zopf S, Walker J, Rankin AL, Schuchmann M, Voehringer D, McKenzie AN, Neurath MF, Pflanz S (2013) Interleukin-33-dependent innate lymphoid cells mediate hepatic fibrosis. Immunity 39(2):357–371

McKenzie AN, Spits H, Eberl G (2014) Innate lymphoid cells in inflammation and immunity. Immunity 41(3):366–374

Mebius RE, Rennert P, Weissman IL (1997) Developing lymph nodes collect CD4+ CD3− LTβ + cells that can differentiate to APC, NK cells, and follicular cells but not T or B cells. Immunity 7(4):493–504

Medici V, Rossaro L, Sturniolo G (2007) Wilson disease—a practical approach to diagnosis, treatment and follow-up. Dig Liver Dis 39(7):601–609

Meng X, Ichim TE, Zhong J, Rogers A, Yin Z, Jackson J, Wang H, Ge W, Bogin V, Chan KW (2007) Endometrial regenerative cells: a novel stem cell population. J Transl Med 5(1):57

Michelotti GA, Machado MV, Diehl AM (2013) NAFLD, NASH and liver cancer. Nat Rev Gastroenterol Hepatol 10(11):656–665

Miki T, Cai H, Strom S (2002) Production of hepatocytes from human amniotic stem cells. Hepatology. pp 171A–171A

Miki T, Lehmann T, Cai H, Stolz DB, Strom SC (2005) Stem cell characteristics of amniotic epithelial cells. Stem Cells 23(10):1549–1559

Miki T, Mitamura K, Ross MA, Stolz DB, Strom SC (2007) Identification of stem cell marker-positive cells by immunofluorescence in term human amnion. J Reprod Immunol 75(2):91–96

Miki T, Marongiu F, Ellis EC, Dorko K, Mitamura K, Ranade A, Gramignoli R, Davila J, Strom SC (2009) Production of hepatocyte-like cells from human amnion. Hepatocyte transplantation. Methods Mol Biol 481:155–168, Springer

Mohamadnejad M, Alimoghaddam K, Mohyeddin-Bonab M, Bagheri M, Bashtar M, Ghanaati H, Baharvand H, Ghavamzadeh A, Malekzadeh R (2007) Phase 1 trial of autologous bone marrow mesenchymal stem cell transplantation in patients with decompensated liver cirrhosis. Arch Iran Med 10(4):459–466

Mohamadnejad M, Alimoghaddam K, Bagheri M, Ashrafi M, Abdollahzadeh L, Akhlaghpoor S, Bashtar M, Ghavamzadeh A, Malekzadeh R (2013) Randomized placebo - controlled trial of mesenchymal stem cell transplantation in decompensated cirrhosis. Liver Int 33(10): 1490–1496

Moriya K, Yoshikawa M, Saito K, Ouji Y, Nishiofuku M, Hayashi N, Ishizaka S, Fukui H (2007) Embryonic stem cells develop into hepatocytes after intrasplenic transplantation in CCl4-treated mice. World J Gastroenterol 13(6):866–873

Moriya K, Yoshikawa M, Ouji Y, Saito K, Nishiofuku M, Matsuda R, Ishizaka S, Fukui H (2008) Embryonic stem cells reduce liver fibrosis in CCl4 - treated mice. Int J Exp Pathol 89(6):401–409

Murphy S, Rosli S, Acharya R, Mathias L, Lim R, Wallace E, Jenkin G (2010) Amnion epithelial cell isolation and characterization for clinical use. Curr Protocol Stem Cell Biol 1E.6.1–1E.6.25

Naito M, Hasegawa G, Takahashi K (1997) Development, differentiation, and maturation of Kupffer cells. Microsc Res Tech 39(4):350–364

Newsome P, Hussain M, Theise N (2004) Hepatic oval cells: helping redefine a paradigm in stem cell biology. Curr Top Dev Biol 61:1–28

Nguyen LN, Furuya MH, Wolfraim LA, Nguyen AP, Holdren MS, Campbell JS, Knight B, Yeoh GC, Fausto N, Parks WT (2007) Transforming growth factor - beta differentially regulates oval cell and hepatocyte proliferation. Hepatology 45(1):31–41

Noort W, Scherjon S, Kleijburg-Van Der Keur C, Kruisselbrink A, Van Bezooijen R, Beekhuizen W, Willemze R, Kanhai H, Fibbe W (2003) Mesenchymal stem cells in human second-trimester bone marrow, liver, lung, and spleen exhibit a similar immunophenotype but a heterogeneous multilineage differentiation potential. Haematologica 88(8):845–852

Noureddin M, Ghany MG (2010) Editorial: treatment of patients with HCV-related cirrhosis with peginterferon and ribavirin: Swinging the pendulum toward treatment. Am J Gastroenterol 105(10):2174–2176

Novo E, Parola M (2008) Redox mechanisms in hepatic chronic wound healing and fibrogenesis. Fibrogenesis & Tissue Repair 1(1):5

Novo E, Busletta C, di Bonzo LV, Povero D, Paternostro C, Mareschi K, Ferrero I, David E, Bertolani C, Caligiuri A (2011) Intracellular reactive oxygen species are required for directional migration of resident and bone marrow-derived hepatic pro-fibrogenic cells. J Hepatol 54(5):964–974

Oswald J, Boxberger S, Jørgensen B, Feldmann S, Ehninger G, Bornhäuser M, Werner C (2004) Mesenchymal stem cells can be differentiated into endothelial cells in vitro. Stem Cells 22(3):377–384

Parola M, Marra F, Pinzani M (2008) Myofibroblast–like cells and liver fibrogenesis: emerging concepts in a rapidly moving scenario. Mol Aspects Med 29(1):58–66

Pellicoro A, Ramachandran P, Iredale JP, Fallowfield JA (2014) Liver fibrosis and repair: immune regulation of wound healing in a solid organ. Nat Rev Immunol 14(3):181–194

Peng L, Xie D, Lin BL, Liu J, Zhu H, Xie C, Zheng Y, Gao Z (2011) Autologous bone marrow mesenchymal stem cell transplantation in liver failure patients caused by hepatitis B: short - term and long - term outcomes. Hepatology 54(3):820–828

Pinzani M, Marra F (2001) Cytokine receptors and signaling in hepatic stellate cells. Semin Liver Dis 21(3):397–416

Poynard T, McHutchison J, Manns M, Trepo C, Lindsay K, Goodman Z, Ling MH, Albrecht J (2002) Impact of pegylated interferon alfa-2b and ribavirin on liver fibrosis in patients with chronic hepatitis C. Gastroenterology 122(5):1303–1313

Pradere JP, Kluwe J, Minicis S, Jiao JJ, Gwak GY, Dapito DH, Jang MK, Guenther ND, Mederacke I, Friedman R (2013) Hepatic macrophages but not dendritic cells contribute to liver fibrosis by promoting the survival of activated hepatic stellate cells in mice. Hepatology 58(4):1461–1473

Ramachandran P, Pellicoro A, Vernon MA, Boulter L, Aucott RL, Ali A, Hartland SN, Snowdon VK, Cappon A, Gordon-Walker TT (2012) Differential Ly-6C expression identifies the recruited macrophage phenotype, which orchestrates the regression of murine liver fibrosis. Proc Natl Acad Sci 109(46):E3186–E3195

Richardson MM, Jonsson JR, Powell EE, Brunt EM, Neuschwander–Tetri BA, Bhathal PS, Dixon JB, Weltman MD, Tilg H, Moschen AR (2007) Progressive fibrosis in nonalcoholic steatohepatitis: association with altered regeneration and a ductular reaction. Gastroenterology 133(1):80–90

Sakuragawa N, Enosawa S, Ishii T, Thangavel R, Tashiro T, Okuyama T, Suzuki S (2000) Human amniotic epithelial cells are promising transgene carriers for allogeneic cell transplantation into liver. J Hum Genet 45(3):171–176

Sancho - Bru P, Altamirano J, Rodrigo - Torres D, Coll M, Millán C, José Lozano J, Miquel R, Arroyo V, Caballería J, Ginès P (2012) Liver progenitor cell markers correlate with liver damage and predict short - term mortality in patients with alcoholic hepatitis. Hepatology 55(6):1931–1941

Sato Y, Araki H, Kato J, Nakamura K, Kawano Y, Kobune M, Sato T, Miyanishi K, Takayama T, Takahashi M (2005) Human mesenchymal stem cells xenografted directly to rat liver are differentiated into human hepatocytes without fusion. Blood 106(2):756–763

Schaub JR, Malato Y, Gormond C, Willenbring H (2014) Evidence against a stem cell origin of new hepatocytes in a common mouse model of chronic liver injury. Cell Rep 8(4):933–939

Scherjon SA, Kleijburg - van der Keur C, de Groot - Swings GM, Claas FH, Fibbe WE, Kanhai HH (2004) Isolation of mesenchymal stem cells of fetal or maternal origin from human placenta. Stem Cells 22(7):1338–1345

Schulz C, Perdiguero EG, Chorro L, Szabo-Rogers H, Cagnard N, Kierdorf K, Prinz M, Wu B, Jacobsen SEW, Pollard JW (2012) A lineage of myeloid cells independent of Myb and hematopoietic stem cells. Science 336(6077):86–90

Schwartz RE, Reyes M, Koodie L, Jiang Y, Blackstad M, Lund T, Lenvik T, Johnson S, Hu W-S, Verfaillie CM (2002) Multipotent adult progenitor cells from bone marrow differentiate into functional hepatocyte-like cells. J Clin Investig 109(10):1291–1302

Shu S-N, Wei L, Wang J-H, Zhan Y-T, Chen H-S, Wang Y (2004) Hepatic differentiation capability of rat bone marrow-derived mesenchymal stem cells and hematopoietic stem cells. World J Gastroenterol 10(19):2818–2822

Silva GF, Villela-Nogueira CA, Mello CEB, Soares EC, Coelho HSM, Ferreira PRA, Ruiz FJG (2014) Peginterferon plus ribavirin and sustained virological response rate in HCV-related advanced fibrosis: a real life study. Braz J Infect Dis 18(1):48–52

Song Z, Cai J, Liu Y, Zhao D, Yong J, Duo S, Song X, Guo Y, Zhao Y, Qin H (2009) Efficient generation of hepatocyte-like cells from human induced pluripotent stem cells. Cell Res 19(11):1233–1242

Spaeth E, Klopp A, Dembinski J, Andreeff M, Marini F (2008) Inflammation and tumor microenvironments: defining the migratory itinerary of mesenchymal stem cells. Gene Ther 15(10):730–738

Spits H, Artis D, Colonna M, Diefenbach A, Di Santo JP, Eberl G, Koyasu S, Locksley RM, McKenzie AN, Mebius RE (2013) Innate lymphoid cells—a proposal for uniform nomenclature. Nat Rev Immunol 13(2):145–149

Su T-H, Kao J-H, Liu C-J (2014) Molecular mechanism and treatment of viral hepatitis-related liver fibrosis. Int J Mol Sci 15(6):10578–10604

Sullivan GJ, Hay DC, Park IH, Fletcher J, Hannoun Z, Payne CM, Dalgetty D, Black JR, Ross JA, Samuel K (2010) Generation of functional human hepatic endoderm from human induced pluripotent stem cells. Hepatology 51(1):329–335

Tacke F, Zimmermann HW (2014) Macrophage heterogeneity in liver injury and fibrosis. J Hepatol 60(5):1090–1096

Takahashi K, Tanabe K, Ohnuki M, Narita M, Ichisaka T, Tomoda K, Yamanaka S (2007) Induction of pluripotent stem cells from adult human fibroblasts by defined factors. Cell 131(5):861–872

Takashima S, Ise H, Zhao P, Akaike T, Nikaido T (2004) Human amniotic epithelial cells possess hepatocyte-like characteristics and functions. Cell Struct Funct 29(3):73–84

Takebe T, Sekine K, Enomura M, Koike H, Kimura M, Ogaeri T, Zhang R-R, Ueno Y, Zheng Y-W, Koike N (2013) Vascularized and functional human liver from an iPSC-derived organ bud transplant. Nature 499(7459):481–484

Takiya CM, Paredes BD, Guintanilha L (2013) Liver resident stem cell. In: Goldenberg RC, de Carvalho AC (eds) Resident stem cells and regenerative therapy. Academic Press, pp 177–203

Taléns-Visconti R, Bonora A, Jover R, Mirabet V, Carbonell F, Castell JV, Gómez-Lechón MJ (2006) Hepatogenic differentiation of human mesenchymal stem cells from adipose tissue in comparison with bone marrow mesenchymal stem cells. World J Gastroenterol 12(36):5834–5845

Tan CY, Lai RC, Wong W, Dan YY, Lim S-K, Ho HK (2014) Mesenchymal stem cell-derived exosomes promote hepatic regeneration in drug-induced liver injury models. Stem Cell Res Ther 5:76

Tanimoto H, Terai S, Taro T, Murata Y, Fujisawa K, Yamamoto N, Sakaida I (2013) Improvement of liver fibrosis by infusion of cultured cells derived from human bone marrow. Cell Tissue Res 354(3):717–728

Than NN, Newsome PN (2015) A concise review of non - alcoholic fatty liver disease. Atherosclerosis 239(1):192–202

Theise ND, Badve S, Saxena R, Henegariu O, Sell S, Crawford JM, Krause DS (2000) Derivation of hepatocytes from bone marrow cells in mice after radiation - induced myeloablation. Hepatology 31(1):235–240

Théry C, Zitvogel L, Amigorena S (2002) Exosomes: composition, biogenesis and function. Nat Rev Immunol 2(8):569–579

Thomas JA, Pope C, Wojtacha D, Robson AJ, Gordon - Walker TT, Hartland S, Ramachandran P, Van Deemter M, Hume DA, Iredale JP (2011) Macrophage therapy for murine liver fibrosis recruits host effector cells improving fibrosis, regeneration, and function. Hepatology 53(6):2003–2015

Torres DM, Williams CD, Harrison SA (2012) Features, diagnosis, and treatment of nonalcoholic fatty liver disease. Clin Gastroenterol Hepatol 10(8):837–858

Touboul T, Hannan NR, Corbineau S, Martinez A, Martinet C, Branchereau S, Mainot S, Strick - Marchand H, Pedersen R, Di Santo J (2010) Generation of functional hepatocytes from human embryonic stem cells under chemically defined conditions that recapitulate liver development. Hepatology 51(5):1754–1765

Tsukada S, Parsons CJ, Rippe RA (2006) Mechanisms of liver fibrosis. Clin Chim Acta 364(1–2):33–60

Vestentoft PS (2014) Adult hepatic progenitor cells. In: Wislet-Gendebien S (ed) Adult stem cell niches. InTech. Available from: http://www.intechopen.com/books/adult-stem-cell-niches/adult-hepatic-progenitor-cells. doi:10.5772/58814.

Viebahn CS, Tirnitz-Parker JE, Olynyk JK, Yeoh GC (2006) Evaluation of the "Cellscreen" system for proliferation studies on liver progenitor cells. Eur J Cell Biol 85(12):1265–1274

Visse R, Nagase H (2003) Matrix metalloproteinases and tissue inhibitors of metalloproteinases structure, function, and biochemistry. Circ Res 92(8):827–839

Wallace K, Burt A, Wright M (2008) Liver fibrosis. Biochem J 411:1–18

Weiskirchen R, Tacke F (2014) Cellular and molecular functions of hepatic stellate cells in inflammatory responses and liver immunology. Hepatobiliary Surg Nutr 3(6):344–363

Williams MJ, Clouston AD, Forbes SJ (2014) Links between hepatic fibrosis, ductular reaction, and progenitor cell expansion. Gastroenterology 146(2):349–356

Wolbank S, Peterbauer A, Fahrner M, Hennerbichler S, Van Griensven M, Stadler G, Redl H, Gabriel C (2007) Dose-dependent immunomodulatory effect of human stem cells from amniotic membrane: a comparison with human mesenchymal stem cells from adipose tissue. Tissue Eng 13(6):1173–1183

Wu N, Liu F, Ma H, Zhu F-X, Liu Z-D, Fei R, Chen H-S, Wang H, Wei L (2008) HBV infection involving in hepatic progenitor cells expansion in HBV-infected end-stage liver disease. Hepatogastroenterology 56(93):964–967

Wynn RF, Hart CA, Corradi-Perini C, O'Neill L, Evans CA, Wraith JE, Fairbairn LJ, Bellantuono I (2004) A small proportion of mesenchymal stem cells strongly expresses functionally active CXCR4 receptor capable of promoting migration to bone marrow. Blood 104(9):2643–2645

Yamamoto H, Quinn G, Asari A, Yamanokuchi H, Teratani T, Terada M, Ochiya T (2003) Differentiation of embryonic stem cells into hepatocytes: biological functions and therapeutic application. Hepatology 37(5):983–993

Yang J, Zhang L, Yu C, Yang X-F, Wang H (2014) Monocyte and macrophage differentiation: circulation inflammatory monocyte as biomarker for inflammatory diseases. Biomark Res 2(1):1

Yin C, Evason KJ, Asahina K, Stainier DY (2013) Hepatic stellate cells in liver development, regeneration, and cancer. J Clin Invest 123:1902–1910

Zekry A, McCaughan G, Stuart K, Adams L, Law M, Sievert W, Nicoll A (2013) The Economic Cost and Health Burden of Liver Disease in Australia www.gesa.org.au

Zhang W, Ge W, Li C, You S, Liao L, Han Q, Deng W, Zhao RC (2004) Effects of mesenchymal stem cells on differentiation, maturation, and function of human monocyte-derived dendritic cells. Stem Cells Dev 13(3):263–271

Zhang D, Jiang M, Miao D (2011a) Transplanted human amniotic membrane-derived mesenchymal stem cells ameliorate carbon tetrachloride-induced liver cirrhosis in mouse. PLoS One 6(2):e16789

Zhang S, Chen S, Li W, Guo X, Zhao P, Xu J, Chen Y, Pan Q, Liu X, Zychlinski D (2011b) Rescue of ATP7B function in hepatocyte-like cells from Wilson's disease induced pluripotent stem cells using gene therapy or the chaperone drug curcumin. Hum Mol Genet 20(16):3176–3187

Zhang Z, Lin H, Shi M, Xu R, Fu J, Lv J, Chen L, Lv S, Li Y, Yu S (2012) Human umbilical cord mesenchymal stem cells improve liver function and ascites in decompensated liver cirrhosis patients. J Gastroenterol Hepatol 27(s2):112–120

Chapter 3
Clinical Applications of Stem Cells in Liver Cirrhosis

Ahmer Irfan

3.1 Introduction

3.1.1 Epidemiology

In the past 50 years there has been a marked increase in the incidence and mortality of liver cirrhosis. As with any complex and advanced disease state, the incidence of severe associated complications has also risen. This has been attributed to an increase in alcohol abuse and non-alcoholic fatty liver disease in western culture and primarily be attributed to viral infection in eastern societies. Consequently, in 2011, liver cirrhosis accounted for over 33,500 deaths each year in the USA. Liver cirrhosis is also a major risk factor for hepatocellular carcinoma, which accounted for 33,000 new cases with 23,000 deaths in USA in the past year (American Cancer Society 2014).

3.1.2 Embryological Development

Liver development begins from the endoderm of the foregut in the third week. Fibroblast growth factors (FGF2) are responsible for allowing the foregut endoderm to differentiate into liver tissue by blocking inhibitors secreted by surrounding tissue. The liver bud penetrates the septum transversum, which will form the central tendon of the diaphragm. The mesoderm of the septum transversum is responsible for the derivation of haematopoietic, Kupffer and connective tissue cells (Sadler and Langman 2010).

A. Irfan (✉)
University of Edinburgh, Edinburgh, UK
e-mail: ahmerirfan@googlemail.com

Although the timeline for the formation of the biliary system remains unclear, once formation has begun the intra-hepatic and extra-hepatic biliary tree remain continuous throughout gestation, allowing a passage for bile into the gastrointestinal tract (Crawford 2002). A narrowing of the connection between the duodenum and the hepatic diverticulum forms the extra-hepatic bile duct (Sadler and Langman 2010). The extra-hepatic bile duct is in a 'solid phase' in the 5th week of gestation and recanalisation occurs to leave it as a hollow tube by the 7th week (Tan et al. 1994). Early intra-hepatic biliary structures appear around week 7 originating from the ductal plate and terminal bile duct formation occurring at the 11th week (Crawford 2002). Bile production in the liver begins around the 12th week when the biliary tree has formed. Failure of correct formation of the bile ducts leads to biliary atresia, which usually presents as jaundice in infants (Sadler and Langman 2010). The growth of the biliary system is closely related to the development of the portal vein system and the hepatic artery vasculature (Collardeau-Frachon and Scoazec 2008; Desmet 1991).

3.1.3 Pathophysiology

Liver cirrhosis occurs as a progression of liver fibrosis with the distortion of the hepatic vasculature (Schuppan and Afdhal 2008). It occurs as a chronic state, when the initial injury continues to persist (Zhang et al. 2012). Liver fibrosis is defined as the initial distortion of hepatic architecture secondary to the accumulation of excessive collagen and extracellular matrix proteins (Bataller and Brenner 2005). This causes hepatocellular hyperplasia (regenerating nodules) and angiogenesis in response to the production of growth regulators (Civan 2013). The angiogenesis produces a low volume, high pressure shunt of (portal and arterial) blood into the central veins. This results in portal hypertension and impaired exchange between sinusoids and hepatocytes (Schuppan and Afdhal 2008; Civan 2013). Histologically, portal tracts are linked to the central veins (and to each other) by vascularised fibrotic septa leading to hepatocyte islands which are devoid of a central vein (Schuppan and Afdhal 2008).

3.1.4 Complications

The earliest complications of a cirrhotic liver relate to impairments in function. These can manifest as coagulopathies, acute kidney injury and hepatic encephalopathy. The inability of the hepatocytes to secrete bile leads to jaundice and fat malabsorption (which has a variety of clinical effects including fat-soluble vitamin deficiencies).

The most common complication is portal hypertension, which presents with its own set of clinical manifestations. It can cause massive bleeding at sites of porto-systemic anastomoses (oesophagus or rectum), ascites (which increases the risk for spontaneous bacterial peritonitis) and pulmonary hypertension (which can present with symptoms of heart failure) (Civan 2013).

Cirrhosis of the liver is also a risk factor for the development of hepatocellular carcinoma, the 6th most common cancer cause of mortality (American Cancer Society 2014).

3.1.5 Current Treatment

Currently the only proven, effective and therefore recommended treatment of end stage liver disease is liver transplantation (Kisseleva et al. 2010), which would require the donation of a healthy organ from either a living or cadaveric donor.

This treatment option presents with its own set of problems; firstly, it is expensive (Khan et al. 2008) 'estimated at $150,000 or more during the first year following transplantation'. The most critical issue, however, is the marked shortage of donor organs available. This problem has been experienced globally and has led to high patient mortality (Kim and Kremers 2008). Post-operatively, lifelong immunosuppression therapy is required to reduce the risk of rejection, compromising the patients' immune system. Long-term renal, cardiovascular and infective complications can occur as well as post-transplant lympho-proliferative diseases (Houlihan and Newsome 2008). The problems aren't just limited to the transplantation itself, as a knock on effect of the long waiting list and the critical condition of many patients, there may be a requirement for intensive supportive care and treatment to be maintained, either as palliative care or as bridging therapy to transplantation.

3.2 Treatment Techniques

3.2.1 Mobilisation of Stem Cells

The use of granulocyte-colony stimulating factor (G-CSF) increases the stem cell population in the blood due to the mobilisation of inherent bone marrow stem cells (BMSCs). The mobilisation of the CD34+ cells in patients with active disease is noted to be less than their healthy counterparts (Gaia et al. 2006); however, quantification and comparisons to clinical outcomes are not known. This intervention has been deemed to be safe and feasible as well as demonstrating a reduction in complications when compared to patients who were undergoing standard therapy (Garg et al. 2012; Lorenzini et al. 2008).

3.2.2 Mobilisation of Stem Cells, Collection and Reinfusion

Once mobilised using G-CSF, stem cells can be harvested using either leukapheresis or via bone marrow extraction from the liver. The extracted HSCs are then re-transplanted through the portal vein or the hepatic artery. A randomised trial showed a higher survival rate at 6 months when compared to patients within the control group (Salama et al. 2010). All other phase I trials of this technique deemed it to be safe and feasible.

3.2.3 Infusion of Stem Cells Extracted from Ileum

All patients undergo bone marrow aspiration from the ileum to harvest BMSCs. There are minor differences in the specified site used and whether the technique was performed uni- or bilaterally. The main adverse effects associated with this technique are associated with the requirement for anaesthetic for the retrieval of the bone marrow.

3.2.4 Novel Approaches

Whilst the methods above are the most commonly utilised for the extraction of BMSCs, there are an increasing number of novel approaches under investigation.

Autologous transplantation of stem cells from the human umbilical cord (Zhang et al. 2012) and a mouse tonsillar tissue (Park et al. 2015) have been found to be effective sources of mesenchymal stem cells (MSCs).

Allogenic transplantation from aborted foetuses has also been trialled (Khan et al. 2010) with relative success.

There are other potential sources that are still under investigation to determine the safest and most effective method of harvesting stem cells that can improve a cirrhotic liver.

3.2.5 Comparison of Techniques

The vast majority of these techniques are still in initial phase testing. Consequently, the results that are published are primarily to assess safety. All these techniques have been proven to be initially safe in their target population. As the treatment options progress and more data becomes available, the comparisons in efficacy will become the focus of the research.

3.3 Mechanism of Action

There are two main theories as to the action of stem cells in the regeneration of liver parenchyma, however, neither has been proven to be fully correct yet.

One theory postulates the trans-differentiation of BMSCs into functional hepatocytes (Terai et al. 2003; Jang et al. 2004). The studies caused liver injury (using carbon tetrachloride [CCl_4]) and BMSCs were transfused into the injured livers. The improvement in albumin levels indicated the functionality of the differentiated hepatocytes from the BMSCs. Both studies also noted that the effects of stem cell transplantation were only present in an injured liver when compared to uninjured controls.

The second model that has been proposed is BMSCs regenerating liver tissue through cell fusion (Wang et al. 2003; Vassilopoulos et al. 2003). It is thought that the repopulation of liver tissue is achieved by the fusion of the host and the donor cells. Vassilopoulos et al predicted that Kupffer cells from the BMSCs are the most likely candidate for fusion. These models utilised fumarylacetoacetate hydrolase (FAH) deficiency, leading to type I tyrosinaemia in mice. The FAH deficiency was achieved by lethal irradiation of the mice and FAH positive BMSCs were transplanted. Analysis of the chromosomes and alleles was the method by which fusion was confirmed.

There was uncertainty in the reporting of both models, with neither being able to rule the other out. It appeared that whilst it was agreed that both models were plausible, the authors felt that the mechanism that they proposed was the main contributor.

3.4 Stem Cell Use in Liver Cirrhosis

3.4.1 Undifferentiated Bone Marrow

In all human trials involving the infusion of stem cells into a cirrhotic patient, the type of stem cell has been clearly identified and defined. A trial utilising unspecified BMSCs is recruiting with view to expand the data in this area.

In some of the earlier trials involving animal models, extracted bone marrow was re-infused back into the mice without any isolation of a particular stem cell type. In these trials, the mice were killed prior to the extraction of the BMSCs. Hepatic fibrosis was induced by administration of carbon tetrachloride (CCl_4) in the mice that were to undergo the treatment. The results showed that up to 25 % of the liver had been repopulated 4 weeks after stem cell administration (Terai et al. 2003). All the trials showed an improvement in liver function and a reduction in liver fibrosis following the therapy (Ali and Masoud 2012; Zheng and Liang 2008; Iwamoto et al. 2013).

3.4.2 Haematopoietic Stem Cells

3.4.2.1 Animal Trials

Studies have shown the migration of haematopoetic stem cells into the foetal liver at day 11. These foetal haematopoetic cells are believed to have a greater proliferative capacity than adult bone marrow cells and it has been concluded that they undergo expansion in the foetal liver (Ema and Nakauchi 2000). There is also evidence to show that bone marrow derived hepatocytes have the capability to convert into fully functioning hepatocytes under the correct conditions (Jang et al. 2004; Wang et al. 2003). This conversion is not without effect, with data supporting an improvement in liver function following infusion of haematopoetic stem cells (Jang et al. 2004; Cho et al. 2011). There is also data to support that the mobilisation of BMSCs (especially with respect to haematopoetic stem cells) can improve liver function, though this was not found to be as effective as direct infusion (Cho et al. 2011).

3.4.2.2 Human Trials

Due to the novel nature of this therapy, the majority of trials that have been carried out focus primarily on the safety of the treatment as opposed to the efficacy.

Mobilisation of haematopoetic stem cells using G-CSF (Gaia et al. 2006) was well tolerated in patients with end stage cirrhosis. There was increase in circulating CD-34+ cells and whilst there may have been potential for regeneration, no change in liver function was reported.

There is limited, conflicting data regarding haematopoetic stem cell infusion (following extraction from ileum). Whilst some data postulates that this type of stem cell is not safe due to significant side effects (hepatorenal syndrome) in a patient following infusion (Mohamadnejad et al. 2007b). It is important to remember that patients with decompensated cirrhosis are at increased risk of complications secondary to their disease and this outcome may not be causal to the haematopoietic stem cell infusion. The rest of the patients that underwent haematopoetic stem cell transplantation did not experience adverse effects and the therapy was deemed safe by that data set (Nikeghbalian et al. 2011).

A combination of the above two methods has also been trialled. BMSCs are mobilised using G-CSF, collected and then re-infused in the human liver. An improvement in liver function and clinical signs was noted with no significant adverse effects from the treatment when compared to the control groups (Salama et al. 2010; Pai et al. 2008).

3.4.3 Mesenchymal Stem Cells

3.4.3.1 Animal Trials

There is a great amount of animal data utilising a variety of novel methods and approaches in the treatment of hepatic fibrosis using MSCs.

Infusion of MSCs into CCl_4-induced fibrotic livers has shown improvement in liver function. There were many variations in the techniques employed.

MSCs treated with hepatic growth factor (HGF) were infused into the liver and improvement was noted by an increase in albumin production (Oyagi et al. 2006) (it was noted that MSCs not treated with HGF did not exert the same effect). Another trial cultured MSCs with HGF and FGF4 (fibroblast growth factor 4) prior to infusion and reported a greater propensity of the treated MSCs to migrate to the liver (when compared to the non-treated group) and this caused a reduction in collagen and alkaline phosphatase, indicating an improvement in liver function (Shams et al. 2015).

Infusion of microencapsulated (but untreated) human MSCs in mice was also found to exert an anti-fibrotic effect on affected livers (Meier et al. 2015) and in one study the administration of nitric oxide following infusion of MSCs was found to augment their ability to repair liver fibrosis (Ali et al. 2012).

Untreated MSCs were found to initiate endogenous hepatic regeneration when administered systemically (as opposed to locally), though this effect was diminished with delayed infusion (Fang et al. 2004).

There are also reports for the utilisation of novel sites to harvest MSCs. One study used MSCs taken from tonsillar tissue and was able to utilise it to ameliorate hepatic fibrosis (Park et al. 2015).

3.4.3.2 Human Trials

Some MSCs trials harvested the cells in the same way as other stem cell lines, extraction from the ileum followed by autologous reinfusion in the liver. These trials provided similar results as compared to other cell lines, with the technique being determined as safe and feasible with some improvements in liver function (Kharaziha et al. 2009; Mohamadnejad et al. 2007a). There was also improvement noted in liver function, but sample numbers were not large enough to make a statistically significant conclusion.

A novel approach that has only been attempted with MSCs is the transplantation of umbilical cord derived MSCs into patients with decompensated cirrhosis. The treatment was found to be safe and feasible with an improvement in the patient's quality of life (Zhang et al. 2012; Lin et al. 2012).

There are currently further trials that are recruiting to investigate this stem cell type further, but there has been no further data released.

3.4.4 Mononuclear Stem Cells

Mononuclear cells isolated from bone marrow extracted from the ileum have also been trialled for liver cirrhosis. The data for the safety of these cells is agreeable, with no significant side effects reported in any patients that have had the autologous infusion. There is, however, conflicting data as to whether these cells cause any alterations in liver function following treatment. Whereas some data shows significant changes (Terai et al. 2006), other trials state no change in any liver parameter (Nikeghbalian et al. 2011).

3.4.5 Foetal Stem Cells

A novel approach infused hepatic progenitor cells taken from foetuses that were aborted between 16 and 20 weeks. There were no adverse affects recorded from the transplantation of foetal stem cells into the hepatic circulation. A very small amount of cells were transplanted (a maximum of 0.5 % of liver mass) but there was a significant improvement in liver function noted (though there was no control group for comparison) (Khan et al. 2010).

3.5 Conclusion

The sporadic nature that this chapter deals with the development and advancement of stem cell therapy in the treatment of liver cirrhosis reflects the data that is currently present on this subject. Whilst there is a push to advance stem cell therapy across all fields, the vast variabilities that exist translate to a slow advancement of data. All the data in this field shows that there is great scope for advancement; there is just a requirement of time until large, statistically strong trials are available to judge the true efficacy of this treatment.

References

Ali G, Masoud MS (2012) Bone marrow cells ameliorate liver fibrosis and express albumin after transplantation in CCl(4)-induced fibrotic liver. Saudi J Gastroenterol 18(4):263–267

Ali G, Mohsin S, Khan M, Nasir GA, Shams S, Khan SN, Riazuddin S (2012) Nitric oxide augments mesenchymal stem cell ability to repair liver fibrosis. J. Transl. Med. 10:75-5876-10-75.

American Cancer Society (2014) Cancer facts and figures 2014. American Cancer Society, Atlanta

Bataller R, Brenner DA (2005) Liver fibrosis. J Clin Invest 115(2):209–218

Cho KA, Lim GW, Joo SY, Woo SY, Seoh JY, Cho SJ, Han HS, Ryu KH (2011) Transplantation of bone marrow cells reduces CCl4 -induced liver fibrosis in mice. Liver Int 31(7):932–939

Civan JM (2013) Cirrhosis. Merck Manual. http://www.merckmanuals.com/professional/hepatic_and_biliary_disorders/fibrosis_and_cirrhosis/cirrhosis.html

Collardeau-Frachon S, Scoazec JY (2008) Vascular development and differentiation during human liver organogenesis. Anat Rec (Hoboken NJ 2007) 291(6):614–627

Crawford JM (2002) Development of the intrahepatic biliary tree. Semin Liver Dis 22(3):213–226

Desmet VJ (1991) Embryology of the liver and intrahepatic biliary tract, and an overview of malformations of the bile duct. In: MacIntyre N, Benhamou JP, Bircher J, Rizzeto M, Rodes J (eds) Oxford textbook of clinical hepatology. Oxford University Press, Oxford, pp 497–517

Ema H, Nakauchi H (2000) Expansion of hematopoietic stem cells in the developing liver of a mouse embryo. Blood 95(7):2284–2288

Fang B, Shi M, Liao L, Yang S, Liu Y, Zhao RC (2004) Systemic infusion of FLK1(+) mesenchymal stem cells ameliorate carbon tetrachloride-induced liver fibrosis in mice. Transplantation 78(1):83–88

Gaia S, Smedile A, Omede P, Olivero A, Sanavio F, Balzola F, Ottobrelli A, Abate ML, Marzano A, Rizzetto M, Tarella C (2006) Feasibility and safety of G-CSF administration to induce bone marrow-derived cells mobilization in patients with end stage liver disease. J Hepatol 45(1):13–19

Garg V, Garg H, Khan A, Trehanpati N, Kumar A, Sharma BC, Sakhuja P, Sarin SK (2012) Granulocyte colony-stimulating factor mobilizes CD34(+) cells and improves survival of patients with acute-on-chronic liver failure. Gastroenterology 142(3):505–512.e1

Houlihan DD, Newsome PN (2008) Critical review of clinical trials of bone marrow stem cells in liver disease. Gastroenterology 135(2):438–450

Iwamoto T, Terai S, Hisanaga T, Takami T, Yamamoto N, Watanabe S, Sakaida I (2013) Bone-marrow-derived cells cultured in serum-free medium reduce liver fibrosis and improve liver function in carbon-tetrachloride-treated cirrhotic mice. Cell Tissue Res 351(3):487–495

Jang YY, Collector MI, Baylin SB, Diehl AM, Sharkis SJ (2004) Hematopoietic stem cells convert into liver cells within days without fusion. Nat Cell Biol 6(6):532–539

Khan AA, Parveen N, Mahaboob VS, Rajendraprasad A, Ravindraprakash HR, Venkateswarlu J, Rao SG, Narusu ML, Khaja MN, Pramila R, Habeeb A, Habibullah CM (2008) Safety and efficacy of autologous bone marrow stem cell transplantation through hepatic artery for the treatment of chronic liver failure: a preliminary study. Transplant Proc 40(4):1140–1144

Khan AA, Shaik MV, Parveen N, Rajendraprasad A, Aleem MA, Habeeb MA, Srinivas G, Raj TA, Tiwari SK, Kumaresan K, Venkateswarlu J, Pande G, Habibullah CM (2010) Human fetal liver-derived stem cell transplantation as supportive modality in the management of end-stage decompensated liver cirrhosis. Cell Transplant 19(4):409–418

Kharaziha P, Hellstrom PM, Noorinayer B, Farzaneh F, Aghajani K, Jafari F, Telkabadi M, Atashi A, Honardoost M, Zali MR, Soleimani M (2009) Improvement of liver function in liver cirrhosis patients after autologous mesenchymal stem cell injection: a phase I-II clinical trial. Eur J Gastroenterol Hepatol 21(10):1199–1205

Kim WR, Kremers WK (2008) Benefits of "the benefit model" in liver transplantation. Hepatology (Baltimore, Md.) 48(3):697–698

Kisseleva T, Gigante E, Brenner DA (2010) Recent advances in liver stem cell therapy. Curr Opin Gastroenterol 26(4):395–402

Lin H, Zhang Z, Shi M, Xu RN, Fu JL, Geng H, Li YY, Yu SJ, Chen LM, Lv S, Wang FS (2012) Prospective controlled trial of safety of human umbilical cord derived-mesenchymal stem cell transplantation in patients with decompensated liver cirrhosis. Zhonghua Gan Zang Bing Za Zhi 20(7):487–491

Lorenzini S, Isidori A, Catani L, Gramenzi A, Talarico S, Bonifazi F, Giudice V, Conte R, Baccarani M, Bernardi M, Forbes SJ, Lemoli RM, Andreone P (2008) Stem cell mobilization and collection in patients with liver cirrhosis. Aliment Pharmacol Ther 27(10):932–939

Meier RP, Mahou R, Morel P, Meyer J, Montanari E, Muller YD, Christofilopoulos P, Wandrey C, Gonelle-Gispert C, Buhler LH (2015) Microencapsulated human mesenchymal stem cells decrease liver fibrosis in mice. J Hepatol 62(3):634–641

Mohamadnejad M, Alimoghaddam K, Mohyeddin-Bonab M, Bagheri M, Bashtar M, Ghanaati H, Baharvand H, Ghavamzadeh A, Malekzadeh R (2007a) Phase 1 trial of autologous bone marrow mesenchymal stem cell transplantation in patients with decompensated liver cirrhosis. Arch Iran Med 10(4):459–466

Mohamadnejad M, Namiri M, Bagheri M, Hashemi SM, Ghanaati H, Zare Mehrjardi N, Kazemi Ashtiani S, Malekzadeh R, Baharvand H (2007b) Phase 1 human trial of autologous bone marrow-hematopoietic stem cell transplantation in patients with decompensated cirrhosis. World J Gastroenterol 13(24):3359–3363

Nikeghbalian S, Pournasr B, Aghdami N, Rasekhi A, Geramizadeh B, Hosseini Asl SM, Ramzi M, Kakaei F, Namiri M, Malekzadeh R, Vosough Dizaj A, Malek-Hosseini SA, Baharvand H (2011) Autologous transplantation of bone marrow-derived mononuclear and CD133(+) cells in patients with decompensated cirrhosis. Arch Iran Med 14(1):12–17

Oyagi S, Hirose M, Kojima M, Okuyama M, Kawase M, Nakamura T, Ohgushi H, Yagi K (2006) Therapeutic effect of transplanting HGF-treated bone marrow mesenchymal cells into CCl4-injured rats. J Hepatol 44(4):742–748

Pai M, Zacharoulis D, Milicevic MN, Helmy S, Jiao LR, Levicar N, Tait P, Scott M, Marley SB, Jestice K, Glibetic M, Bansi D, Khan SA, Kyriakou D, Rountas C, Thillainayagam A, Nicholls JP, Jensen S, Apperley JF, Gordon MY, Habib NA (2008) Autologous infusion of expanded mobilized adult bone marrow-derived CD34+ cells into patients with alcoholic liver cirrhosis. Am J Gastroenterol 103(8):1952–1958

Park M, Kim YH, Woo SY, Lee HJ, Yu Y, Kim HS, Park YS, Jo I, Park JW, Jung SC, Lee H, Jeong B, Ryu KH (2015) Tonsil-derived Mesenchymal Stem Cells Ameliorate CCl4-induced Liver Fibrosis in Mice via Autophagy Activation. Sci Rep 5:8616

Sadler TW, Langman J (2010) Langman's medical embryology, 11th edn. Wolters Kluwer, Lippincott Williams & Wilkins, Philadelphia

Salama H, Zekri AR, Bahnassy AA, Medhat E, Halim HA, Ahmed OS, Mohamed G, Al Alim SA, Sherif GM (2010) Autologous CD34+ and CD133+ stem cells transplantation in patients with end stage liver disease. World J Gastroenterol 16(42):5297–5305

Schuppan D, Afdhal NH (2008) Liver cirrhosis. Lancet 371(9615):838–851

Shams S, Mohsin S, Nasir GA, Khan M, Khan SN (2015) Mesenchymal stem cells pretreated with HGF and FGF4 can reduce liver fibrosis in mice. Stem Cells Int 2015:747245

Tan CE, Driver M, Howard ER, Moscoso GJ (1994) Extrahepatic biliary atresia: a first-trimester event? Clues from light microscopy and immunohistochemistry. J Pediatr Surg 29(6):808–814

Terai S, Sakaida I, Yamamoto N, Omori K, Watanabe T, Ohata S, Katada T, Miyamoto K, Shinoda K, Nishina H, Okita K (2003) An in vivo model for monitoring trans-differentiation of bone marrow cells into functional hepatocytes. J Biochem 134(4):551–558

Terai S, Ishikawa T, Omori K, Aoyama K, Marumoto Y, Urata Y, Yokoyama Y, Uchida K, Yamasaki T, Fujii Y, Okita K, Sakaida I (2006) Improved liver function in patients with liver cirrhosis after autologous bone marrow cell infusion therapy. Stem Cells (Dayton, Ohio) 24(10):2292–2298

Vassilopoulos G, Wang PR, Russell DW (2003) Transplanted bone marrow regenerates liver by cell fusion. Nature 422(6934):901–904

Wang X, Willenbring H, Akkari Y, Torimaru Y, Foster M, Al-Dhalimy M, Lagasse E, Finegold M, Olson S, Grompe M (2003) Cell fusion is the principal source of bone-marrow-derived hepatocytes. Nature 422(6934):897–901

Zhang Z, Lin H, Shi M, Xu R, Fu J, Lv J, Chen L, Lv S, Li Y, Yu S, Geng H, Jin L, Lau GK, Wang FS (2012) Human umbilical cord mesenchymal stem cells improve liver function and ascites in decompensated liver cirrhosis patients. J Gastroenterol Hepatol 27(Suppl 2):112–120

Zheng JF, Liang LJ (2008) Intra-portal transplantation of bone marrow stromal cells ameliorates liver fibrosis in mice. Hepatobiliary Pancreat Dis Int 7(3):264–270

Chapter 4
Mesenchymal Stem Cell Therapy for Liver Cirrhosis Treatment: Mechanisms and Bioeffects

Nhung Hai Truong and Phuc Van Pham

4.1 Introduction

The liver plays a vital role as a metabolism machinery with the following main functions: absorption of nutrients, elimination of toxins, energy storage, control of blood sugar, and production of protein (e.g., bile and blood clotting factors, transport proteins, etc.). The liver is the only organ in the body that can easily regenerate damaged cells. Notably, the liver shows limited cell turnover, but as soon as cell damage occurs, the regenerative process is enhanced rapidly to recover and maintain organ function (Alison et al. 2009). Although the liver compensatory regeneration is a rapid and efficient process, sometimes liver cells may not able to be restored, as in the case with massive hepatocyte damage, in which case liver failure occurs. Though there are many causes of chronic liver disease, the consequences are the same. Liver injury may cause liver cirrhosis that leads to reduction of liver function and induction hepatocellular carcinoma (HCC) (Forbes 2009; Stefan et al. 2015).

Cirrhosis is defined as the replacement of healthy liver tissues by fibrosis and regenerative nodule formation. Cirrhosis ranks 14th in the world and 4th in Central Europe as cause of death (Tsochatzis et al. 2014). Currently, cirrhosis is listed among the 25 diseases with the highest mortality worldwide. There are many different causes leading to liver cirrhosis; these include viral hepatitis, alcohol, cholestatic phenomenon, and toxic chemicals. In particular, hepatitis B virus (HBV)-associated disease causes the highest mortality rate, up to 4.8/100,000 people (Moon et al. 2009; Murray Christopher et al. 2012).

N.H. Truong • P. Van Pham (✉)
Laboratory of Stem Cell Research and Application, University of Science, Vietnam National University, Ho Chi Minh City, Vietnam
e-mail: pvphuc@hcmuns.edu.vn

© Springer International Publishing AG 2017
P.V. Pham (ed.), *Liver, Lung and Heart Regeneration*,
Stem Cells in Clinical Applications, DOI 10.1007/978-3-319-46693-4_4

Cirrhosis is characterized by inflammation and necrosis induced by Kupffer cells and activation of hepatic satellite cells (HeSCs). Kupffer cells and HeSCs are the main types of cells which cause degeneration of liver tissue (GI-PPEUM et al. 2005). Chronic liver injury leads to the activation of HeSCs, the production of extracellular matrix components, and the secretion of proinflammatory cytokines, chemokines and growth factors, such as transforming growth factor beta (TGF-ß) (Friedman 2008). Metalloproteinase and its regulators (tissue inhibitors of metalloproteinases; TIMPs) control the deposition and degradation of the extracellular matrix (Stefan et al. 2015).

Presently, orthotopic liver transplantation (OLT) is the priority treatment for decompensated liver cirrhosis. In most clinical studies, for many years, the percentage of surviving patients ranged from 70 to 90 % after 1 year of OLT (Abbasoglu et al. 1997; Botha et al. 2000; Busuttil et al. 1994; Sudan et al. 1998). However, patients face many obstacles, such as the high costs of OLT, high risk from invasive surgery, limitation of donor tissue, and lifelong immunosuppressive treatment (Lorenzini and Andreone 2007; Zheng and Wang 2013). Therefore, hepatocyte transplantation has emerged as a more promising alternative therapy to OLT. Hepatocyte transplantation therapy is a non-invasive method, does not require high immunity in conformity, and has full function of mature hepatocyte which accelerates recover of liver. Hepatocyte transplantation has the potential to overcome obstacles of OLT (Nussler et al. 2006). Although hepatocyte transplantation began in the 1980s, the use of therapeutic stem cells for liver cell transplantation is a recent shift.

4.2 Liver Regeneration: Underlying Fundamentals of Stem Cell Application in Liver Disease Treatment

Studies of liver regeneration capacity were first demonstrated by partial hepatectomy (PH) in rats and mice (Zhao et al. 2009). After 10 days of PH, one-third of rat/mouse liver tissue that remained after the PH procedure could regenerate and fill the empty liver mass. The PH model demonstrated the mechanism of liver generation (Alison et al. 2009; Kang et al. 2012). It is possible that many cells are involved in liver regeneration after PH. However, scientists suggest two main mechanisms related directly to liver regeneration and restoration of liver mass. Firstly, although hepatocytes are differentiated cells, during liver injury they can activate themselves; the mechanism is called self-proliferation (Zhao et al. 2009). Many studies have shown that liver stem cells can be elevated in liver damage. These stem cells can differentiate into liver cells and biliary epithelial cells and are called bipotent progenitor cells or liver progenitor cells (Thorgeirsson 1996; Zhao et al. 2009). In rodents, these progenitor cells are called "oval cells." The repopulation of liver from liver progenitor cells is a process known as trans-differentiation (Kang et al. 2012; Zhao et al. 2009). Secondly, in addition to cells/stem cells, molecular signals play an important role in liver generation (Kang et al. 2012). Experiments in rats demonstrated the important role of circulating growth factors, such as hepatocyte growth factor (HGF) and transforming growth factor alpha (TGF-α), which are involved in hepatocyte proliferation (Kang et al. 2012). Together with circulating growth factors, cell–cell interaction and cell-matrix reorganization also play vital roles in liver regeneration (Kang et al. 2012). For instance, urokinase-type plasminogen activator (uPA) is released soon after PH and induces the reorganization of extracellular matrix via activation of metalloproteinases (MMP) and HGF (Kang et al. 2012; Kim et al. 2000). Overall, the liver regeneration process is dependent on mechanisms that relate to liver cells/progenitor cells and molecular signals (Alison et al. 2009; Kang et al. 2012).

Along with the understanding of liver regeneration, a greater understanding of stem cell biology could advance the application of stem cells for liver cirrhosis treatment. Stem cells are defined as cells with self-renewal capability and potential to differentiate into specific cell types. Because of these features, stem cells can provide considerably beneficial for use in liver cirrhosis therapy. First, self-stem cells can generate numerous cells. In treating liver disease, a major problem is limited cell quantity; use of stem cells could overcome this obstacle. Second, many studies have revealed that stem cells can differentiate into hepatocyte like cells (Anna et al. 2010; Forbes and Newsome 2012) (Bagher et al. 2012; Zhang et al. 2012; Zhao et al. 2009). In vivo studies have also demonstrated the great potential of stem cells to ameliorate liver disease. Stem cells not only can retain their ability to renew and rejuvenate new cells in response to injury, but they also play a key role in many important physiological processes (Anna et al. 2010). In the case of liver cirrhosis, the use of targeted drugs can help reduce injury but cell therapy can overturn damage by supplying new functional cells. Compared to the mechanism of liver regeneration in PH, stem cell therapy can cure liver disease by similar means. The consideration of therapeutic mechanism of MSCs in liver cirrhosis, the kinds of stem cells that should be used, and strategies for liver cirrhosis treatment continue to be debated and controversial.

4.3 Mesenchymal Stem Cells

Stromal bone marrow cells were first described by Friedenstein in 1968. This cell population was fibroblast like in shape, adherent to plastic, and able to differentiate into bone. Friedenstein considered these cells to be osteogenic precursor cells (Friedenstein et al. 1968). Thereafter, many studies indicated that these cells could differentiate into many cell types of mesoderm lineage, including cartilage, tendon, and adipocyte. Based on their differentiation potential, in 1991, Caplan defined the term of mesenchymal stem cells (MSC) (AI 1991). Recently, the term of mesenchymal stromal cells was suggested to describe the population of cells capable of plastic adherence in vitro and showing resemblance to fibroblasts (Horwitz et al. 2005). Bonnet et al. have demonstrated that single cells from mouse bone marrow MSC population express embryonic specific antigens (Anjos-Afonso and Bonnet 2007), therefore these cells have properties of embryonic stem cells.

Although the origin of MSCs was from bone marrow (Pittenger et al. 1999), recently MSCs have been isolated also from adipose tissue, umbilical cord blood, placenta, amniotic fluid, etc. (In't Anker et al. 2003a, b, 2004; Zuk et al. 2001). Due to the variety of sources, different laboratories have different protocols for the isolation and characterization of MSCs, based on the source of cells and on specific properties. In 2006, the International Society of Cellular Therapy introduced the criteria for MSC characterization including: (1) MSCs must have capacity of plastic adherence in vitro; (2) MSCs must express specific surface markers, such as CD105 (SH2 or endoglin), CD73 (SH3 or SH4), CD90, CD166, CD44, and CD29 (Deans and Moseley 2000; Pittenger et al. 1999), but lack expression of hematopoietic and endothelial markers, such as CD11b, CD14, CD31, and CD45 (Pittenger et al. 1999); and (3) MSCs must have differentiation capacity into mesoderm lineage cells, such as osteoblasts, chondroblasts, and adipocytes.

Recent studies show that the MSCs differentiate not only into mesoderm lineage but also into endoderm and neural ectoderm lineage, which include nerve cells, liver cells, and endothelial cells (AI 1991; Sanchez-Ramos et al. 2000; Schwartz et al. 2002). In addition to their differentiation capability, MSCs can migrate to injury sites in response to homing signals. The mechanism of MSC migration resembles that of leukocyte migration. In brief, surface markers of MSCs, such as CD44 receptors, CCR1, c-met, and CXCR4, have affinity to chemokines, growth factors (MIP1a, HGF, and SDF-1), and hypoxic signals at injury sites. MSCs migrate in response to chemotaxis. In the blood vessels, MSCs attach to VCAM-1 receptors via VLA-4 receptors. The transmigration is facilitated by MT1-MMP factors and MMP2. The homing capacity is very essential in regenerative medicine since physicians have a variety options for transplantation routes. For instance, MSCs can be transplanted into liver via intravenous, portal venous, intrahepatic, or intrasplenic injections.

Furthermore, other important characteristics of MSC are their immunosuppressive and anti-inflammatory properties, which have been reported from in vitro and in vivo studies. MSCs have low expression of major histocompatibility complex (MHC) class I and non-expression of MHC class II, Fas ligand, and co-stimulatory molecules (CD40, CD80, and CD86). MSCs avoid identification and rejection by the immune system, and they also inhibit immune responses through the regulation of immune cells. Regarding their effects on the innate immune system, MSCs can inhibit the release of tumor necrosis factor (TNF-α) and interleukin (IL)-12, as well enhancing IL-10 secretion by dendritic cells (DCs). MSCs also prohibit nature killer (NK) cells and cause the reduction of interferon gamma (IFN-γ) secretion by NK cells (Aggarwal and Pittenger 2005; Bartholomew et al. 2002; Beyth et al. 2005; Le Blanc and Ringdén 2007; Tse et al. 2003). With regard to the adaptive immune system, MSCs can inhibit the proliferation of T cells; both naïve T cells and memory CD4+ and CD8+ T cells are inhibited by MSCs. B cell proliferation is also inhibited by MSCs. Additionally, MSCs are known to release variety of soluble factors, which are important for the immunomodulatory properties of MSCs. These factors include HGF, TFG-β1, IL-10, prostaglandin-E2, nitric oxide, human leukocyte antigen-G5, and indoleamine 2,3-dioxygenase. Other intrinsic properties of MSCs are that they are tractable cells, making them easy to modify in vitro to accelerate protein expression, to promote cell proliferation, and to orient differentiation towards a specific type of cell. Their broad properties make them ideal for use in regenerative medicine (Hodgkinson et al. 2010; Soland et al. 2012; Phillips and Tang 2008).

4.4 Mechanism of Liver Cirrhosis Treatment Using Mesenchymal Stem Cells

4.4.1 MSCs Differentiate into Hepatocytes In Vitro and In Vivo

Studies have shown that MSCs can be derived from bone marrow (Li et al. 2010b; Wang et al. 2010; Mohsin et al. 2011; Pournasr et al. 2011; Shi et al. 2008), cord blood (Moon et al. 2009), and adipose tissue (Abbas Sahebghadam 2012; Harn et al. 2012; Lue et al. 2010; Puglisi et al. 2011a; Ruiz 2011; Sgodda et al. 2007; Sterodimas et al. 2010), and are able to differentiate into hepatocytes in vitro. The use of growth factors, such as epithelial growth factor (EGF) and HGF, and other factors, such as oncostatin M (OSM), dexamethasone, and insulin-transferrin-selenite (ITS), can induce MSCs to become hepatocytes (Ruiz 2011; Shi et al. 2008). Talens-Visconti et al. showed that MSCs derived from bone marrow and adipose tissue can be differentiated into hepatocytes with the provision of HGF, FGF, dexamethasone, ITS, and OSM (Talens-Visconti et al. 2006). MSCs alter morphology and function after 21 days of differentiation (Talens-Visconti et al. 2006). Moreover, Agnieszka Banas et al. also differentiated MSCs into hepatocytes using dexamethasone and ITS (Banas et al. 2007) (Fig. 4.1).

In vivo studies, such as those as reported by Kyung Hee Jung et al. (2009), demonstrate that MSCs (e.g., umbilical cord blood derived) can be labeled with CM-Dil dye and injected in vivo to study liver fibrosis; cells were injected in a mouse model of liver fibrosis via tail vein and grafted in the liver after 4 weeks. Significantly, the labeled cells were positive for albumin and alpha-fetoprotein (AFP) (Jung et al.

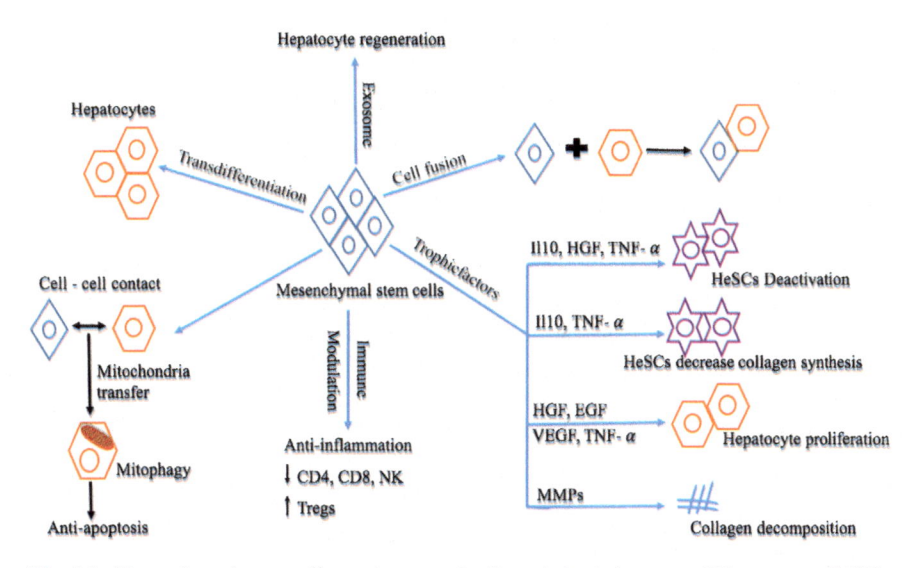

Fig. 4.1 Mesenchymal stem cells can improve the liver cirrhosis by some different ways. MSCs can be differentiated into hepatocytes, fused with hepatocytes, inhibited the apoptosis of hepatocytes and the inflammation, and stimulated the regenerative process by some trophic factors

2009). It is claimed that human MSCs from bone marrow and umbilical cord blood can integrate in mouse liver and express human albumin (Ren et al. 2010) (Wang et al. 2003). Similarly, bone marrow derived MSCs engrafted in the liver and showed expression of alpha-1 antitrypsin (AAT) and albumin protein after 14 days of transplantation (Li et al. 2010a). Overall, the outcomes of in vivo studies have indicated that MSCs express several liver markers, including AAT, AFP, and albumin (Jung et al. 2009; Li et al. 2010a; Ren et al. 2010; Wang et al. 2003). However, the trans-differentiation of MSCs into hepatocytes in vivo has not correlated with in vitro data. In fact, the quantity of trans-differentiated MSCs (into hepatocytes) was less than 1 % of the total liver mass in animal models. Therefore, the principal therapeutic properties of MSCs that are relevant to liver cirrhosis may be associated with their paracrine/trophic effects, secretory properties, immunomodulatory properties, anti-fibrotic features, ability to deactivate hepatic stellate cells, and role in exosome/cell fusion and mitochondrial transfer (Tan et al. 2014; Eom et al. 2015; Liang et al. 2014).

4.4.2 *Immune Modulation of MSCs Ameliorates Inflammation*

Liver injury is caused by persistent inflammation associated with T cells, B cells, and monocyte infiltration to the liver (Kisseleva et al. 2006). It has been demonstrated that immunosuppressive therapy can prevent the recurrence of liver diseases, both before and after OLT (Mehdi et al. 2005; Pinelopi et al. 2010). Moreover, the reduction of inflammation may be beneficial for liver regeneration in acute liver failure (Antonino et al. 2011; Zhen Fan et al. 2007). In this approach, the immunomodulatory and immunosuppressive properties of MSC may play an important role in liver diseases. MSCs are not immunogenic cells; indeed, MSCs show low MHC-I expression and no expression of MHC-II and co-stimulatory molecules (CD80, CD86, or CD40) on their surface, and therefore do not trigger T cell responses (Sandra et al. 2013). Moreover, MSCs have immunosuppressive abilities; they can inhibit naïve, memory, and cytotoxic T cells (Antonio et al. 2008). In vitro studies by Nicola Di et al. lend support for their immunosuppressive properties; MSCs were shown to reduce cell proliferation of CD4 and CD8 T cells (Di Nicola et al. 2002). Additionally, Aggarwal and Pittenger demonstrated that MSCs secrete prostaglandin E2 to trigger IL-10 secretion by DCs (Sudeepta and Mark 2005). Moreover, MSCs mediate the induction of regulatory T cells (Tregs) and decrease TNF-α (produced by dendritic cells), interferon-γ (IFN-γ) (produced by T helper 1 cells), and IL-4 (produced by T helper 2 cells) (Sudeepta and Mark 2005). The immunosuppressive properties of MSCs have also been confirmed in preclinical studies (Bartholomew et al. 2002). In clinical studies, MSC-mediated immune suppression led to successful treatment of graft versus host disease (Le Blanc and Ringdén 2005). A dose of 1.4×10^6 MSCs was injected in 55 patients; improvement was observed in 71 % of the patients (Deeg 2007). Studies have also shown that MSC can participate in reducing the leukocyte penetration in acute and chronic liver disease, following liver transplantation.

4.4.3 Trophic Factor Secretion/Paracrine Effects of MSCs

MSCs produce a variety of trophic factors, including cytokines, growth factors, and chemokines, which have biological effects in liver regeneration and anti-fibrosis. These trophic factors are known not only to reduce collagen synthesis, apoptosis, and fibrosis, but also to deactivate HeSCs. MSCs can release trophic factors, such as HGF, IL-6, IL-10, TGF-β3, and TNF-α (Li et al. 2013). These factors play an important role in inhibiting the proliferation of HeSCs and reducing collagen synthesis in the liver through inhibition of cell signaling related to cirrhosis. For instance, HGF can inhibit alpha-actin formation of smooth muscle cells (Kim et al. 2005; Li et al. 2013).

Moreover, MSCs also secrete potential mitogens, such as EGF, TGF-α, and VEGF, which accelerate the hepatocyte proliferation. In fact, it has recently claimed that MSCs transplantation is associated with an increase of insulin-like growth factor 1 (IGF-1) and HGF in the fibrotic liver, and followed by anti-inflammation. MSCs can also secrete matrix metalloproteinases (MMPs) and tissue inhibitors of MMPs (TIMPs) (Li et al. 2013). MMPs (e.g., MMP-2, -9, -13, and -14) play an important role in the reconstruction of the extracellular matrix (Kang et al. 2012). MMPs and TIMPs cause collagen decomposition directly through the resolution of the extracellular matrix.

4.4.4 Cell–Cell Contact, Mitochondria Transfer, and Exosome Secretion Properties of MSCs

Recently, some studies have investigated the fusion of MSCs with neighboring cells. It is thought that cell–cell contact mechanisms may cause cell fusion between MSCs and hepatocytes, thereby contributing to the effectiveness of MSC-based therapy. Although, no in vivo studies have proven MSC fusion with hepatocytes after liver fibrosis, Acquistapace et al. predict that the fusion mechanism process might resemble that for MSC/cardiomyocyte fusion (Acquistapace et al. 2011). From in vitro studies, direct coculture of hepatocytes with MSCs (from umbilical cord blood or adipose tissue) improved hepatocyte viability and function; after coculture with MSCs, hepatocytes showed improved production of urea and albumin (Fitzpatrick et al. 2015; Isoda et al. 2004). Therefore, cell–cell contact may have a role in supporting hepatocyte growth (Fitzpatrick et al. 2015).

In addition, analysis of conditioned medium of MSCs revealed that microparticles or exosomes may have beneficial effects on tissue regeneration in the heart, kidney, and liver. It is reported that exosome treatment attenuates liver recovery in mice after toxicant-induced injuries (Tan et al. 2014). Interestingly, exosome of MSCs may transfer RNA to target cells. MicroRNA in these exosomes may have a role in the therapeutic effect. Moreover, bioactive molecules in exosomes have been identified; these include IL-10, TNF-α-stimulated gene/protein 6 (TSG-6), stanniocalcin-1, and TNF-α. These proteins may promote tissue repair by anti-inflammatory and anti-oxidant signals (Fung and Thebaud 2014).

Recently, it was uncovered that MSCs may transfer their mitochondria into target cells via tunneling nanotubes. This mechanism stimulates endogenous stem cells to "wake up" and amplify their response to tissue injury (Fung and Thebaud 2014). By transferring mitochondria DNA to cells that have been damaged or that lack functional mitochondria, MSCs can rescue injured cells. In fact, in a report presented in the 2015 Joint Congress of IPITA, IXA, and CTS, Raquel Fernandez Dacosta et al. (2015) revealed that mitochondria transfer from human MSCs to human hepatocytes via tunneling nanotubes plays a major role for hepatocyte survival through direct cell–cell contact. The mitochondria transfer from MSCs to hepatocytes is facilitated by actin-based tunneling nanotubes. More in vivo studies are underway to further elucidate these mechanisms.

4.5 MSC-Based Therapy for Liver Cirrhosis: From Bench to Bedside

Theoretically, MSCs can contribute to fibrogenesis since MSCs highly respond to TGF-β and platelet-derived growth factor (PDGF), which are both highly expressed in fibrotic liver (Aquino et al. 2010). TGF-β accelerates myofibroblast differentiation in MSCs while PDGF enhances migration of MSCs, serving as a chemoattractant for MSCs (Dennis and Charbord 2002; Ozaki et al. 2007; Ponte et al. 2007). In fact, MSCs recruited to the fibrotic liver change into myofibroblasts, as revealed by studies by di Bonzo et al. (2008) and Li et al. (2009a). In addition, secreted protein acidic and rich in cysteine (SPARC), which is overexpressed in liver cirrhosis, is specifically expressed in HeSCs in both animal models and patients, also appear in MSCs with significant level (Kulterer et al. 2007; Silva et al. 2003). It is suggested that MSCs might play a role in fibrogenesis mechanism.

Contrarily to what could be anticipated, MSCs transplantation revealed antifibrotic in animal models (Table 4.1). As a matter of fact, many sources of MSCs have been studied for the liver cirrhosis, including MSCs from bone marrow (Nasir et al. 2013; Guilherme et al. 2011; Hiroyuki et al. 2011; Park et al. 2011; Lian et al. 2014; Linhua Zheng et al. 2013; Meier et al. 2015; Mohamadnejad et al. 2007; Puglisi et al. 2011b; Volarevic et al. 2014; Jang et al. 2014), adipose tissue stem cells (Koellensperger et al. 2013; Puglisi et al. 2011a; Saito et al. 2014; Wang et al. 2012; Wilson et al. 2011), umbilical cord blood (Briquet et al. 2014; Hong et al. 2014; Jung et al. 2009; Li et al. 2009b), etc., aimed to reduce the demand of liver transplantation. MSC transplantation induced the recover faster and more effectively in animal models of liver cirrhosis. MSCs restore AST, ALT, albumin, bilirubin level, the expression of the fibrotic genes, and liver function. The underlying mechanism and the effect of MSCs in liver cirrhosis have been elucidated. Preclinical studies are kept increasing day by day.

Despite promising outcomes from preclinical studies, clinical trials are still limited. The results from 7 clinical trials, during 2011–2014, which used autologous MSCs for end-stage liver cirrhosis revealed that MSC therapy is safe and efficient,

Table 4.1 Preclinical studies using MSC-based therapy for liver cirrhosis treatment

Source of MSC	Model	Biological effects	References
Mouse BM	CCl_4	BMP7 secretion	Li et al. (2015)
Mouse BM	CCl_4	Fibrogenesis- and ECM-related genes down-regulation	Nhung et al. (2016)
Human BM	CCl_4/BDL	IL-6, IGFBP-2, IL-1Ra releasing	Meier et al. (2015)
Mouse BM	CCl_4	iNOS activation in MSC	Chen et al. (2014)
Mouse BM	CCl_4	EGFR, MMP9, IL-1Ra production	Huang et al. (2013)
Mouse BM	CCl_4	Dlk1 secretion	Pan et al. (2011)
Rat adipose tissue	CCl_4	VEGF down-regulation	Wang et al. (2012)
Human adipose tissue	CCl_4	Secretion of interleukin 1 receptor α (IL-1Rα), IL-6, IL-8, granulocyte colony-stimulating factor (G-CSF), granulocyte-macrophage colony-stimulating factor (GM-CSF), monocyte chemotactic protein 1, nerve growth factor, and hepatocyte growth factor	Banas et al. (2008)
Human umbilical cord blood	Diethylnitrosamine (DEN)	Breakdown of collagen fibers	Hong et al. (2014)

and restores liver function. In the aforementioned trials, a total of 256 patients were treated with a dose of 10^6–8×10^8 cells/kg MSCs and a variety of transplantation routes were used, including intravenous, hepatic artery, portal venous, etc. A number of clinical trials using MSCs to evaluate their potential efficacy for treatment of liver cirrhosis disease have been conducted worldwide (Table 4.2). Although recent studies have shown positive effects of MSC therapy for end-stage liver disease, many controversial issues remain, such as the optimal types of stem cells and optimal transplantation route. Moreover, it is unclear whether MSC is best administered alone, or in combination with growth factors.

4.6 Conclusion

MCSs have great potential for the treatment of liver cirrhosis. MSCs can improve cirrhotic liver by four different mechanisms. These include: (1) MSC differentiation into hepatocytes in vitro and in vivo, (2) immune modulation of MSCs which ameliorate inflammation, (3) trophic factor secretion and paracrine effects of MSCs, and

Table 4.2 Clinical trials using MSC-based therapy for liver cirrhosis treatment

Source of MSC	Diseases	Phase	Number of patients	Country	ID
Adipose tissue	Liver cirrhosis	–	4	Japan	NCT01062750
	Liver cirrhosis	1	6	Taiwan	NCT02297867
Bone marrow	Alcoholic liver cirrhosis	2	11	Korea	NCT01741090
	Liver failure caused by chronic hepatitis B virus	1/3	Treatment: 53 Control: 105	China	NCT00956891
	Liver cirrhosis	1/2	8	Iran	NCT00420134
Umbilical cord blood	Liver cirrhosis	1/2	Treatment: 30 Control: 15	China	NCT01220492
	Liver failure	1/2	Treatment: 24 Treatment: 19	China	NCT01218464
	HBV related liver cirrhosis	1/2	240	China	NCT01728727
	Decompensated liver cirrhosis	1/2	20	China	NCT01342250
	Hepatic cirrhosis	1	20	China	NCT02652351

(4) cell–cell contact, mitochondria transfer, and exosome secretion properties of MSCs. Paracrine effects, mitochondria transfer, and exosome secretion may be main role of MSCs in liver disease treatment. Concern of liver cirrhosis, it has been believed that effective therapies cause a change of liver micro-milieu which plays crucial role in treatment. Scientists should concentrate on paracrine effects of MSCs to learn more about secretion products of these cells. Many evidences suggest that secretion products of MSC can be enhanced by preconditioning and genetic induction. Paracrine secretion of MSCs implies using of cocktail trophic factors for treatment. Furthermore, mitochondria transfer mechanism is attractive issue to further elucidation. Likewise, preclinical and clinical trials using MSC transplantation should be elevated to prove the potential of MSCs for cirrhosis treatment.

References

Abbas Sahebghadam L (2012) High yield generation of hepatocyte like cells from adipose derived stem cells. Sci Res Essays 7:1141–1147

Abbasoglu O, Levy MF, Brkic BB, Testa G, Jeyarajah DR, Goldstein RM, Husberg BS, Gonwa TA, Klintmalm GB (1997) Ten years of liver transplantation: an evolving understanding of late graft Loss1. Transplantation 64:1801–1807

Acquistapace A, Bru T, Lesault PF, Figeac F, Coudert AE, le Coz O, Christov C, Baudin X, Auber F, Yiou R, Dubois-Randé JL, Rodriguez AM (2011) Human mesenchymal stem cells reprogram adult cardiomyocytes toward a progenitor-like state through partial cell fusion and mitochondria transfer. Stem Cells 29:812–824

Aggarwal S, Pittenger MF (2005) Human mesenchymal stem cells modulate allogeneic immune cell responses. Blood 105:1815–1822

Ai C (1991) Mesenchymal stem cells. J Orthop Res 9:641–650
Alison MR, Islam S, Lim S (2009) Stem cells in liver regeneration, fibrosis and cancer: the good, the bad and the ugly. J Pathol 217:282–298
Anjos-Afonso F, Bonnet D (2007) Non-hematopoietic/endothelial SSEA-1pos cells defines the most primitive progenitors in the adult murine bone marrow mesenchymal compartment. Blood 109:1298–1306
Anna CP, Campanale M, Gasbarrini A, Gasbarrini G (2010) Stem cell-based therapies for liver diseases: state of the art and new perspectives. Stem Cells Int 2010:10
Antonino S, Carmen G-G, Philippe M, Baertschiger RM, Niclauss N, Mentha G, Majno P, Serre-Beinier V, Buhler L (2011) Interleukin-1 receptor antagonist modulates the early phase of liver regeneration after partial hepatectomy in mice. PloS One 6:e25442
Antonio U, Lorenzo M, Vito P (2008) Mesenchymal stem cells in health and disease. Nat Rev Immunol 8:726–736
Aquino JB, Bolontrade MF, Garcia MG, Podhajcer OL, Mazzolini G (2010) Mesenchymal stem cells as therapeutic tools and gene carriers in liver fibrosis and hepatocellular carcinoma. Gene Ther 17:692–708
Bagher L, Ensieh NE, Peyvand A (2012) Stem cell therapy in treatment of different diseases. Acta Med Iran 50:79–96
Banas A, Teratani T, Yamamoto Y, Tokuhara M, Takeshita F, Quinn G, Okochi H, Ochiya T (2007) Adipose tissue-derived mesenchymal stem cells as a source of human hepatocytes. Hepatology 46:219–228
Banas A, Teratani T, Yamamoto Y, Tokuhara M, Takeshita F, Osaki M, Kawamata M, Kato T, Okochi H, Ochiya T (2008) IFATS collection: in vivo therapeutic potential of human adipose tissue mesenchymal stem cells after transplantation into mice with liver injury. Stem Cells 26:2705–2712
Bartholomew A, Sturgeon C, Siatskas M, Ferrer K, McIntosh K, Patil S, Hardy W, Devine S, Ucker D, Deans R, Moseley A, Hoffman R (2002) Mesenchymal stem cells suppress lymphocyte proliferation in vitro and prolong skin graft survival in vivo. Exp Hematol 30:42–48
Beyth S, Borovsky Z, Mevorach D, Liebergall M, Gazit Z, Aslan H, Galun E, Rachmilewitz J (2005) Human mesenchymal stem cells alter antigen-presenting cell maturation and induce T-cell unresponsiveness. Blood 105:2214–2219
Botha J, Spearman C, Millar A, Michell L, Gordon P, Lopez T, Butt A, Thomas J, McCulloch M, James M (2000) Ten years of liver transplantation at Groote Schuur Hospital. SAMJ 90:880–883
Briquet A, Gregoire C, Comblain F, Servais L, Zeddou M, Lechanteur C, Beguin Y (2014) Human bone marrow, umbilical cord or liver mesenchymal stromal cells fail to improve liver function in a model of CCl4-induced liver damage in NOD/SCID/IL-2Rgamma(null) mice. Cytotherapy 16:1511–1518
Busuttil RW, Shaked A, Millis JM, Jurim O, Colquhoun SD, Shackleton CR, Nuesse BJ, Csete M, Goldstein LI, McDiarmid SV (1994) One thousand liver transplants. The lessons learned. Ann Surg 219:490
Chen X, Gan Y, Li W, Su J, Zhang Y, Huang Y, Roberts AI, Han Y, Li J, Wang Y, Shi Y (2014) The interaction between mesenchymal stem cells and steroids during inflammation. Cell Death Dis 5:e1009
Dacosta RF, Lee C, Walker S, Lehec S, Fitzpatrick E, Dhawan A, Filippi C (2015) Investigation of the mitochondrial transfer from human mesenchymal stem/stromal cells to human primary hepatocytes through tunnelling nanotubes as a potential mechanism for cell survival and function enhancement during hepatocyte transplantation. Xenotransplantation 22(S1):S47
Deans RJ, Moseley AB (2000) Mesenchymal stem cells: biology and potential clinical uses. Exp Hematol 28:875–884
Deeg H (2007) How I treat refractory acute GVHD. Blood 109:4119–4126
Dennis JE, Charbord P (2002) Origin and differentiation of human and murine stroma. Stem Cells 20:205–214

di Bonzo LV, Ferrero I, Cravanzola C, Mareschi K, Rustichell D, Novo E (2008) Human mesenchymal stem cells as a two-edged sword in hepatic regenerative medicine: engraftment and hepatocyte differentiation versus profibrogenic potential. Gut 57:223–231

Di Nicola M, Carlo-stella C, Magni M, Milanesi M, Longoni PD, Matteucci P, Grisanti S, Gianni AM (2002) Human bone marrow stromal cells suppress T-lymphocyte proliferation induced by cellular or nonspecific mitogenic stimuli. Blood 99:3838–3843

Eom YW, Shim KY, Baik SK (2015) Mesenchymal stem cell therapy for liver fibrosis. Korean J Intern Med 30:580–589

Fitzpatrick E, Wu Y, Dhadda P, Hughes RD, Mitry RR, Qin H, Lehec SC, Heaton ND, Dhawan A (2015) Coculture with mesenchymal stem cells results in improved viability and function of human hepatocytes. Cell Transplant 24:73–83

Forbes SJ (2009) Stem cell therapy for liver disease. World Stem Cell Report (http://www.worldstemcellsummit.com)

Forbes SJ, Newsome PN (2012) New horizons for stem cell therapy in liver disease. J Hepatol 56:496–499

Friedenstein AJ, Petrakova KV, Kurolesova AI, Frolova GP (1968) Heterotopic of bone marrow. Analysis of precursor cells for osteogenic and hematopoietic tissues. Transplantation 6:230–247

Friedman SL (2008) Mechanisms of hepatic fibrogenesis. Gastroenterology 134:1655–1669

Fung ME, Thebaud B (2014) Stem cell-based therapy for neonatal lung disease: it is in the juice. Pediatr Res 75:2–7

Gi-Ppeum L, Jeong W-I, Jeong D-H, Do S-H, Kim T-H, Jeong K-S (2005) Diagnostic evaluation of carbon tetrachloride-induced rat hepatic cirrhosis model. Anticancer Res 25:1029–1038

Guilherme B, Kretzmann NA, Tieppo J, Filho G.P, Cruz CU, Meurer L, Silveira TRD, Santos JLD, Marroni CUA, Marroni NP et al (2011) Bone marrow cells reduce collagen deposition in the rat model of common bile duct ligation. J Cell Sci Ther 2:112

Harn HJ, Lin SZ, Hung SH, Subeq YM, Li YS, Syu WS, Ding DC, Lee RP, Hsieh DK, Lin PC et al (2012) Adipose-derived stem cells can abrogate chemical-induced liver fibrosis and facilitate recovery of liver function. Cell Transplant 21:2753–2764

Hiroyuki K, Yasuhiro F, Takumi T, Junji I, Naoya K, Kouji N, Tsuruyama T, Uemoto S, Kobayashi E (2011) Bone marrow-derived mesenchymal stem cells ameliorate hepatic ischemia reperfusion injury in a rat model. PloS one 6(4):e19195

Hodgkinson CP et al (2010) Genetic engineering of mesenchymal stem cells and its application in human disease therapy. Hum Gene Ther 21:1513–1526

Hong J, Jin H, Han J, Hu H, Liu J, Li L, Huang Y, Wang D, Wu M, Qiu L et al (2014) Infusion of human umbilical cord derived mesenchymal stem cells effectively relieves liver cirrhosis in DEN induced rats. Mol Med Rep 9:1103–1111

Horwitz EM, Le Blanc K, Dominici M (2005) Clarification of the nomenclature for MSC: the international society for cellular therapy position statement. Cytotherapy 7:393–395

Huang CK, Lee SO, Lai KP, Ma WL, Lin TH, Tsai MY, Luo J, Chang C (2013) Targeting androgen receptor in bone marrow mesenchymal stem cells leads to better transplantation therapy efficacy in liver cirrhosis. Hepatology 57:1550–1563

In't Anker PS, Noort WA, Scherjon SA (2003a) Mesenchymal stem cells in human second trimester bone marrow, liver, lung, and spleen exhibit a similar immunophenotype but a heterogeneous multilineage differentiation potential. Haematologica 88:845–852

In't Anker PS, Scherjon SA, Kleijburg-van der Keur C (2003b) Amniotic fluid as a novel source of mesenchymal stem cells for therapeutic transplantation. Blood 102:1548–1549

In't Anker PS, Scherjon SA, Kleijburg-van der Keur C (2004) Isolation of mesenchymal stem cells of fetal or maternal origin from human placenta. Stem Cells 22:1338–1345

Isoda K, Kojima M, Takeda M, Higashiyama S, Kawase M, Yagi K (2004) Maintenance of hepatocyte functions by coculture with bone marrow stromal cells. J Biosci Bioeng 97:343–346

Jang YO, Kim MY, Cho MY, Baik SK, Cho YZ, Kwon SO (2014) Effect of bone marrow-derived mesenchymal stem cells on hepatic fibrosis in a thioacetamide-induced cirrhotic rat model. BMC Gastroenterol 14:198. http://www.biomedcentral.com/1471-1230X/1414/1198

Jung KH, Shin HP, Lee S, Lim YJ, Hwang SH, Han H, Park HK, Chung JH, Yim SV (2009) Effect of human umbilical cord blood-derived mesenchymal stem cells in a cirrhotic rat model. Liver Int 29:898–909

Kang L-I, Mars WM, Michalopoulos GK (2012) Signals and cells involved in regulating liver regeneration. Cell 1:1261–1292

Kim TH, Mars WM, Stolz DB, Michalopoulos GK (2000) Expression and activation of pro-MMP-2 and pro-MMP-9 during rat liver regeneration. Hepatology 31(1):75–82

Kim W, Matsumoto K, Bessho K, Nakamura T (2005) Growth Inhibition and apoptosis in liver myofibroblast promoted by hepatocyte growth factor leads to resolution from liver cirrhosis. Am J Pathol 166:1017–1028

Kisseleva T, Uchinami H, Feirt N, Quintana-Bustamante O, Segovia J, Schwabe RF, Brenner DA (2006) Bone marrow-derived fibrocytes participate in pathogenesis of liver fibrosis. J Hepatol 45:429–438

Koellensperger E, Niesen W, Kolbenschlag J, Gramley F, Germann G, Leimer U (2013) Human adipose tissue derived stem cells promote liver regeneration in a rat model of toxic injury. Stem Cells Int 2013:534263

Kulterer B, Friedl G, Jandrositz A, Sanchez-Cabo F, Prokesch A, Paar C, Scheideler M, Windhager R, Preisegger K-H, Trajanoski Z (2007) Gene expression profiling of human mesenchymal stem cells derived from bone marrow during expansion and osteoblast differentiation. BMC Genomics 8:1–15

Le Blanc K, Ringdén O (2005) Immunobiology of human mesenchymal stem cells and future use in hematopoietic stem cell transplantation. Biol Blood Marrow Transplant 11:321–334

Le Blanc K, Ringdén O (2007) Immunomodulation by mesenchymal stem cells and clinical experience. J Int Med 262:509–525

Li C, Kong Y, Wang H, Wang S, Yu H, Liu X (2009) Homing of bone marrow mesenchymal stem cells mediated by sphingo- sine 1-phosphate contributes to liver fibrosis. J Hepatol 50:1174–1183

Li S, Sun Z, Lv G, Guo X, Zhang Y et al (2009) Microencapsulated UCB cells repair hepatic injure by intraperitoneal transplantation. Cytotherapy 11:1032–1040

Li H, Lu Y, Witek RP, Chang LJ, Campbell-Thompson M, Jorgensen M, Petersen B, Song S (2010a) Ex vivo transduction and transplantation of bone marrow cells for liver gene delivery of alpha1-antitrypsin. Mol Ther 18:1553–1558

Li J, Tao R, Wu W, Cao H, Xin J, Li J, Guo J, Jiang L, Gao C, Demetriou AA et al (2010b) 3D PLGA scaffolds improve differentiation and function of bone marrow mesenchymal stem cell-derived hepatocytes. Stem Cells Dev 19:1427–1436

Li Z, He C, Xiao J, Chen Z (2013) Treating end-stage liver diseases with mesenchymal stem cells: an oak is not felled at one stroke. OA Tissue Eng 1(1):3

Li B, Shao Q, Ji D, Li F, Chen G (2015) Mesenchymal stem cells mitigate cirrhosis through BMP7. Cell Physiol Biochem 35:433–440

Lian J, Lu Y, Xu P, Ai A, Zhou G, Liu W, Cao Y, Zhang WJ (2014) Prevention of liver fibrosis by intrasplenic injection of high-density cultured bone marrow cells in a rat chronic liver injury model. PLoS One 9:e103603

Liang X, Ding Y, Zhang Y, Tse HF, Lian Q (2014) Paracrine mechanisms of mesenchymal stem cell-based therapy: current status and perspectives. Cell Transplant 23:1045–1059

Linhua Zheng JC, Yongquan S, Xinmin J, Ling T, Qiang L, Lina C, Zheyi H, Ying H, Daiming F (2013) Bone marrow-derived stem cells ameliorate hepatic fibrosis by down-regulating interleukin-17. Cell Biosci 3:46

Lorenzini S, Andreone P (2007) Stem cell therapy for human liver cirrhosis: a cautious analysis of the results. Stem Cells 25:2383–2384

Lue J, Lin G, Ning H, Xiong A, Lin CS, Glenn JS (2010) Transdifferentiation of adipose-derived stem cells into hepatocytes: a new approach. Liver Int 30:913–922

Mehdi M, Reza M, Siavosh N-M, Sepideh H-A, Nasser R et al (2005) Impact of immunosuppressive treatment on liver fibrosis in autoimmune hepatitis. Dig Dis Sci 50:547–551

Meier RP, Mahou R, Morel P, Meyer J, Montanari E, Muller YD, Christofilopoulos P, Wandrey C, Gonelle-Gispert C, Buhler LH (2015) Microencapsulated human mesenchymal stem cells decrease liver fibrosis in mice. J Hepatol 62:634–641

Mohamadnejad M, Alimoghaddam K, Mohyeddin-Bonab M, Bagheri M, Bashtar M, Ghanaati H, Baharvand H, Ghavamzadeh A, Malekzadeh R (2007) Phase 1 trial of autologous bone marrow mesenchymal stem cell transplantation in patients with decompensated liver cirrhosis. Arch Iran Med 10:459–466

Mohsin S, Shams S, Ali Nasir G, Khan M, Javaid Awan S, Khan SN, Riazuddin S (2011) Enhanced hepatic differentiation of mesenchymal stem cells after pretreatment with injured liver tissue. Differentiation 81:42–48

Moon YJ, Yoon HH, Lee MW, Jang IK, Lee DH, Lee JH, Lee SK, Lee KH, Kim YJ, Eom YW (2009) Multipotent progenitor cells derived from human umbilical cord blood can differentiate into hepatocyte-like cells in a liver injury rat model. Transplant Proc 41:4357–4360

Murray Christopher JL, Vos T et al (2012) Disability-adjusted life years (DALYs) for 291 diseases and injuries in 21 regions, 1990–2010: a systematic analysis for the Global Burden of Disease Study 2010. Lancet 380:2197–2223

Nasir GA, Mohsin S, Khan M, Shams S, Ali G, Khan SN, Riazuddin S (2013) Mesenchymal stem cells and Interleukin-6 attenuate liver fibrosis in mice. J Transl Med 11:78. http://www.translational-medicine.com/content/11/11/78

Nhung HT, Nam HN, Trinh VL, Ngoc BV, Nghia H, Thanh VN, Huy ML, Ngoc KP, Phuc VP (2016). Comparison of the treatment efficiency of bone marrow-derived mesenchymal stem cell transplantation via tail and portal veins in CCl4-induced mouse liver fibrosis. Stem Cells Int 2016:13

Nussler A et al (2006) Present status and perspectives of cell-based therapies for liver diseases. J Hepatol 45:144–159

Ozaki Y, Nishimura M, Sekiya K, Suehiro F, Kanawa M, Nikawa H (2007) Comprehensive analysis of chemotactic factors for bone marrow mesenchymal stem cells. Stem Cells Dev 16:119–129

Pan RL, Wang P, Xiang LX, Shao JZ (2011) Delta-like 1 serves as a new target and contributor to liver fibrosis down-regulated by mesenchymal stem cell transplantation. J Biol Chem 286:12340–12348

Park JA, Kim GD, Cha JH, Kim HL, Choi ES, Jung ES, Yoon SK, Oh SH, Bae SH (2011) Therapeutic potential of human mesenchymal stem cells derived from amnion and bone marrow in a rat model of acute liver injury and fibrosis. Tissue Eng Regen Med 8:422–431

Phillips MI, Tang YL (2008) Genetic modification of stem cells for transplantation. Adv Drug Deliv Rev 60:160–172

Pinelopi M, Vasiliki A, Emmanuel T, Graziella I, Kate J, Graham S, Amar PD, James OB, David P, Andrew KB (2010) Primary biliary cirrhosis after liver transplantation: influence of immunosuppression and human leukocyte antigen locus disparity. Liver Transpl 16:64–73

Pittenger MF, Mackay AM, Beck SC (1999) Multilineage potential of adult human mesenchymal stem cells. Science 284:143–147

Ponte AL, Marais E, Gallay N, Langonné A, Delorme B, Hérault O (2007) The in vitro migration capacity of human bone marrow mesenchymal stem cells: comparison of chemokine and growth factor chemotactic activities. Stem Cells 25:1737–1745

Pournasr B, Mohamadnejad M, Bagheri M, Aghdami N, Shahsavani M, Malekzadeh R, Baharvand H (2011) In vitro differentiation of human bone marrow mesenchymal stem cells into hepatocyte-like cells. Arch Iran Med 14:244–249

Puglisi MA, Saulnier N, Piscaglia AC, Tondi P, Agnes S, Gasbarrini A (2011a) Adipose tissue-derived mesenchymal stem cells and hepatic differentiation: old concepts and future perspectives. Eur Rev Med Pharmacol Sci 15:355–364

Puglisi MA, Tesori V, Lattanzi W, Piscaglia AC, Gasbarrini GB, D'Ugo DM, Gasbarrini A (2011b) Therapeutic implications of mesenchymal stem cells in liver injury. J Biomed Biotechnol 2011:860578

Ren H, Zhao Q, Cheng T, Lu S, Chen Z, Meng L, Zhu X, Yang S, Xing W, Xiao Y et al (2010) No contribution of umbilical cord mesenchymal stromal cells to capillarization and venularization of hepatic sinusoids accompanied by hepatic differentiation in carbon tetrachloride-induced mouse liver fibrosis. Cytotherapy 12:371–383

Ruiz JC (2011) Generation of adipose stromal cell-derived hepatic cells. Methods Mol Biol 702:249–260

Saito Y, Shimada M, Utsunomiya T, Ikemoto T, Yamada S, Morine Y, Imura S, Mori H, Arakawa Y, Kanamoto M, Iwahashi S, Takasu C (2014) Homing effect of adipose-derived stem cells to the injured liver: the shift of stromal cell-derived factor 1 expressions. J Hepatobiliary Pancreat Sci. doi:10.1002/jhbp.147

Sanchez-Ramos J, Song S, Cardozo-Pelaez F (2000) Adult bone marrow stromal cells differentiate into neural cells in vitro. Exp Neurol 164:247–256

Sandra AJ, Valerie DR, Catherine MV, Stefaan WVG (2013) Immunological characteristics of human mesenchymal stem cells and multipotent adult progenitor cells. Immunol Cell Biol 91:32–39

Schwartz RE, Reyes M, Koodie L (2002) Multipotent adult progenitor cells from bone marrow differentiate into functional hepatocyte-like cells. J Clin Invest 109:1291–1302

Sgodda M, Aurich H, Kleist S, Aurich I, Konig S, Dollinger MM, Fleig WE, Christ B (2007) Hepatocyte differentiation of mesenchymal stem cells from rat peritoneal adipose tissue in vitro and in vivo. Exp Cell Res 313:2875–2886

Shi XL, Mao L, Xu BY, Xie T, Zhu ZH, Chen JH, Li L, Ding YT (2008) Optimization of an effective directed differentiation medium for differentiating mouse bone marrow mesenchymal stem cells into hepatocytes in vitro. Cell Biol Int 32:959–965

Silva Jr WA, Covas D, Panepucci RA, Proto-Siqueira R, Siufi JL, Zanette DL (2003) The profile of gene expression of human marrow mesenchymal stem cells. Stem Cells 21:661–669

Soland MA, Bego MG, Colletti E, Porada CD, Zanjani ED, St Jeor S, Almeida-Porada G. (2012). Modulation of human mesenchymal stem cell immunogenicity through forced expression of human cytomegalovirus US proteins. PLoS One 7:e36163

Stefan M, Thomas B, Jürgen R, Christoph S, Heiner W (2015) Hepatology, 6th edn. Flying Publisher, Germany

Sterodimas A, de Faria J, Nicaretta B, Pitanguy I (2010) Tissue engineering with adipose-derived stem cells (ADSCs): current and future applications. J Plast Reconstr Aesthet Surg 63:1886–1892

Sudan DL, Shaw BW Jr, Langnas AN (1998) Causes of late mortality in pediatric liver transplant recipients. Ann Surg 227:289

Sudeepta A, Mark FP (2005) Human mesenchymal stem cells modulate allogeneic immune cell responses. Blood 105:1815–1822

Taléns-Visconti R, Bonora A, Jover R, Mirabet V, Carbonell F, Castell JV, Gómez-Lechón MJ (2006) Hepatogenic differentiation of human mesenchymal stem cells from adipose tissue in comparison with bone marrow mesenchymal stem cells. World J Gastroenterol 12(36):5834–5845

Tan CY, Lai RC, Wong W, Dan YY, Lim S-K, Ho HK (2014) Mesenchymal stem cell-derived exosomes promote hepatic regeneration in drug-induced liver injury models. Stem Cell Res Ther 5(3):76

Thorgeirsson SS (1996) Hepatic stem cells in liver regeneration. FASEB J 10:1249–1256

Tse WT, Pendleton JD, Beyer WM, Egalka MC, Guinan EC (2003). Suppression of allogeneic T-cell proliferation by human marrow stromal cells: implications in transplantation. Transplantation 75:389–397

Tsochatzis EA, Bosch J, Burroughs AK (2014) Liver cirrhosis. Lancet 383:1749–1761

Volarevic V, Nurkovic J, Arsenijevic N, Stojkovic M (2014) Concise review: therapeutic potential of mesenchymal stem cells for the treatment of acute liver failure and cirrhosis. Stem Cells 32:2818–2823

Wang X, Ge S, McNamara G, Hao QL, Crooks GM, Nolta JA (2003) Albumin-expressing hepatocyte-like cells develop in the livers of immune-deficient mice that received transplants of highly purified human hematopoietic stem cells. Blood 101:4201–4208

Wang M, Pei H, Zhang L, Guan L, Zhang R, Jia Y, Li B, Yue W, Wang Y, Pei X (2010) Hepatogenesis of adipose-derived stem cells on poly-lactide-co-glycolide scaffolds: in vitro and in vivo studies. Tissue Eng 16:1041–1050

Wang Y, Lian F, Li J, Fan W, Xu H, Yang X, Liang L, Chen W, Yang J (2012) Adipose derived mesenchymal stem cells transplantation via portal vein improves microcirculation and ameliorates liver fibrosis induced by CCl4 in rats. J Transl Med 10:133

Wilson A, Butler PE, Seifalian AM (2011) Adipose-derived stem cells for clinical applications: a review. Cell Prolif 44:86–98

Zhang Z, Lin H, Shi M, Xu R, Fu J, Lv J, Chen L, Lv S, Li Y, Yu S et al (2012) Human umbilical cord mesenchymal stem cells improve liver function and ascites in decompensated liver cirrhosis patients. J Gastroenterol Hepatol 27(Suppl 2):112–120

Zhao Q, Ren H, Zhu D, Han Z (2009) Stem/progenitor cells in liver injury repair and regeneration. Biol Cell 101:557–571

Zhen Fan Y, Ho DWY, Ngai P, Lau CK, Zhao Y, Poon RTP, Fan ST (2007) Antiinflammatory properties of IL-10 rescue small-for-size liver grafts. Liver Transpl 13:558–565

Zheng Z, Wang F-S (2013) Stem cell therapies for liver failures and cirrhosis. J Hepatol 59:183–185

Zuk PA, Zhu M, Mizuno H (2001) Multilineage cells from human adipose tissue: implications for cell based therapies. Tissue Eng 7:211–228

Part II
Lung Regeneration

Chapter 5
Mesenchymal Stem Cell Therapy for Airway Restoration Following Surgery

Francesco Petrella, Stefania Rizzo, Fabio Acocella, Stefano Brizzola, and Lorenzo Spaggiari

Abbreviations

BLI	Bioluminescence imaging
BMMSC	Bone marrow-derived mesenchymal stem cell
BPF	Bronchopleural fistula
CT	Computed tomography
FDA	Food and Drugs Administration
FLI	Fluorescence imaging
MRI	Magnetic resonance imaging
MSC	Mesenchymal stem cell
PET	Positron emission tomography
SPECT	Single-photon emission computed tomography
SPIO	Super paramagnetic iron oxide

F. Petrella (✉)
Department of Thoracic Surgery, European Institute of Oncology,
Via Ripamonti, 435, 20141 Milan, Italy
e-mail: francesco.petrella@ieo.it

S. Rizzo
Department of Radiology, European Institute of Oncology, Milan, Italy

F. Acocella • S. Brizzola
Department of Health, Animal Science and Public Health, University of Milan, Milan, Italy

L. Spaggiari
Department of Thoracic Surgery, European Institute of Oncology,
Via Ripamonti, 435, 20141 Milan, Italy

School of Medicine, University of Milan, Milan, Italy

© Springer International Publishing AG 2017
P.V. Pham (ed.), *Liver, Lung and Heart Regeneration*,
Stem Cells in Clinical Applications, DOI 10.1007/978-3-319-46693-4_5

5.1 Introduction

Stem cells can be used to repair defects in the airway wall, resulting from tumors, trauma, and diseases associated with epithelial damage (Roomans 2010). Two groups of stem cells can be used for this purpose: endogenous progenitor cells present in the respiratory tract or exogenous stem cells derived from other tissues in the body; in this second group of stem cells there are embryonic stem cells, induced pluripotent stem cells, amniotic fluid stem cells, and fat-derived or bone marrow-derived mesenchymal stem cells (BMMSCs) (Roomans 2010).

The aim of this chapter is to focus on autologous BMMSCs for closing post-resectional airway tissue defects, potentially developing after lung resections.

5.2 Clinical Background of Airway Diseases Potentially Benefitting from MSC Therapy

Large airway defects and tracheobronchial dehiscence following curative lung resection present a major problem for clinicians because no effective methods of treatment are available (Macchiarini et al. 2008; Sonobe et al. 2000).

Post-resectional BPF is a pathological connection between the airway (bronchus) and the pleural space that may develop after lung resection. The incidence of BPF in pleuropulmonary surgery ranges from 1 to 4 %, but its mortality rate ranges from 12.5 to 71.2 % (Sonobe et al. 2000). BPF may be caused by incomplete bronchial closure, impediment of bronchial stump wound healing, or stump destruction by residual neoplastic tissue (Sonobe et al. 2000).

The clinical effect of impaired bronchial stump healing after anatomic lung resection may culminate in a life-threatening septic and ventilatory catastrophe (Shields 2005). For many patients with empyema, the presence or absence of a fistula makes the difference between recovery, chronicity, or death (Patterson 2016).

5.3 Mesenchymal Stem Cells and Airway Problems: Current Approaches and Future Perspectives

Development of cell therapies and bioengineering approaches for lung diseases has rapidly progressed over the past decade (Weiss 2014). A number of early reports initially suggested that bone marrow-derived cells, including MSC and other populations, could structurally engraft as mature differentiated airway and alveolar epithelial cells or as pulmonary vascular or interstitial cells (Kassmer and Krause 2010). Some recent reports continue to suggest that engraftment of the donor-derived airway can occur with several different types of bone marrow-derived cells (Wong et al. 2009).

MSC from bone marrow, adipose and placental tissue, and other origins have been widely investigated for their immunomodulatory effects in a broad range of inflammatory and immune diseases (Keating 2012). However, the mechanisms of MSC actions are only partially understood. In addition to the paracrine actions of soluble peptide and other mediators, a growing body of data suggests that release of episomal or microsomal particles by MSC can influence the behavior of both surrounding structural and inflammatory cells (Weiss 2014). A recent report suggests that MSC may also promote repair by activation of endogenous distal lung airway progenitor cell populations in mouse models (Tropea et al. 2012). Administration of MSC of either bone marrow or placental origin has also been demonstrated to decrease injury and inflammation in endotoxin or bacterially injured human lung explants (Lee et al. 2013).

MSC can also exert effects on lung inflammation and injury through primary interactions with the immune system rather than through direct actions in lung, in particular when the cells are systemically delivered (Weiss 2014).

In the light of more recent findings, chronic persistent lung diseases with low level or smoldering inflammation—such as chronic obstructive pulmonary disease or idiopathic pulmonary fibrosis—may not represent the best therapeutic targets for MSC intervention. On the contrary, more acute inflammatory lung or systemic diseases such as adult respiratory distress syndrome or sepsis/septic shock may be better targets (Bishop et al. 2012). Consequently, the scenario of an acute postresectional BPF, where the acute inflammation component is prominent along with infection, may represent an ideal target for MSC clinical use in thoracic surgery.

The use of autologous MSC eliminates the patient's need for lifelong immunosuppressant therapy and avoids the risk of infection. Bone marrow remains one of the best sites for MSC harvesting, although even adipose tissue is a viable option and may be further explored in the future.

There may be some concern that MSC may have an undesirable effect on tumor growth, but the in vivo evidence collected so far remains inconclusive. A recent clinical study in which autologous BMMSC from cancer patients was locally administered at the site of malignant primary bone tumor resection showed no increase in the cancer local recurrence risk in patients treated with the cell-based therapy after an average follow-up of 15 years (Hernigou et al. 2014).

5.4 Airway Restoration: Animal Models and Experimental Results

To evaluate the safety and efficacy of the MSC local treatment during BPF an ad hoc large animal model has been developed. Even though a clinical BPF is multifactorial we planned to create a simple fistula between the airways and the pleura cavity and consequent pleural empyema. The choice of a large animal model was driven essentially by two needs: the evaluation of the efficacy of the cell treatment and the evaluation of the mesenchymal stem cells (MSCs) local administration procedure through the respiratory tree (Gomez-de-Antonio et al. 2010). The choice of the goat model has been based on its anatomical similarities of the pleural cavity and mediastinum with those human beings, essentially because they do not have communication between the two pleural cavities thus limiting pneumothorax and empyema to one side. To reduce animals' discomfort, 10 % of the entire lungs volume has been removed by lobectomy. Animals were treated according to the requirements of the European Union Directive 86/609 regarding the protection of animals used for experimental or other scientific purpose and the Council of Europe Convention for the protection of vertebrate animals used for experimental and other scientific purpose (ET 123). A total of 9 goats underwent the same surgical procedure to create an artificial BPF and were randomized in two groups: (a) experimental group ($n=4$) and (b) control group ($n=5$). Standard right upper tracheal lobectomy, simulating human right upper fissureless lobectomy, was performed under general anesthesia without the need for single lung ventilation. Bronchial stump closure was achieved using single interrupted stitches leaving the medial edge of the stump open so as to obtain a uniform size of the fistula (30 % of the internal bronchial diameter). Water submersion test under standard airway pressure of 15 cm H_2O was then performed to confirm the bronchial fistula and exclude other sites of air leakage. Mean whole bronchial stump caliber was 10.1 ± 1 mm. Autologous MSCs already isolated from the bone marrow (Pittenger et al. 1999) and before administration were submitted to infection with a lentiviral vector codifying for a report gene, the galactosidase gene (LacZ). In the experimental group, 2 millions of MSCs/ml in 5 ml of modified fibrin glue (Evicel; Ethicon; Somerville, NJ) were injected bronchoscopically into the BPF either in the submucosal layer around the fistula and into the bronchial lumen. The control group followed the same procedure using Evicel alone. A mid term (15 days) evaluation was performed using bronchoscopy. After 1 month the animals were sacrificed and specimens of the bronchus and the surrounding tissues were collected and immediately frozen. The samples were submitted to histological, immunohistochemical, and imaging analyses computed tomography(CT) and magnetic resonance (MRI) (Fig. 5.1). The disease model of this study proved to be feasible and the surgical technique showed neither major discomfort nor complication for all animals enrolled. The MCS endoscopical transplantation was always effective even if the reproducibility of cells injection has to be improved. Differently from the control group, microscopical and imaging analyses highlighted the full closure of the fistula one month after MCS transplantation by fibroblast proliferation and collagenous matrix development. Given the extremely high mortality rate of BPF (up to 71.2 %), the proposed technique could drastically improve post-BPF survival rate, providing the patients an effective therapeutic chance in an otherwise highly fatal situation.

5.5 MSC Imaging and Tracking in the Airways

Successful implementation of cell therapies requires a better understanding of cell fate after transplantation, which can be achieved by the application of molecular imaging. Indeed, questions dealing with survival on the site of injection, and possible patterns of migration after injection of MSCs, are still to be answered. Cell

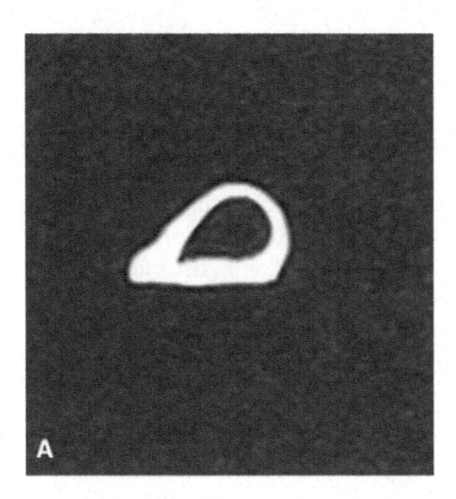

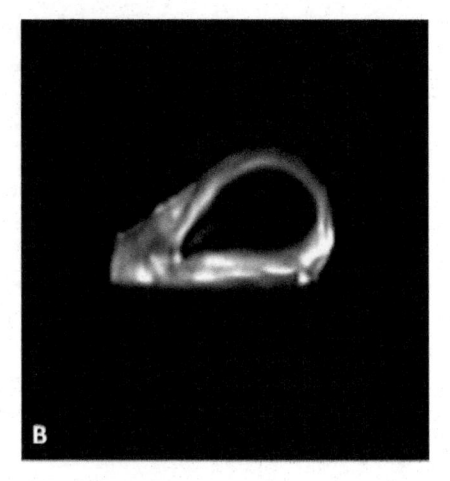

Fig. 5.1 MR discloses the black right bronchial lumen, externally closed by hyperintense solid tissue in the axial T1 image (**a**), and by inhomogeneous slightly hypointense tissue in the axial T2 image (**b**). Notably, the intensity of the regenerated tissue is comparable to the normal intensity of the tracheal and bronchial wall, where the stem cells were not present

tracking can be performed either by labeling cells with molecular probes that enter the cell by active/passive transport and are trapped intracellularly (direct labeling), or by inducing overexpression of specific reporter genes.

Signal generated from cells labeled by either technique can then be visualized using imaging systems such as fluorescence imaging, bioluminescence imaging, single-photon emission computed tomography (SPECT), positron emission tomography (PET), or magnetic resonance (MR) (Nguyen et al. 2014).

Optical methods usually require sacrifice of the animal: moreover, long-term cell tracking would be desirable. SPECT and PET are by far the most frequently applied techniques in clinical biodistribution studies. However, radioactive tracers with short half-lives, that are more commonly used in clinical stem cell imaging studies, are not suitable to assess cell migration due to a short tracking period and limited resolution.

In the last few years, super paramagnetic iron oxide (SPIO) nanoparticles, visible at MR, have been used to label MSCs, thanking the significant field distortions caused on the $T2^*$ relaxation time. These contrast media produce a "negative" contrast, greater than the area occupied by the nanoparticles (the so-called blooming artifact). However, the iron oxide particles may be retained in a tissue, even if the grafted (stem) cell dies, hence leading to false positive signals (Reagan and Kaplan 2011). Moreover, most FDA approved SPIOs have now been discontinued from the market, so moving to clinics with SPIO-labeled cells will be difficult in the near future.

In the specific field of molecular imaging applied to the airways, there is still lack of experimental studies. Because of the high spatial resolution, the interest on a labeling method visible at MR may be considered desirable. However the SPIO appearance (hypointensity on all sequences), identical to the appearance of the air, makes difficult the use of this specific category of contrast medium in the airways.

Other MR imaging techniques are currently under evaluation in other research contests, such as imaging based on perfluorocarbon formulations. The fluorine signal may be quantified from the MR images by a comparison with a standard reference. Then the images of the fluorine signal can be fused with the standard anatomical images (Ribot et al. 2014).

Regardless of model systems and components used, it is clear that the best clinical and basic research results derive from multimodal imaging systems provide functional and anatomical data. Molecular imaging may play in the future a more extensive role by evaluating not only site and fate of the cells, but also strategies to improve the viable cell concentration at the target site. As demonstrated in a growing number of preclinical studies, cell imaging can help define the optimal cell type, delivery method (Hofmann et al. 2005), timing of delivery (Rosenblum et al. 2012; Swijnenburg et al. 2010), and host microenvironment (Pearl et al. 2012).

5.6 Conclusion

Experimental and clinical evidence exists regarding MSC efficacy in airway defects restoration; although clinical MSC use—in the daily practice—is not yet reached for airway diseases, we can argue that MSC do not represent any more merely an experimental approach to airway tissue defects restoration but they can considered as a "salvage" therapeutic tool in very selected patients and diseases (Petrella and Spaggiari 2015; Petrella et al. 2014, 2015).

References

Bishop JA, Teruya-Feldstein J, Westra WH, Pelosi G, Travis WD, Rekhtman N (2012) p40 (ΔNp63) is superior to p63 for the diagnosis of pulmonary squamous cell carcinoma. Mod Pathol 25(3):405–415

Gomez-de-Antonio D, Zurita M, Santos M, Salas I, Vaquero J, Varela A (2010) Stem cells and bronchial stump healing. J Thorac Cardiovasc Surg 140(6):1397–1401. doi:10.1016/j.jtcvs.2010.03.009, Epub 2010 Apr 18

Hernigou P, Flouzat Lachaniette CH, Delambre J, Chevallier N, Rouard H (2014) Regenerative therapy with mesenchymal stem cells at the site of malignant primary bone tumour resection: what are the risks of early or late local recurrence? Int Orthop 38(9):1825–1835

Hofmann M, Wollert KC, Meyer GP, Menke A, Arseniev L, Hertenstein B, Ganser A, Knapp WH, Drexler H (2005) Monitoring of bone marrow cell homing into the infracted human myocardium. Circulation 111:2198–2202

Kassmer SH, Krause DS (2010) Detection of bone marrow-derived lung epithelial cells. Exp Hematol 38:564–573

Keating A (2012) Mesenchymal stromal cells: new directions. Cell Stem Cell 10:709–716

Lee JW, Krasnodembskaya A, McKenna DH, Song Y, Abbott J, Matthay MA (2013) Therapeutic effects of human mesenchymal stem cells in ex vivo human lungs injured with live bacteria. Am J Respir Crit Care Med 187:751–760

Macchiarini P, Jungebluth P, Go T et al (2008) Clinical transplantation of a tissue-engineered airway. Lancet 372(9655):2023–2030

Nguyen PK, Riegler J, Wu JC (2014) Stem cell imaging: from bench to bedside. Cell Stem Cell 14(4):431–444. doi:10.1016/j.stem.2014.03.009

Patterson GA, Cooper JD, Deslauires J, Lerut AEM, Luketich JD, Rice TW (2016) Pearsons's thoracic and esophageal surgery, 3rd edn. Churchill Livingstone, London

Pearl JI, Kean LS, Davis MM, Wu JC (2012) Pluripotent stem cells: immune to the immune system? Sci Transl Med 4:164ps125

Petrella F, Spaggiari L (2015) Bronchopleural fistula treatment: from the archetype of surgery to the future of stem cell therapy. Lung India 32(2):100–101

Petrella F, Toffalorio F, Brizzola S, De Pas TM, Rizzo S, Barberis M, Pelicci P, Spaggiari L, Acocella F (2014) Stem cell transplantation effectively occludes bronchopleural fistula in an animal model. Ann Thorac Surg 97(2):480–483

Petrella F, Spaggiari L, Acocella F, Barberis M, Bellomi M, Brizzola S, Donghi S, Giardina G, Giordano R, Guarize J, Lazzari L, Montemurro T, Pastano R, Rizzo S, Toffalorio F, Tosoni A, Zanotti M (2015) Airway fistula closure after stem-cell infusion. N Engl J Med 372(1):96–97

Pittenger MF, Mackay AM, Beck SC, Jaiswal RK, Douglas R, Mosca JD, Moorman MA, Simonetti DW, Craig S, Marshak DR (1999) Multilineage potential of adult human mesenchymal stem cells. Science 284(5411):143–147

Reagan MR, Kaplan DL (2011) Concise review: mesenchymal stem cell tumor-homing: detection methods in disease model systems. Stem Cells 29(6):920–927

Ribot EJ, Gaudet JM, Chen Y, Gilbert KM, Foster PJ (2014) In vivo MR detection of fluorine-labeled human MSC using the bSSFP sequence. Int J Nanomed 9:1731–1739

Roomans GM (2010) Tissue engineering and the use of stem/progenitor cells for airway epithelium repair. Eur Cells Mater 19:284–299

Rosenblum S, Wang N, Smith TN, Pendharkar AV, Chua JY, Birk H, Guzman R (2012) Timing of intra-arterial neural stem cell transplantation after hypoxia-ischemia influences cell engraftment, survival, and differentiation. Stroke 43:1624–1631

Shields TW, Locicero J III, Ponn RB et al (2005) General thoracic surgery. Complications of pulmonary resection, 6th edn. Lippincott Williams & Wilkins, Philadelphia, 554–586

Sonobe M, Nakagawa M, Ichinose M et al (2000) Analysis of risk factors in bronchopleural fistula after pulmonary resection for primary lung cancer. Eur J Cardiothorac Surg 18:519–523

Swijnenburg RJ, Govaert JA, van der Bogt KE, Pearl JI, Huang M, Stein W, Hoyt G, Vogel H, Contag CH, Robbins RC, Wu JC (2010) Timing of bone marrow cell delivery has minimal effects on cell viability and cardiac recovery after myocardial infarction. Circ Cardiovasc Imaging 3:77–85

Tropea KA, Leder E, Aslam M, Lau AN, Raiser DM, Lee JH, Balasubramaniam V, Fredenburgh LE, Alex Mitsialis S, Kourembanas S, Kim CF (2012) Bronchoalveolar stem cells increase after mesenchymal stromal cell treatment in a mouse model of bronchopulmonary dysplasia. Am J Physiol Lung Cell Mol Physiol 302:L829–L837

Weiss DJ (2014) Current status of stem cells and regenerative medicine in lung biology and diseases. Stem Cells 32:16–25

Wong AP, Keating A, Lu WY, Duchesneau P, Wang X, Sacher A, Hu J, Waddel TK (2009) Identification of a bone marrow-derived epithelial-like population capable of repopulating injured mouse airway epithelium. J Clin Inv 119:336–348

Chapter 6
Regenerative Potential of Mesenchymal Stem Cells: Therapeutic Applications in Lung Disorders

Kavita Sharma, Syed Yawer Husain, Pragnya Das, Mohammad Hussain, and Mansoor Ali Syed

6.1 Introduction

Despite substantial clinical advances over the past decades, inflammatory lung diseases are a major cause of morbidity and mortality and have become one of the major public health issues across the world. The World Health Organization positions lung diseases second in epidemiology, mortality, and cost and predicts that about one-fifth deaths will be attributed to lung diseases by 2020. Currently, there are no therapeutic ways to inhibit or reverse the pathobiology of many destructive lung diseases that results in dysfunctional lung renovation. However, recent studies indicate that lung mesenchymal stem cells (MSCs) are triggered by local factors to differentiate into myofibroblasts that contribute to disease progression. Therefore, it is of critical significance to understand the molecular and cellular basis of endogenous lung MSCs participation in lung injury and repair. Interestingly, while the lung exhibits tremendous regenerative capacity, restoration of pulmonary function does

---

K. Sharma
Department of Pathology, AIIMS, New Delhi, India

S.Y. Husain
Stem Cell Research Lab, Rajasthan University of Health Sciences,
Subhash Nagar, Jaipur 302016, Rajasthan, India

P. Das
Drexel University School of Medicine, Philadelphia, PA, USA

M. Hussain
Department of Biotechnology, Jamia Millia Islamia (Central University), New Delhi, India

M.A. Syed (✉)
Stem Cell Research Lab, Rajasthan University of Health Sciences,
Subhash Nagar, Jaipur 302016, Rajasthan, India

Department of Biotechnology, Jamia Millia Islamia (Central University), New Delhi, India
e-mail: smansoor@jmi.ac.in

not occur in many adult lung diseases. To address this condition emphasis has been increased on the development of cell-based therapies, but due to diverse cell types and functions lung is considered as a recalcitrant candidate for these strategies. Currently, origins and contributions of stem cells are under intense investigation for cell-based therapy and pulmonary remodelling. Specialized microenvironments for resident multipotent MSCs have been identified in many adult tissues and normal differentiation processes of these may be disrupted by pathologic microenvironmental stimuli during disease, epigenetic changes, or genetic alteration, which program their contribution to pathologic expansion at the expense of functional tissue regeneration. MSCs are most widely investigated and clinically tested type of stem cell because of their regenerative capacity to mesoderm/non-mesoderm-derived tissues and they also display immune-enhancing as well as immunosuppressive properties. MSCs especially from human are widely studied as compared to other cells, different early stage clinical and scientific studies show potential for repair and renewal of lung tissues and offer great ability for the treatment of several devastating and incurable lung diseases. Bone marrow derived MSCs (BM-MSCs) are currently tested in clinical trials as a potential therapy in patients with such inflammatory lung diseases. Here, we will review the biology of MSCs, their interaction with molecular and cellular pathways, and their modulation of immune responses in lung disorders. Additionally, we discuss what stem cell therapy offers in specific acute and chronic lung disorders which includes respiratory distress syndrome (RDS), chronic obstructive pulmonary disease (COPD), asthma, fibrosis, bronchopulmonary dysplasia (BPD), and pulmonary hypertension.

6.2 Stem Cell Biology

Stem cells are a population of undifferentiated cells characterized by their ability to divide asymmetrically, clonality arising from a single cell, and potency to differentiate into different type of cells or tissues (Kolios and Moodley 2013). Lung stem cell research slightly lags behind studies on other organs because of their limited knowledge about the endogenous progenitor cells. Moodley and colleagues demonstrated reduced inflammation and fibrosis by performing a xenograft implantation in which they injected human amnion epithelial cells parenterally into bleomycin treated severe combined immunodeficiency mice as pulmonary fibrosis model (Moodley et al. 2010). More interestingly, Kajstura and coworkers identified stem cells from adult human lung tissue using c-kit stem cell antigen marker (Kajstura et al. 2011). After administration into an injured mouse lung, the cells demonstrated pluripotent capability of producing human bronchioles, alveoli, and pulmonary vessels (Anversa et al. 2011). These evidences collectively suggest that isolated adult lung stem cells are potentially of great therapeutic importance for cell-based therapies in pulmonary diseases. MSCs are a prototypical adult stem cell with capacity for self-renewal and differentiation with a broad tissue distribution. Firstly, defined in bone marrow, MSCs have the capacity to differentiate into mesoderm- and non-mesoderm-derived tissues. The endogenous role for MSCs is maintenance of stem cell niches (classically the hematopoietic), organ homeostasis, wound healing, and successful aging. Considering their ease of preparation and immunologic privilege, MSCs are emerging as a tremendously promising therapeutic agent for tissue regeneration and thus can be considered as a potential therapeutic candidate for clinical studies in relation to lung diseases (Dimmeler et al. 2011).

6.3 Historical Background of MSCs

Consideration of MSC discovery and expansion is an important aspect to design future therapeutic studies. MSCs were first isolated from bone marrow and defined by Friedenstein in 1968 as a distinct entity having plastic adherent fibroblast like appearance that distinguish them from other hematopoietic stem cells (HSCs) (Friedenstein et al. 1968). Also, he was the first investigator to demonstrate the potential of MSCs to differentiate into mesoderm-derived tissue and to further identify their importance in controlling the hematopoietic niche. Numerous laboratories have confirmed Friedenstein's protocol and expanded these findings by showing that these cells are also present in the human bone marrow, and can be sub-passaged and differentiated in vitro into a variety of cells of the mesenchymal lineages such as osteoblasts, chondrocytes, adipocytes, and myoblasts (Bianco et al. 2008; Caplan 2007; Kolf et al. 2007; Pittenger et al. 1999; Prockop 1997) (Fig. 6.1). Caplan and colleagues later renamed the cells isolated from bone marrow as "mesenchymal stem cell" or MSC (Caplan 2007). These MSCs or MSC-like cells are also found in tissues such as fat, umbilical cord blood (UCB), amniotic fluid, placenta, dental pulp, tendons, synovial membrane, and skeletal muscle, though the complete equivalency of such populations has not been formally demonstrated using robust scientific methods (Rogers and Casper 2004; Bieback and Klüter 2007; Xu et al. 2005; Gronthos et al. 2003; Tsai et al. 2004; Igura et al. 2004; De Bari et al. 2001; Crisan et al. 2008). Several studies are conducted on MSCs to phenotypically characterize, expand, and identify factors that keep them in undifferentiated state in order to then transplant them back in vivo for the purpose of repairing specific tissues (Tsutsumi et al. 2001; Kulterer et al. 2007; Pochampally et al. 2004). Therefore, our knowledge of MSCs is based entirely on the characterization of cultured cells, in vitro. Interestingly, no evidence of asymmetric cell division, a property of self-renewing cells in some settings (Wu et al. 2008), has yet been provided for MSCs. In vivo knowledge of MSCs is modest in relation to their phenotypic characteristics, origin of development, organogenesis role, postnatal tissue homeostasis, and anatomical localization. Thus, a faithful assay that would rigorously test for their ability to self-renew in vivo is required to prove their missing identity. An in vivo characterization of MSCs could allow for either pharmacological or genetic manipulations to provide new possibilities for isolation of more enriched population for tissue engineering applications with potentially better capabilities of self-renewal upon in vitro expansion and in vivo transplants with appropriate scaffolds. Moreover, identification of cell-specific markers/antigens on MSCs would be very useful for a

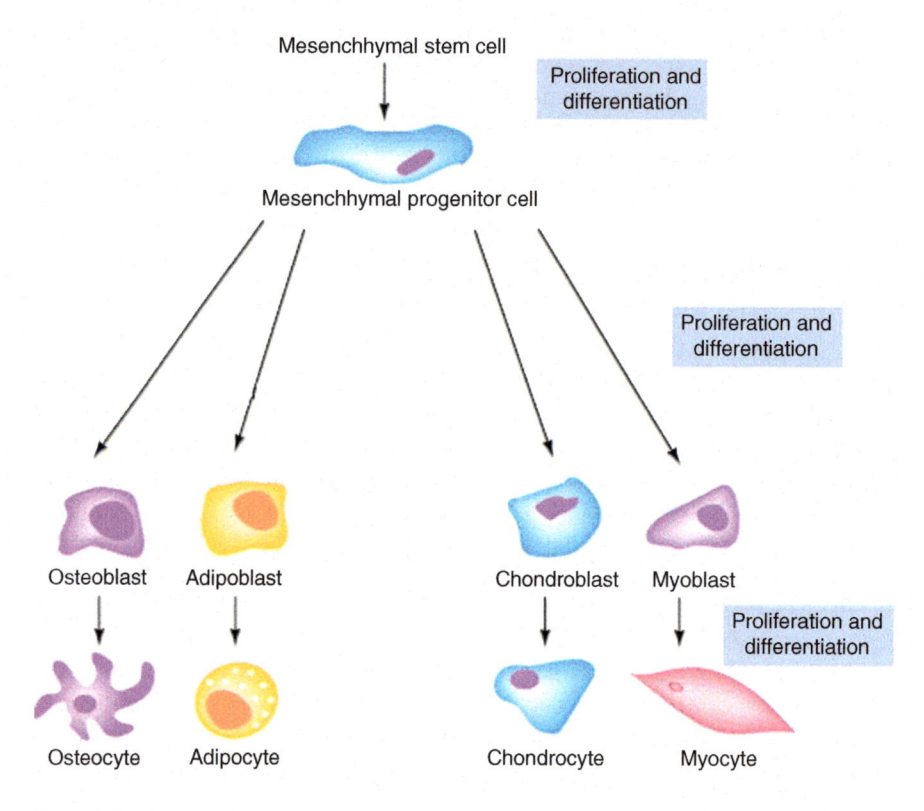

Fig. 6.1 Mesenchymal stem cells go through a series of steps to help accomplish repair

phenotypic separation from HSCs (Schipani and Kronenberg 2008). Over the years, numerous cells-surface antigens expressed by MSCs have been cultured and sub-passaged in vitro, but only a few laboratories have attempted a phenotypic characterization of MSCs in vivo (Ratajczak et al. 2007). Histological localization of MSCs in vivo is impossible in absence of unique markers. The findings by Bianco and colleagues represent the first rigorous attempt to histologically localize and phenotypically define MSC-like cells, or at least a subpool of this population. Crisan and colleagues suggest that MSCs with perivascular localization exist in numerous human organs (Crisan et al. 2008), although it remains to be established whether the vascular setting provides a true niche for pericytic MSC-like cells or not and is still the main source of MSCs in vivo (Schipani and Kronenberg 2008). Thereafter, MSCs have been isolated from many other tissues, including UCB, placenta, adipose tissue, amniotic fluid, and Wharton's jelly. Control of stem cell niches by MSCs is emerging as a key role played by MSCs in a broad array of tissues, including hair follicles and the gut. It is also shown that ablation of MSCs results in hematopoiesis disruption. During the early twenty-first century, in vivo studies demonstrated that human MSCs trans-differentiate into endoderm-derived cells and cardiomyocytes while in vitro co-culturing of ventricular myocytes with MSCs induced trans-differentiation into cardiomyocytes. Also, MSCs were demonstrated to suppress T-lymphocyte proliferation, which paved the way for the application of MSC therapy for allogeneic transplantation and as a potential immunomodulatory therapy (Williams and Hare 2011).

6.4 Identification and Characterization of MSCs

Adult tissues have some capacity to regenerate through the replication and differentiation of stem cells. While lung resident stem cells are an obvious candidate cell therapy for lung diseases, limitations exist regarding knowledge of their biology. Recently, there is considerable interest in the therapeutic application of exogenous cells, especially MSCs for lung diseases (Antunes et al. 2014). These cells are identified according to the following consensus criteria (Dominici et al. 2006): adherence to plastic under standard culture conditions; expression of CD105, CD73, and CD90 and lack of surface expression of CD45, CD34, CD14, CD11b, CD79, CD19, and HLA-DR; and ability to differentiate into adipocytes, chondrocytes, and osteocytes in vitro. Multipotent MSCs have the potential to differentiate into cells of various tissues like bone, cartilage, muscle, liver, and lung (Anversa et al. 2011). However, as explained previously, there are no consistent cell surface markers to isolate them from different tissues. Mesenchymal and Tissue Stem Cell International Committee has proposed minimal criteria to define human MSCs that can be taken into consideration during their identification and isolation (Dominici et al. 2006). Since MSCs have mesodermal origin they should ideally include blood and connective; however, in practice connective tissue is described as the derivative of MSCs.

Studies suggest that it is possible for MSCs and HSCs to share a common precursor. However, this statement is controversial and the physiological relevance of these cells remains to be demonstrated. Some of the inconsistencies surrounding the identification of MSC arise from the fact that specific markers have not yet been agreed on. In the absence of a universal antigenic indication and a universal assay, MSCs are often identified simply by testing a culture's differentiation potential into colony forming units indicative of proliferative capacity (Pittenger et al. 1999). Presently, adherence to tissue culture plastic and a fibroblast like morphology are considered as the characteristic features for MSCs (Prockop 1997). Recent studies performed on MSCs establish the association of other markers including D7fib (Jones et al. 2002), Stro1 (Stenderup et al. 2001), CD45 and glycophorin A (Pittenger et al. 1999; Jones et al. 2004), and BMPR1a (Zvaifler et al. 2000). Lineage tracing studies with fibroblast growth factor (FGF-10) have disclosed the status of two distinct populations of MSCs—one residing at the trachea and the other at the branching tip of the epithelium along the airway. Studies demonstrated that in vitro MSCs can be enriched with $CD45^{neg}$, $CD31^{neg}$, $EpCAM^{neg}$, and $Sca-1^{pos}$ cell surface markers to support proliferation and differentiation in the epithelial progenitor cell niche (McQualter et al. 2010). Lama and colleagues successfully isolated MSCs from the bronchoalveolar lavage of adult human lung allografts which expressed common mesenchymal markers (CD73, CD90, and CD105) that are capable of differentiating into adipocytes, chondrocytes, and osteocytes. Although, adult lung contains some MSC lineages, in human disease models more studies are focused on the bone marrow derived MSCs (BM-MSCs) (Lama et al. 2007). Gazdhar and colleagues isolated a population of human hepatocyte growth factor (HGF)-expressing stem cells that are co-expressed with MSC markers such as CD44, CD29, CD105, CD90, and CXCR4 at fibrotic area in usual interstitial pneumonia patients (Gazdhar et al. 2013). In order to understand whether they have antifibrotic property, they instilled BM-MSCs transfected with HGF into rat lung for 7 days after bleomycin treatment and observed an attenuated fibrosis. Although several lines of evidence support a paracrine effect to repair injured cells, the exact mechanisms of how MSCs contribute to beneficial effects against diseases are not yet fully understood. However, in vitro studies show that conditioned media (CM) of human MSCs facilitates wound repair in A549 cell lines and primary human small airway epithelial cells by producing secretome containing proteins such as fibronectin, lumican, periostin, and IGFBP7 (Akram et al. 2013). Also, studies show that MSCs CM inhibited cardiac fibroblast proliferation by up-regulating cell-cycle arresting genes such as elastin, myocardin, and DNA-damage inducible transcripts (Li et al. 2009; Ohnishi et al. 2007). Extracellular vesicles (EVs) that are released from the endosomal compartment contain a cargo that includes miRNA, mRNA, and proteins that may have role in modulation of MSCs paracrine functions. Recent, animal-based studies suggest that EVs have a significant potential as a novel alternative to whole cell therapies (Foronjy and Majka 2012). Tissue resident MSCs are considered to be the important regulators of lung repair, fibrosis, inflammation, angiogenesis, and tumor formation. By using flow cytometry lung MSCs have been identified and characterized to detect Hoechst 33342 vital dye efflux by lung cells in combination with absence of the hematopoietic marker, CD45. These Hoechst33342^{dim}CD45^{neg} cells demonstrate multilineage mesenchymal differentiation potential to osteocyte, adipocyte, and chondrocytes lineages and express the characteristic mesenchymal cell surface determinants ABCG2, CD90, CD105, CD106, CD73, CD44, and Sca-I, in vitro. The expression of high levels of telomerase activity in lung MSCs gives them the ability to replicate and contribute significantly towards tissue regeneration and proliferative diseases (Lama et al. 2007; Martin et al. 2008; Summer et al. 2007; Jun et al. 2011). Lung MSCs have also shown to mediate pathogenic changes in lung when subjected to certain conditions as their behavior is considered highly sensitive to the microenvironment to which they are exposed. As an example, it was recently shown that TGF-β expression within the lungs of premature infants stimulates MSCs to differentiate into myofibroblasts that may induce dysfunctional matrix remodelling resulting in the development of the chronic lung disease, BPD. Similar findings were observed in lung allografts from transplanted patients where lung derived MSCs isolated from the airways had increased expression of type I collagen and α-smooth muscle actin and readily differentiated into myofibroblasts upon treatment with IL-13 or TGF-β. Thus, these findings indicate that MSCs are a critical factor in the development of dysfunctional lung remodelling in these diseases (Foronjy and Majka 2012). Different populations of MSCs have also been isolated from other adult tissues—brain, spleen, liver, kidney, lung, bone marrow, muscle, thymus, pancreas, fat, and fetal tissues and they are shown to exhibit different immunophenotypes, secreted cytokine profiles, and proteome analyses (Antunes et al. 2014). From a therapeutic perspective, MSCs are emerging as an extremely promising therapeutic agent for tissue regeneration (Williams and Hare 2011) owing to their ease of preparation and immunologic privilege.

6.5 MSC: Mechanism of Action

Immunomodulation and paracrine functions make MSCs to be the eligible candidate for allogeneic transplantation. In this regard, MSCs home specifically to injured tissue and exert their immunomodulatory activity with the secretion of vascular endothelial growth factor (Guan et al. 2013), antiapoptotic (Bcl-2) (Zhen et al. 2008), and anti-inflammatory factors (Guan et al. 2013; Rojas et al. 2005; Shigemura et al. 2006; Gupta et al. 2007; Németh et al. 2009) thus stimulating angiogenesis, promoting host cell recovery, and repairing the injured tissue. Exogenously administered MSCs modulate the function of host cells within the injury by cell contact-dependent and paracrine mechanisms and by the transfer of cellular materials via microvesicles (Islam et al. 2012; Zhu et al. 2014). Preclinical studies using MSCs provide promising results for lung disorders, including emphysema (Guan et al. 2013; Zhen et al. 2010), BPD (Chang et al. 2011; van Haaften et al. 2009), fibrosis (Cargnoni et al. 2009; Moodley et al. 2009), and acute RDS (Gupta et al. 2007; Németh et al. 2009; Mei et al. 2010). However, development of stem cell therapy for pulmonary diseases remains a major challenge. Further, understanding of lung stem cell fate during disease may prove to be of critical importance for drug intervention and autologous therapies. Niches for resident MSC have been identified in many adult tissues and more recently in the lungs (Martin et al. 2008). Lung regeneration involves activation of progenitor cells as well as cell replacement through proliferation of remaining undamaged cells. The pathways and factors that control this process and its role in disease are only now being explored.

6.6 Pathways Involved in the Pathophysiology of the Lung Diseases

Platelet-derived growth factor subunit B (PDGF-BB) and Wnt are the two key targets involved in the signalling of lung diseases through MSCs. Lung MSCs respond to increased PDGF-BB with proliferation and collagen production (Heldin and Westermark 1999; Barst 2005) and its expression is up-regulated in lung parenchyma following exposure to low oxygen tension, hypobaric hypoxia, and mechanical stress (Ghofrani et al. 2005; Schermuly et al. 2005; Berg et al. 2010). Significant increase in the levels of PDGF are reported in diseases such as lung fibrosis, bronchiolitis obliterans pneumonia, post-transplant obliterative bronchiolitis, histiocytosis X, pneumoconiosis, and fibrosis associated with pulmonary arterial hypertension (PAH) (Heldin and Westermark 1999). Studies using in vivo and in vitro models of vascular neo-intimal thickening with various injuries have substantiated a role for PDGF-BB signalling. Inhibition of PDGFB signalling with multiple inhibitors attenuates pulmonary fibrosis and PAH by blocking fibroblast proliferation, muscularization of the vasculature, intimal thickening, and matrix deposition (Abdollahi et al. 2005; Wang et al. 2005). Studies established the association of PDGF-BB signalling to link the transition of mesenchyme to a profibrotic phenotype that disrupts the normal pulmonary architecture. MSC derived from bone marrow constitutively express PDGF-B and it is shown that the addition of exogenous protein promotes their growth and differentiation. PDGF-BB receptor and PDGFRE are pivotal in the adhesion and migration of MSC through regulation of integrin binding to fibronectin suggesting the role of fibronectin rich matrix in wound healing (Veevers-Lowe et al. 2011). Taken together these studies provide a basis to investigate PDGF dependent mechanisms that define endogenous lung MSCs and the differentiation of other multipotent stem cells during disease in order to identify interventions and facilitate repair. On the other hand, family of Wnt proteins are a highly conserved group of signalling molecules important in fundamental biological process including development and repair. Recently, Wnt pathway has been linked to the pathogenesis of important lung diseases, such as lung cancer and pulmonary fibrosis. Wnt signalling is known to regulate mesenchymal cell biology in an autocrine and paracrine fashion. β-catenin via co-activation of TCF/LEF transcription factors acts as a central mediator of canonical Wnt signalling pathway and protein regulation occurs at the level of synthesis, activation/translocation to the nucleus, and proteasomal degradation through a destruction complex (CKI and GSK kinases) (Baksh et al. 2007; Etheridge et al. 2004; Kirton et al. 2007; Reya and Clevers 2005). Hypercellularity and deregulated self-renewal of tissues in BM-derived MSCs can occur due to sustained activation of β-catenin and its downstream genes that regulate processes such as proliferation, differentiation, migration, and survival (Berg et al. 2010). On the other hand, non-canonical Wnt pathways regulate the proliferation and differentiation of MSCs via Dvl or Ca++ dependent processes (Ling et al. 2009). However, the effects of Wnt signalling on MSCs are quite complex. Not only canonical and non-canonical pathways antagonize each other but it is also suggested that low dose Wnt treatment stimulates MSC proliferation while high dose treatment inhibits its proliferation (De Boer et al. 2014). Studies have reported that Wnt inhibitory factor (WIF-1) and secreted frizzled related proteins (SFRPs) bind soluble Wnts and inhibit interaction with the frizzled receptor thereby antagonizing their action (Uren et al. 2000) while Dickkopfs (Dkks) and Sclerostins target the LRP receptors and prevent the propagation of intracellular signals through these receptors (Leyns et al. 1997; Semënov et al. 2001). Studies have shown that β-catenin also functions in processes such as cell adhesion, thus its disruption may lead to distorted Wnt signalling, followed by dysregulated cell adhesion. DKK1 treatment of mouse lung organ cultures has led to impaired branching morphogenesis and defects in the formation of the pulmonary vascular network while addition of exogenous fibronectin rescued the DKK1-mediated phenotype. However, epithelium-specific overexpression of WNT5a led to an inhibition of epithelial branching morphogenesis (Cho et al. 2008). The fact that some of the above-mentioned studies have resulted in apparent opposite results further underlines the complexity of the Wnt signalling system, which is subjected to a tightly regulated spatiotemporal expression pattern in the lung. These studies have demonstrated a critical role of properly regulated Wnt signalling for normal epithelial–mesenchymal interaction during lung morphogenesis, and emphasized the deleterious impact of distorted Wnt signalling on proper lung development. Wnt pathway also plays an important role in the fibrotic changes that occur in the development proliferative myofibroblast lesions during lung fibrosis (Königshoff et al. 2009). Further studies are needed to understand how Wnt signalling and MSCs interact to modulate the development of pathogenic changes in the lung. Since effects of Wnts depend on the specific cell type, disease process, and expression of counter regulatory inhibitory molecules, it will be really challenging to decipher this complex system (Foronjy and Majka 2012).

6.7 Regenerative Significance of MSCs

No treatments using MSCs are yet available. However, several possibilities for their use in the clinic are currently being explored.

6.7.1 MSCs in Tissue Engineering

Adult lung is architecturally complex as a hierarchical model of homoeostasis; therefore, it is impractical to build a whole functional lung organ with the current knowledge and technology. Interestingly, it is still possible to produce part of the upper, lower airway, or the alveolar tissue. Significant progress has been made using de-cellularized or synthetic scaffolds to generate tracheal cartilage as well as tendon tissue in diaphragm (Badylak et al. 2012). In a study, restoration of trachea function in the recipient was observed on implantation of MSC-derived chondrocytes to the de-cellularized donor trachea (Macchiarini et al. 2008). Studies suggest that seeding somatic lung progenitor cells onto synthetic polymer scaffold in vitro or implanted in vivo promoted cell differentiation. However, the in vivo transplantation caused an inflammatory response which disrupted lung development. Peterson and colleagues demonstrated that rat's bioengineered lung achieved gas exchange upon re-implantation suggesting a bright future for bioengineered lung tissues in regenerative medicine (Petersen et al. 2010). These mainly include the optimal origin of progenitor stem cells, extracellular matrix (ECM) components, potential immunogenicity, ideal in vitro culture condition, and other implantation-related dosage, order, route, and function monitoring system (Weiss et al. 2014; Nichols et al. 2012). Investigators have identified efficient modalities of expanding MSCs isolated from bone marrow or adipose tissue aspirates, while maintaining their multipotency (Caplan 2007). These cells have been used with appropriate scaffolds to form tissues such as bone and cartilage, upon transplantation at specific sites in experimental animal models. However, no human MSC-based technology is currently available (Schipani and Kronenberg 2008).

6.7.2 MSCs in Tissue Repair

Upon exogenous administration, MSCs have been demonstrated to serve as effective therapeutic agents in a variety of experimental models of tissue injuries (Ortiz et al. 2003; Kunter et al. 2006; Minguell and Erices 2006; Phinney and Isakova 2005). However, many of the studies suggest that therapeutic efficacy did not correlate with the efficiency of engraftment, which was low in general. The ability to repair was very likely secondary not to trans-differentiation of MSCs into the appropriate cell phenotype or to cell fusion, but rather to the secretion by MSCs of soluble factors that altered the tissue microenvironment (Prockop and Oh 2009). In this regard, extensive proteomic analyses have shown that MSCs produce a variety of factors in vitro that may influence a broad range of biological functions (Caplan 2007). Additionally it is suggested that chemokines such as SDF1 and its cognate receptor CXCR4; cytokine receptor CCR2, and its intracellular adapter FROUNT may have an essential role (Chamberlain et al. 2007) in homing MSCs to the sites of lung injury (Belema-Bedada et al. 2008). Further, MSCs are involved in the release of proteins that stimulate the growth of endothelial precursor cells and guide the assembly of new blood vessels in a process called neovascularization from preexisting endothelial cells. Thus, the above-mentioned studies that elucidate the function of MSCs have led researchers to hope that MSCs may provide a way to repair the blood vessel damage linked critical diseases (Schipani and Kronenberg 2008). A number of early stage clinical trials using MSCs in patients are currently underway that may prove to be effective.

6.7.3 Immunomodulatory Properties

MSCs have significant clinical implications in the immune modulation as they exert potent immune-suppressive and anti-inflammatory effects through the interactions between the lymphocytes associated with both the innate and adaptive immune systems. MSCs function to suppress T cell proliferation (Glennie et al. 2005; Di Nicola et al. 2002), B cell functions (Augello et al. 2005; Corcione et al. 2006), natural killer cell proliferation, and cytokine production, and prevent the differentiation, maturation, and activation of dendritic cells (Aggarwal and Pittenger 2005; Nauta et al. 2006; Beyth et al. 2010) (Fig. 6.2). Also, MSCs have the ability to suppress

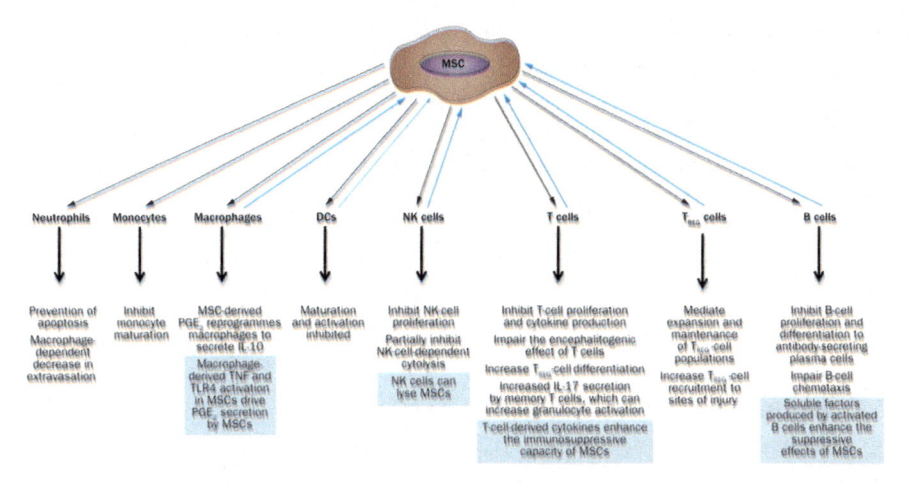

Fig. 6.2 Each arrow represents functional interaction between the indicated cell types. In many cases bi-directional crosstalk exists, which influences the final outcome of the interaction, and thus the effects of MSC-based therapy on disease. *Black arrows* and text pertain to responses driven by MSCs, whereas the effects of immune cells on MSCs are indicated by *blue arrows* and boxes. In general, the effects of MSCs on cells of the immune system are anti-inflammatory. *DCs* dendritic cells, *MSCs* mesenchymal stem cells, *NK* natural killer (cell), *PGE2* prostaglandin E2, *TLR4* Toll-like receptor 4, *TREG* regulatory T (cell)

cells independently of the major histocompatibility complex (MHC) individuality between donor and recipient due to low expression of their MHC-II and other costimulatory molecules (Stagg et al. 2006). Primary mechanism of MSCs involves the production of soluble factors, including transforming growth factor-β (Keating 2008), HGF (Di Nicola et al. 2002), nitric oxide (Ren et al. 2008), and indoleamine 2,3-dioxygenase (Meisel et al. 2004) while cell to cell contact modulates the immunosuppressive effects. Furthermore, through cell to cell contact and production of soluble factors, the property of MSCs to induce an immunosuppressive environment by generation of regulatory T cells (Tregs) has been observed in both in vitro (Di Ianni et al. 2008) and in vivo conditions (Patel et al. 2010; Madec et al. 2009). These observations suggest that MSCs are key regulators in immunomodulation and act by directly suppressing activated immune cells and indirectly recruiting Tregs. Microenvironmental stimuli represent an additional variable that can greatly influence MSCs immunosuppressive activity which in turn depends on the process of activation. Under acute inflammatory conditions polarized by M1 macrophages and helper T lymphocyte (Th)-type-1 cytokines, especially the pro-inflammatory cytokine interferon (IFN)-g, the immunosuppressive capacity of MSCs is enhanced through increased production of ICAM-1, CXCL-10, CCL-8, and IDO (Marigo and Dazzi 2011; Dazzi and Marelli-Berg 2008; Krampera et al. 2006) while under chronic inflammatory conditions when MSCs are polarized by M2 macrophages and Th2 cytokines, MSCs can be recruited into the fibrotic process. Thus, inflammatory microenvironment is the key component which should be taken into consideration when MSCs are used for therapy. Studies suggest that MSCs are able to slow down the multiplication of immune cells to reduce inflammation and can help treat transplant rejection or autoimmune diseases (Yang and Jia 2014). Various studies have analyzed the inhibitory effect exerted by MSCs on cells belonging to acquired or to innate immunity. Studies are conducted to monitor the inhibitory effect of MSCs and it is shown that through type I immune responses they are able to inhibit proliferation and function of natural killer (NK) cells and hinder the generation of dendritic cells and macrophages. Further investigation on the biology of MSC and on the regulatory events involved in their functional activities can help to optimize their use in clinical practice (Kim and Cho 2013).

6.8 Acute Respiratory Distress Syndrome

Acute respiratory distress syndrome (ARDS) is a devastating disease that continues to have a high mortality and for which currently there is no treatment. It is considered as a major cause of acute respiratory failure and is often found to be associated with multiple organ failure. Also, conditions such as pneumonia, sepsis, aspiration of gastric contents, and major trauma may stimulate ARDS and even with the current advances in lung-protective ventilation and fluid management, patient mortality rate remains high. The diagnostic criteria for ARDS have recently been revised and the new definition of the criteria include the following parameters: timing (within 1 week of a known clinical insult or new or worsening respiratory symptoms); chest imaging (bilateral opacities consistent with pulmonary edema); origin of lung edema: respiratory failure not fully explained by cardiac failure or fluid overload; and oxygenation (different categories according to the degree of hypoxemia severity—mild, moderate, and severe). ARDS is a major public health problem worldwide, with more than 100,000 cases per year, and carries a high economic burden associated with long intensive care unit and hospital stays. ARDS mortality remains high (30–40 %), and no therapy exists for this devastating disease, underlining the urgent need for the development of new approaches. Regardless of the cause, pathogenesis of ARDS involves lung endothelial injury, alveolar epithelial injury leading to increased permeability and extravasations of protein-rich fluid into the alveolar space, which compromises surfactant synthesis. Further, the alveolar damage is worsened by alveolar neutrophil influx, formation of hyaline membranes, and the injury process may be exacerbated by mechanical ventilation (Definition Task Force et al. 2012) (Fig. 6.3). In severe cases, the proliferative phase results in an increased number of type II alveolar cells, fibroblasts, myofibroblasts and matrix deposition, fibrosis, and multiple organ failure, this is the leading cause of death in ARDS. The immunomodulatory and reparative potential of MSCs makes them potential therapeutic tools for the acute inflammatory response to infection and pulmonary injury seen in ARDS. It is reported that biomarkers that reflect inflammation (IL-6, IL-8) (Parsons et al. 2005), coagulation (plasminogen activator inhibitor-1, protein C, and thrombomodulin) (Ware et al. 2007), endothelial cell injury (von Willebrand factor) (Parsons et al. 2005), and epithelial cell injury [SP-D and receptor for advanced

Fig. 6.3 Inflammation in alveoli after fluid accumulation

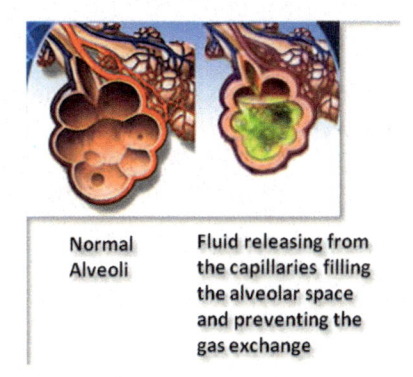

Normal Alveoli

Fluid releasing from the capillaries filling the alveolar space and preventing the gas exchange

glycosylation end products (RAGE)] (Parsons et al. 2005) have all been linked to increased disease severity and poorer clinical outcomes in patients with ARDS. The effects of MSC therapy in relevant preclinical models of ARDS further underline their therapeutic potential.

6.9 MSC Therapy in ARDS

Several preclinical studies of ARDS have demonstrated that MSCs may improve the pulmonary and systemic inflammation characteristic of the disease (Rojas et al. 2005; Gupta et al. 2007; Németh et al. 2009). MSC treatment in mouse models of lipopolysaccharide-induced ARDS not only attenuates inflammation by decreasing several inflammatory mediators, including TNF-α, MIP-2, IFN-γ, IL-1β, MIP-1α, IL-6, IL-8, and keratinocyte-derived cytokine in plasma and bronchoalveolar lavage fluid, but also increases secretion of the antibacterial protein lipocalin-2 (Gupta et al. 2012) and is able to rescue epithelial cells with mitochondrial dysfunction by mitochondria transfer (Islam et al. 2012; Spees et al. 2006). Thus, administration of MSCs favorably influences the host response to bacterial infections which is considered to be the commonest and most severe cause of ARDS. Studies have shown that MSC therapy can reduce bacterial counts via increased antimicrobial peptide secretion (Gupta et al. 2012) and enhanced macrophage phagocytosis (Németh et al. 2009). It is suggested that MSC therapy restores lung function following ventilator-induced lung injury via a keratinocyte growth factor (KGF) dependent mechanism. Based on these promising preclinical findings, a number of early phase clinical trials have begun to investigate the potential of MSC therapy for severe ARDS and presently, two studies of MSC therapy in patients with ARDS are ongoing (Ref?). Adipose-derived MSCs have also been used in preclinical studies in ARDS and in one such study, Martinez-González and colleagues demonstrated that genetically engineered adipose MSCs overexpressing soluble IL-1 receptor-like-1 decreased lung airspace inflammation, vascular leakage, reduced protein content, differential neutrophil counts, and pro-inflammatory cytokine (TNF-a, IL-6, and macrophage inflammatory protein 2) concentrations in bronchoalveolar lavage fluid (Martínez-González et al. 2013). In the randomized placebo-controlled phase I clinical trial, the primary goal of the study was to evaluate the safety and feasibility of systemic administration of allogeneic adipose-derived MSCs in ARDS patients (Joo et al. 2010). These studies have laid the foundation for ARDS clinical trials with MSCs. BM-MSCs reside in the bone marrow stroma in relatively small quantities and it has been estimated that they comprise about 0.001–0.01% of the total marrow nucleated cells (Bernardo et al. 2009), whereas the proportion of adipose-derived MSCs is approximately 2% of all nucleated cells of adipose tissue (Qayyum et al. 2012). Thus, this difference is particularly relevant for making adipose-derived MSCs more suited for clinical applications due to their ease of accessibility. Moreover, the findings demonstrated that infusion of allogeneic adipose-derived MSCs was safe and there were no significant adverse events related to the MSCs in ARDS. The change in ARDS biomarker, SP-D, after treatment may suggest the protective effect of MSCs. Additional large studies with a long follow-up period are necessary to confirm the safety and efficacy profile of MSCs in ARDS and to establish the best strategy for their administration, including concomitant medication and dosage (Zheng et al. 2014) (Table 6.1).

Table 6.1 Clinical trial status of mesenchymal stem cell therapy in ARDS as per the information reported in ClinicalTrials.gov

Title	Identifier	Status	Link
Human Mesenchymal Stem Cells for Acute Respiratory Distress Syndrome	NCT01775774	Recruiting patients	https://ClinicalTrials.gov/show/NCT01775774
Human Mesenchymal Stem Cells for Acute Respiratory Distress Syndrome (START)	NCT02097641	Unknown[a]	https://ClinicalTrials.gov/show/NCT02097641
Human Umbilical-Cord-Derived Mesenchymal Stem Cell Therapy in Acute Lung Injury	NCT02444455	Recruiting patients	https://ClinicalTrials.gov/show/NCT02444455
Adipose-Derived Mesenchymal Stem Cells in Acute Respiratory Distress Syndrome	NCT01902082	Recruiting patients	https://ClinicalTrials.gov/show/NCT01902082
Mesenchymal Stem Cell in Patients with Acute Severe Respiratory Failure	NCT02112500	Recruiting patients	https://ClinicalTrials.gov/show/NCT02112500

[a]Status has not been verified in more than 2 years

6.10 Chronic Obstructive Pulmonary Disease

COPD is characterized by persistent airflow limitation that is usually progressive and driven by an enhanced chronic inflammatory response in the airways and lung tissue in response to noxious particles or gases, especially cigarette smoke exposure, as well as genetic predisposition (α1-antitrypsin deficiency) (Global Strategy for the Diagnosis, Management and Prevention of COPD, 2014). Cigarette smoking is the commonest preventable cause of COPD and even after smoking cessation; inflammation, apoptosis, and oxidative stress can persist and continue to contribute towards diseased state. It is the fourth leading cause of chronic morbidity and mortality in the USA, and is projected to rank fifth in 2020 in burden of disease caused worldwide, according to a study published by the World Bank/World Health Organization. Furthermore, although COPD has received increasing attention from the medical community in recent years, awareness among the people is still very less. Although current therapies based on anti-inflammatory drugs such as corticosteroids, theophylline, and bronchodilators have the potential to reduce airway obstruction, limit exacerbation, and improve the patient's health-related quality of life; none can prevent disease progression or reduce mortality (Jin et al. 2015). Destruction of lung parenchyma by inflammation leads to the loss of the alveolar attachments to the small airways and decreases elastic recoil, which diminishes the ability of the airways to remain open, resulting in their collapse during expiration (Fig. 6.4) (Global Strategy for the Diagnosis, Management and Prevention of COPD, 2014). Inflammation in COPD involves an increased number of CD8+ cytotoxic lymphocytes, neutrophils, and macrophages, which release inflammatory mediators and enzymes and interact with structural cells of the airways (epithelial cells and fibroblasts), lung parenchyma (alveolar epithelial cells), and pulmonary

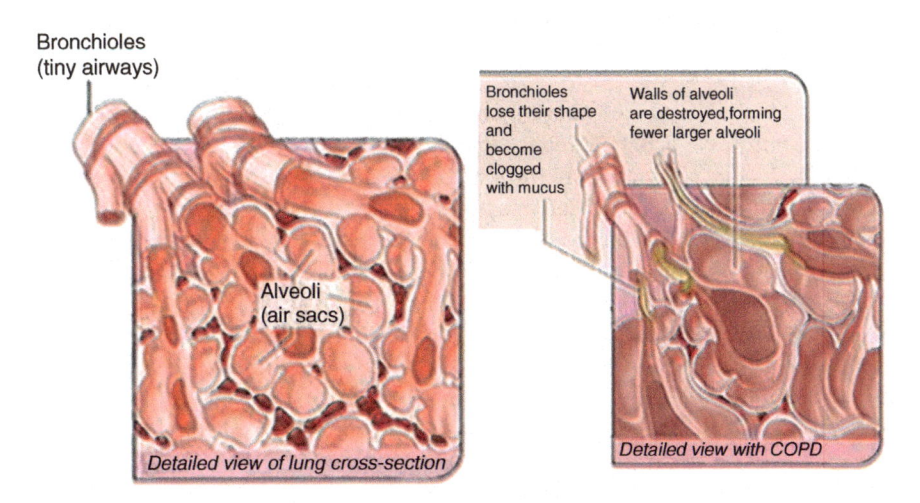

Fig. 6.4 Comparative view of a normal lung and a lung with COPD

vasculature (endothelial cells), leading to further inflammation and destruction of the pulmonary tissue (Global Strategy for the Diagnosis, Management and Prevention of COPD, 2014). In severe cases, COPD is considered a systemic disorder triggered by the primary pulmonary injury. Moreover, hypoxic vasoconstriction results in intimal hyperplasia and smooth muscle hypertrophy/hyperplasia, which in severe cases may lead to PAH (Antunes et al. 2014). Repairing the destroyed lung structure with stem cell therapies becomes an extremely attractive choice to treat COPD and MSC is the most extensively studied candidates in clinical trials not only for COPD, but also for other chronic diseases.

6.11 Current Therapeutic Progress with MSC

Current therapies aim to control symptoms, limit inflammation, and enhance functional capacity in patients with COPD. However, no available therapy has been able to reconstitute the alveolar architecture or halt the fibrogenic process. MSC treatment has been demonstrated to attenuate inflammation by decreasing levels of inflammatory mediators, such as IL-1 b, TNF- a, IL-8, as well as decrease apoptosis (Zhen et al. 2010; Huh et al. 2011), improve parenchymal repair, and increase lung perfusion (Guan et al. 2013; Shigemura et al. 2006). Based on these preclinical findings, several groups are investigating the therapeutic potential of MSC therapy in COPD patients and to date, two early phase clinical trials have been finished. The first safety trial is registered in ClinicalTrials.gov (NCT01110252) using bone marrow mononuclear cells, which encompasses the whole fraction of HSCs and MSCs from bone marrow, was carried out in Brazil. On 12-month follow-up, four patients/volunteers with advanced COPD (stage IV dyspnea) exhibited no adverse effects and experienced a significant improvement in the quality of life consistent with a more clinical stable condition (Ribeiro-Paes et al. 2011). The only published trial to date was carried out in the USA (NCT00683722), using intravenous allogeneic MSCs (PROCHYMAL1; Osiris Therapeutics Inc). Sixty-two patients were randomized to double-blinded intravenous infusions of either allogeneic MSCs or vehicle control. Patients received 4 monthly infusions and were subsequently followed for 2 years after the first infusion. This trial demonstrated that use of MSCs administration in COPD patients may be considered safe. Interestingly, the subgroup of patients who had increased circulating C-reactive protein levels at baseline demonstrated a significant decrease following MSC infusion (Weiss et al. 2013). One of the mechanisms postulated for MSC protection against emphysema is suppression of the inflammatory response by modulating the release of soluble anti-inflammatory molecules and activation of cellular anti-inflammatory pathways. Intrapulmonary administration of MSCs in a rat model of cigarette smoke-induced emphysema has been shown to ameliorate emphysematous pathology in these animals, in part via downregulation of pro-inflammatory mediators such as TNF-a, IL-1 b, IL-6, and MCP-1 (Guan et al. 2013). Pulmonary administration of MSCs has been shown to reverse the induction of matrix metalloproteinase (MMP)-9 and

MMP-12 in the lungs of rats with cigarette smoke-induced emphysema, at both the mRNA and protein levels. Although the mechanistic basis of this effect is not completely understood, it has been attributed in part to the inhibition (by MSCs) of a positive feedback loop, involving the release of proteases by inflammatory and structural cells activated by cigarette smoke, and the recruitment by these proteases of additional inflammatory cells (Churg et al. 2003). Since, studies have shown where MSCs stimulate VEGF secretion (Wong et al. 2007) and VEGFR2 induction, amelioration by MSC transplantation of alveolar cell apoptosis in the lungs of papain (Zhen et al. 2010) or cigarette smoke-induced rat models of emphysema has therefore been postulated to involve reversal of the effects of cigarette smoke exposure on the VEGF signalling pathway (Kanazawa and Yoshikawa 2005). An alternative mechanism by which MSCs suppress alveolar cell apoptosis has been suggested to involve alterations in the expression of apoptotic or antiapoptotic genes in these cells (Zhen et al. 2008). It has been reported, for example, that the apoptotic gene Bax and the antiapoptotic gene Bcl-2 are induced and repressed, respectively, after pulmonary administration of MSCs in a papain induced model of emphysema in rats. A third mechanism for MSC amelioration of alveolar apoptosis is suggested by the suppression by MSCs of alveolar levels of cleaved-caspase a key player in the apoptotic programme in these cells (Kim et al. 2012). Modulation by MSCs of the redox environment is a rapidly emerging area of interest. For example, the increased survival rate of lipopolysaccharide-induced lung injury rats after transplantation of bone marrow MSCs has been shown to be accompanied by decreased oxidative stress (Li et al. 2012). Moreover, reduction by MSCs of pulmonary levels of malondialdehyde occurs in parallel with increased synthesis of heme oxygenase-1, an enzyme with strong antioxidative stress and cytoprotective effects (Fredenburgh et al. 2007). In addition, transplantation of bone marrow MSCs is known to decrease oxidative stress in the brain of a rat model of spontaneous stroke, suggesting that MSCs may decrease oxidative stress in cigarette smoke-induced emphysema (Calió et al. 2014). Further studies are needed to understand the effects of MSCs on oxidative stress in emphysema, and the antioxidative mechanism of its action in alveolar cells. Differentiation of MSCs into type I and/or type II alveolar epithelial cells has been reported in rat models of lipopolysaccharide and cigarette smoke-induced emphysema and bleomycin-induced lung injury (Rojas et al. 2005; Kotton et al. 2001). On a mechanistic level, it is shown that differentiation of MSCs into type II alveolar epithelial cells in a co-culture system is associated with activation of the canonical Wnt signalling pathway (Liu et al. 2013). At present, there are no therapies that can reduce the disease progression or mortality associated with COPD. Transplantation of MSCs represents a potentially promising therapy for COPD, and may involve modulation of inflammation, protease/antiprotease balance, apoptosis and oxidative stress, or the differentiation of MSCs into lung parenchyma cells. A major obstacle to the clinical application of MSCs in COPD is the dearth of data on the long-term safety of MSCs in patients with COPD. Further larger scale clinical trials will be necessary to more fully assess the efficacy and long-term safety of MSCs in patients with COPD. A second major challenge to the clinical application of MSCs in COPD is that the therapeutic schedule is not clear, and additional studies are warranted to ascertain the appropriate cellular dose, infusion rate, and route of administration. A final challenge is the poor survival of MSCs and the low level of engraftment in host organs (Parekkadan and Milwid 2010); therefore, a pressing need exists for the development of approaches that increase survival and engraftment of MSCs in host organs. In summary, although several challenges exist, transplantation of MSCs represents a potentially promising therapy for COPD (Table 6.2).

6.12 Asthma

Asthma is a chronic disease that involves inflammation of the pulmonary airways and bronchial hyper-responsiveness affecting over 24 million people in the USA (Moorman et al. 2007). The disease results in the clinical expression of a lower

Table 6.2 Clinical trial status of mesenchymal stem cell therapy in COPD as per the information reported in ClinicalTrials.gov

Title	Identifier	Status	Link
PROCHYMAL™ (Human Adult Stem Cells) for the Treatment of Moderate to Severe Chronic Obstructive Pulmonary Disease (COPD)	NCT00683722	Completed	https://ClinicalTrials.gov/show/NCT00683722
Safety Study of Endobronchial Transplantation of Autologous Mesenchymal Stem Cells (MSCs) in Emphysema Patients	NCT01758055	Unknown[a]	https://ClinicalTrials.gov/show/NCT01758055
Clinical Study of the Efficacy and Safety of the Application of Allogeneic Mesenchymal (Stromal) Cells of Bone Marrow, Cultured Under the Hypoxia in the Treatment of Patients with Severe Pulmonary Emphysema	NCT01849159	Recruiting patients	https://ClinicalTrials.gov/show/NCT01849159
Safety and Feasibility Study of Administration of Mesenchymal Stem Cells for Treatment of Emphysema	NCT01306513	Completed	https://ClinicalTrials.gov/show/NCT01306513
Adipose-Derived Stem Cells Transplantation for Chronic Obstructive Pulmonary Disease	NCT02645305	Recruiting patients	https://ClinicalTrials.gov/show/NCT02645305

[a]Status has not been verified in more than 2 years

airway obstruction that usually is reversible. Bronchial hyper-responsiveness is documented by decreased bronchial airflow after broncho-provocation with methacholine or histamine. Broncho-provocation with allergen induces a prompt early phase immunoglobulin E (IgE) mediated decrease in bronchial airflow followed in many patients by a latephase IgE mediated reaction with a decrease in bronchial airflow for 4–8 h. Additionally, the gross pathology of asthmatic airways displays lung hyperinflation, smooth muscle hypertrophy, lamina reticularis thickening, mucosal edema, epithelial cell sloughing, cilia cell disruption, and mucus gland hypersecretion. Also, asthma is characterized by the presence of increased numbers of eosinophils, neutrophils, lymphocytes, and plasma cells in the bronchial tissues, bronchial secretions, and mucus (Fig. 6.5). Initially, there is recruitment of leukocytes from the bloodstream to the airway by activated CD4 Tlymphocytes that directs the release of inflammatory mediators from eosinophils, mast cells, and lymphocytes. Additionally, the subclass 2 helper Tlymphocytes subset of activated Tlymphocytes produces interleukin (IL)-4, IL5, and IL-13. IL4 in conjunction with IL13 signals the switch from IgM to IgE antibodies. Further, crosslinkage of two IgE molecules by allergen causes mast cells to degranulate that leads to histamine release, leukotrienes, and other mediators that increase the obstruction in airways (Dimmeler et al. 2011). Long-term structural changes of the airways may arise due to repeated cycles of inflammation and injury to the pulmonary tissues. These structural and inflammatory changes throughout the airway wall may lead to bronchial thickening and edema as well as increased mucus production and broncho-constriction. The type of inflammatory cell predominately found in the airway denotes a specific phenotype in asthma; therefore, finding targeting antibodies, inflammatory cytokines, and inflammatory cells will be helpful for the treatment of asthma (Hanania 2008; Murphy and O'Byrne 2010; Robinson 2010). Although airway smooth muscle (ASM) has been implicated in constrictor hyper-responsiveness in asthma, other important roles of ASM have also been identified such as impaired airway relaxation, mediation of structural changes, and inflammatory signalling (Doeing and Solway 2013). It had been recognized that patients who die from acute asthma attacks have grossly inflamed airways and the airway lumen is occluded by a tenacious mucus plug composed of plasma

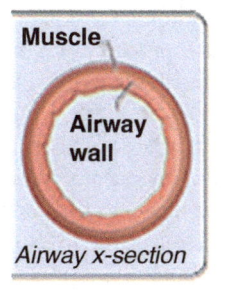

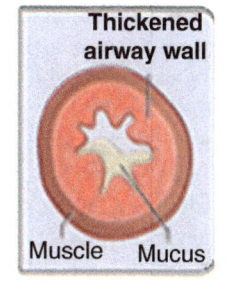

Fig. 6.5 Cross-section of a lung airway during asthma symptoms

proteins exuded from airway vessels and mucus glycoproteins secreted from surface epithelial cells. The airway epithelium is invariably shed in a patchy manner and clumps of epithelial cells are found in the airway lumen. Although, several therapeutic strategies are currently available to reduce airway inflammation, no treatment has so far been able to hasten repair of the damaged lung.

6.13 Possible Therapeutic Role of MSCs in Asthma

Several studies have reported that MSCs reduce lung inflammation and remodelling in experimental allergic asthma (Abreu et al. 2013; Goodwin et al. 2011; Lathrop et al. 2014). Extracellular vesicles (EVs) released from several cells are shown to be involved in allergies through mast cells, DCs (what is DC?), T cells, and bronchial epithelial cells (BECs) in the lungs. For instance, mast cell-derived EVs induce DC maturation, and DC-derived EVs can transport allergens and activate allergen-specific T-helper (Th) type 2 cells (Admyre et al. 2008). Earlier studies had shown that when BECs are exposed to compressive stress they simulate the bronchoconstriction that is one of the characteristic features seen in asthma and produce EVs bearing tissue factor which may participate in promotion of sub-epithelial fibrosis and angiogenesis (Park et al. 2012). Additionally, findings that target the inhibition of EVs secretion by these cells may lead to the development of future treatments for asthma patients. Therapeutic effects of EVs derived from human MSCs (hMSCs) and mouse MSCs (mMSCs) were investigated in experimental asthma and it was observed that systemic administration of EVs from either hMSCs or mMSCs were each more effective than the administration of hMSCs or mMSCs themselves in the process of mitigating allergic airway hyper-responsiveness and lung inflammation (Cruz et al. 2015). Also, they were able to alter the antigen-specific CD4+ T cells phenotype in an immunocompetent mice model of severe, acute, mixed Th2/Th17-mediated eosinophilic and neutrophilic airway allergic inflammation. Further, blocking EVs release led to an absence of protective effects associated with both hMSCs and mMSCs. Role of MSCs in asthma MSC immunoregulatory effects and allogenic tolerance (García-Castro et al. 2008) led to the treatment of severe immune-mediated diseases which in turn attracted interest on anti-inflammatory MSCs as a potential candidate for asthma treatment (Le Blanc et al. 2008; Trounson et al. 2011). Evidence suggests that airway remodelling is an important feature of asthma pathophysiology and MSCs administered for therapeutic purposes in asthma may bear a dual property of reducing inflammation and serving as building blocks for airway remodelling (Brewster et al. 1990; Gizycki et al. 1997; Schmidt et al. 2003; Nihlberg et al. 2006; Dolgachev et al. 2009; Saunders et al. 2009; Ramos-Barbón et al. 2010). Pardo et al. showed that MSCs homed to the lungs and rapidly downregulated airway inflammation in association with raised T-helper-1 lung cytokines; however, such an effect declined under sustained allergen challenge despite persistent presence of MSCs (Mariñas-Pardo et al. 2014). Their study concluded that therapeutic MSC infusion in murine experimental asthma is free of unwanted pro-remodelling effects and ameliorates airway hyperresponsiveness and contractile tissue remodelling (Cubillo et al. 2014). Airway dendritic cells (DCs) have an important role in the pathogenesis of allergic asthma, and disrupting their function may be a novel therapeutic approach. To prove this hypothesis, Zeng et al. used a mouse model of asthma to demonstrate that transplantation of MSCs suppressed features of asthma by targeting the function of lung myeloid DCs where MSCs suppressed the maturation and migration of lung DCs to the mediastinal lymph nodes, and thereby reducing the allergen-specific T helper type 2 (Th2) responses in the nodes. In addition, MSC-treated DCs were less potent in activating naive and effector Th2 cells and the capacity of producing chemokine (C-C motif) ligand 17 (CCL17) and CCL22, which are chemokines attracting Th2 cells, to the airways was reduced. These results supported that MSCs may be used as a potential treatment for asthma (Zeng et al. 2015). Another study done by Braza and colleagues indicates the role of lung macrophages in attenuating asthma by phagocytosis of MSCs (Braza et al. 2016). Delivery of compact bone derived MSCs to the injured lungs is an alternative treatment strategy for chronic asthma. In the study conducted by Ogulur et al. mouse compact bone MSCs migrated to the lung tissue and suppressed histological changes in murine model of asthma (Ogulur et al. 2014). These results provide evidence that mouse compact bone MSCs may be used as an alternative strategy for the treatment of remodelling and inflammation associated with chronic asthma (Table 6.3).

6.14 Idiopathic Pulmonary Fibrosis

Idiopathic pulmonary fibrosis (IPF) is defined as a chronic, progressive fibrosing interstitial pneumonia of unknown cause. The disease is characterized by the ECM deposition and scar tissue formation in the interstitial lungs over time (Fig. 6.6). It is seen primarily in older adults, is a condition that is limited to the lungs. The incidence of IPF is 13–20 cases per 100,000 people (Raghu et al. 2006). Typically the disease is found in old adults and the median survival time is 3–5 years. The clinical presentation includes exertional dyspnea, cough, functional and exercise limitation, impaired quality of life, and risk for acute respiratory failure and death. Unfortunately,

Table 6.3 Clinical trial status of mesenchymal stem cell therapy in asthma as per the information reported in ClinicalTrials.gov

Title	Identifier	Status	Link
Safety and Feasibility Study of Intranasal Mesenchymal Trophic Factor (MTF) for Treatment of Asthma	NCT02192736	Active, not recruiting patients	https://ClinicalTrials.gov/show/NCT02192736

Fig. 6.6 Cross-section of the lungs airways and air sacs in IPF

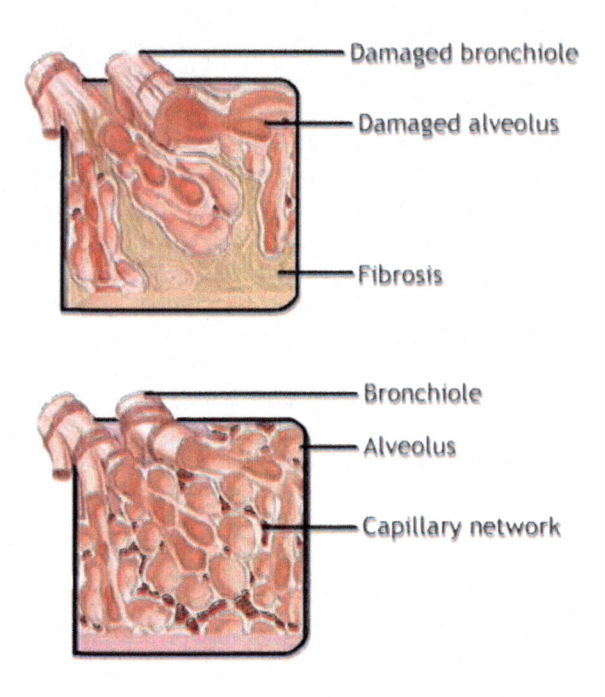

there is no FDA approved treatment or cure for IPF patients. Pirfenidone is an approved antifibrotic and anti-inflammatory drug in Europe and Japan, but still in clinical trials in North America (Taniguchi et al. 2010; Noble et al. 2011). Presently, accumulation of activated myofibroblasts is thought to be the source of interstitial collagens and there are at least four proposed cellular origins of myofibroblasts including expansion of lung residential fibroblasts, pericytes, recruitment of BM-derived fibrocytes, and alveolar epithelial cells undergoing epithelial-mesenchymal transition (EMT) (Hung et al. 2013; Yang et al. 2013). Among these cellular origins, the circulating fibrocytes have been proposed as a clinical marker to assess the progress of disease condition (Moeller et al. 2009). Importantly, other known causes of interstitial pneumonia (idiopathic interstitial pneumonias and interstitial lung disease associated with environmental exposure, medication, or systemic disease) must have been ruled out, making IPF a diagnosis of exclusion (Raghu et al. 2011). The hallmark of IPF is a pattern of fibrosis area with scarring and honeycomb changes alternating with areas of less affected or normal parenchyma. These alterations often affect the subpleural and paraseptal parenchyma most intensely, while inflammation is usually mild and is constituted by an irregular interstitial infiltrate of lymphocytes and plasma cells associated with hyperplasia of type 2 alveolar epithelial cells and bronchiolar epithelium. While there are no definitive studies of the incidence or prevalence of IPF, prevalence estimates have ranged from 2 to 29 cases per 100,000 in the general population (Raghu et al. 2011). While treatment of conditions such as pulmonary hypertension, gastroesophageal reflux disease, obesity, emphysema, and obstructive sleep apnea may improve respiratory symptoms in IPF patients, the Committee on Idiopathic Pulmonary Fibrosis does not support the use of any direct pharmacologic therapy for IPF. Conversely, the use of several nonpharmacologic therapies, including oxygen therapy, lung transplantation, mechanical ventilation, and pulmonary rehabilitation, is recommended in appropriate patients (Antunes et al. 2014). Further larger scale trials are necessary to examine the potential effects of MSCs on clinical outcome in patients with IPF.

6.15 Treating IPF with MSCs

Murine bleomycin model has been considered as one of the best characterized animal models for human IPF disease. With this model, several groups have shown that administration of allogeneic bone marrow MSCs reduced inflammation and collagen deposition (Rojas et al. 2005; Ortiz et al. 2003, 2007). Stem cells from other sources such as placenta and human umbilical cord also demonstrated reduced lung tissue damage in the mouse bleomycin models (Cargnoni et al. 2009; Moodley et al. 2009). Interestingly, HSCs and MSCs have been genetically manipulated as vehicles to deliver KGF into the lung injured sites. In a recent clinical trial allogeneic MSCs were found to be safe and well tolerated to treat refractory lupus erythematosus (Liang et al. 2010). Considering the potential beneficial role of MSCs in preclinical models, a few clinical trials on IPF patients have been approved. In addition, FDA has approved a clinical trial phase I study of autologous MSCs to treat IPF patients (initiated in March 2013, ClinicalTrials.gov Identifier: NCT01919827) that will evaluate the safety and feasibility of the endobronchial administration of autologous BM-MSCs in patients with mild to moderate IPF. On the other hand, two separate groups found that MSCs may play a role in the fibrogenic process of the lung. Antoniou and colleagues reported an increased expression of the axis stromal-cell-derived factor-1 (SDF-1)/CXCR4 in BM-MSCs from IPF patients, suggesting the BM-MSCs may probably implicate in the pathogenesis of IPF by recruiting MSCs to lung injury sites (Antoniou et al. 2010). Walker and colleagues recently found that allograft derived local MSCs contain a profibrotic phenotype by up-regulation of FOXF1, α-SMA, and collagen I in human lung transplant recipients with BOS (Walker et al. 2011). Preclinical studies using a bleomycin-induced IPF model (Bitencourt et al. 2011) demonstrated that stem cells decreased inflammation, with reductions in neutrophil infiltration, fibrosis, and collagen deposition and an increase in epithelial repair (Moodley et al. 2009; Ortiz et al. 2007). Due to these positive effects, MSCs are currently being explored in the clinical studies for IPF patients. Currently, there are three trials officially registered with NIH/FDA to evaluate the safety and feasibility of MSC therapy in IPF patients. In the USA, a phase I, randomized, blinded, and placebo-controlled trial is recruiting 25 IPF patients to investigate the safety, tolerability, and potential efficacy of intravenous infusion of allogeneic human MSCs (NCT02013700). The migration of MSCs to the injured tissues seems to be central to the efficacy of this cell therapy, raising the possibility that localized inflammatory processes may induce MSCs to release chemokines and growth factors, stimulating tissue repair. Considering that the lung offers the intratracheal/endobronchial route as a direct pathway for drug/cell delivery, this route of administration may potentiate MSC efficacy. Another ongoing trial is being conducted at Navarra University in Spain to examine this issue. This phase I, open-label, multicenter, non-randomized study will evaluate the safety and feasibility of the endobronchial infusion of autologous bone marrow MSCs at escalating doses in patients with mild-to-moderate IPF (NCT01919827). Thus, MSCs may have therapeutic potential for patients with IPF (Antunes et al. 2014) (Table 6.4).

Table 6.4 Clinical trial status of mesenchymal stem cell therapy in IPF as per the information reported in ClinicalTrials.gov

Title	Identifier	Status	Link
Study of Autologous Mesenchymal Stem Cells to Treat Idiopathic Pulmonary Fibrosis	NCT01919827	Recruiting patients	https://ClinicalTrials.gov/show/NCT01919827
Allogeneic Human Cells (hMSC) in Patients with Idiopathic Pulmonary Fibrosis Via Intravenous Delivery (AETHER)	NCT02013700	Active, not recruiting patients	https://ClinicalTrials.gov/show/NCT02013700
A Study on Radiation-induced Pulmonary Fibrosis Treated with Clinical Grade Umbilical Cord Mesenchymal Stem Cells	NCT02277145	Recruiting patients	https://ClinicalTrials.gov/show/NCT02277145
A Study to Evaluate the Potential Role of Mesenchymal Stem Cells in the Treatment of Idiopathic Pulmonary Fibrosis	NCT01385644	Completed, has results	https://ClinicalTrials.gov/show/NCT01385644
Safety and Efficacy of Allogeneic Mesenchymal Stem Cells in Patients with Rapidly Progressive Interstitial Lung Disease	NCT02594839	Recruiting patients	https://ClinicalTrials.gov/show/NCT02594839
Evaluate Safety and Efficacy of Intravenous Autologous ADMSc for Treatment of Idiopathic Pulmonary Fibrosis	NCT02135380	Unknown[a]	https://ClinicalTrials.gov/show/NCT02135380

[a]Status has not been verified in more than 2 years

6.16 Bronchopulmonary Dysplasia

BPD is a chronic respiratory disease that results from complications related to the treatment of RDS in low-birth-weight premature infants, or when abnormal lung development occurs in older infants (Walsh et al. 2006). BPD is assessed at a postmenstrual age of 36 weeks and clinically diagnosed if there is higher than 21 % oxygen dependency for 28 days or more. It occurs almost exclusively in very preterm infants born before 30 weeks of gestation, when around one in three infants develop BPD (Smith et al. 2005). The incidence of BPD appears to be rising in parallel with the increased survival rate of very-low-birth-weight infants who are treated for and recover from RDS. Histologically, BPD is characterized by diffuse pulmonary inflammation, with alveolar and vascular simplification and an arrest of lung development at the late canalicular to early saccular stages. This results in a reduced surface area for gas exchange and chronic pulmonary dysfunction (Thibeault et al. 2003) (Fig. 6.7). Although its etiology is not completely understood, mechanisms that contribute to the pathogenesis of BPD include oxidative and inflammation-mediated lung injury due to chorioamnionitis, hyperoxia, infection, ventilator-induced injury, and other adverse stimuli. Inflammatory cells and cytokines in tracheal fluid from premature newborns are strongly associated with high risk for the subsequent development of BPD. Injury disrupts normal growth factor and cellular signalling during key stages of lung maturation, which impairs alveolar and vascular growth and results in the abnormal lung structure of BPD. Data from clinical and laboratory studies suggest that early interventions with anti-inflammatory agents and pharmacologic strategies that directly preserve lung cell survival and function should prevent the development of BPD. The immature lung does not support the increase in respiratory system requirements, requiring supplemental oxygen and mechanical ventilation, triggering inflammation, exacerbating structural deficits, inducing alveolar arrest, and ultimately leading to BPD. The pulmonary damage induced by BPD is irreversible for many children, and the respiratory impairment initiated during neonatal life may continue into adolescence and adulthood. The most effective therapy for BPD has been its prevention, but pharmacologic approaches have limited efficacy, particularly in more extreme infant prematurity. Despite a better understanding of the pathophysiology of BPD and significant research effort into its management, there remains today no effective treatment. Cell-based therapy is a novel approach that offers much promise in the prevention and treatment of BPD. Recent research supports a therapeutic role for cell transplantation in the management of a variety of acute and chronic adult and childhood lung diseases, with potential of such therapy to reduce inflammation and prevent acute lung injury. However, considerable uncertainties remain regarding cell therapies before they can be established as safe and effective clinical treatments for BPD. Over recent decades, improvements in the pre-and post-natal care of the pre term neonate have seen significant reductions in mortality and acute morbidities such as neonatal RDS. However, longer term sequelae of preterm birth remain a concern in these infants. In particular, lung immaturity and consequent BPD remain a major cause of morbidity and death.

Fig. 6.7 Normally, air flows easily through the bronchioles (small airways) in and out of the alveoli (air sacs). Scarring can constrict the airways and keep air sacs from opening fully

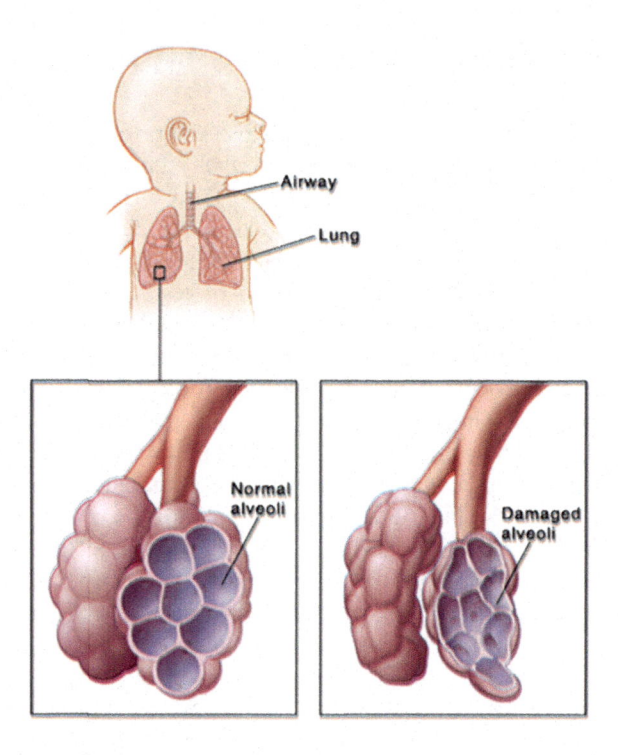

6.17 Therapeutic Dimensions of MSCs in BPD

MSC therapy may have therapeutic potential for infants with BPD, given their immunomodulatory effects and regenerative capacity. Interestingly, preclinical studies using hyperoxia-induced BPD models have demonstrated that bone marrow MSC therapy improves alveolar and vascular repair, lung function, and survival in preterm mouse pups (van Haaften et al. 2009; Aslam et al. 2009; Hansmann et al. 2012). Studies have shown that MSCs may also be isolated from UCB, and evidence from a model of neonatal hyperoxia-induced lung injury in rats suggests that they may contribute to alveolar repair through paracrine mechanisms (Chang et al. 2011). All five ongoing clinical trials of MSC therapy for BPD that have been registered at ClinicalTrials.gov are using PNEUMOSTEM1 MSCs, which are a human UCB-MSC preparation developed commercially as a potential therapy for premature infants with BPD. Only one trial has been completed to date which was conducted in South Korea and the study was an open-label, single-center, phase I clinical study, which evaluated the safety and the efficacy of PNEUMOSTEM1 for BPD treatment in premature infants (NCT01297205) (Antunes et al. 2014). Interestingly, Tropea and colleagues showed bronchoalveolar stem cells increase after mesenchymal stromal cell treatment in a mouse model of BPD (Tropea et al. 2012). Reported follow-up studies of children who had BPD as a neonate show that they have higher than average rates of chronic cough and wheeze, airway hyper-responsiveness, and lung function abnormalities throughout childhood. Unfortunately, not all studies of MSCs in BPD models have been so promising. In one study of hyperoxia-exposed mouse pups, the administration of whole bone marrow isolate resulted in substantial lung engraftment of MSCs, but the cells did not differentiate into epithelial cells instead, they unexpectedly undertook a chimeric change generating alveolar macrophages. To conclude, they worsened the lung injury thus, although some studies suggest promise for selected MSCs as a therapy for BPD, uncertainty remains about the safety of these cells. Importantly, the unexpected generation of alveolar macrophages from whole bone marrow transplant is a concerning factor and requires additional consideration before considered in clinical therapy (Vosdoganes et al. 2012). Two important and remarkably complementary studies report that MSC therapy reduces acute lung inflammation and improves late lung structure in experimental rodent models of hyperoxia-induced BPD. In this study it has been shown that hyperoxia reduces MSCs in lung and blood, suggesting that decreased endogenous MSC numbers may be a pathogenetic mechanism of BPD. Although engrafted MSCs co-expressed the alveolar type II cell marker and surfactant protein C, the numbers of engrafted cells were extremely low, making cell differentiation and replacement as the primary mechanism of lung protection unlikely. Parallel in vitro studies demonstrate the protective effects of conditioned media (CM) from MSC on epithelial and endothelial functions during hyperoxia, supporting the concept that mediators from the MSC secretome, and not cell replacement itself, is the most likely mechanism of lung protection in this model. Study demonstrated that intravenous delivery of MSC shortly after birth protected neonatal mice from hyperoxia-induced lung injury, as reflected by decreased inflammation, improved alveolar structure and less pulmonary hypertension. Importantly, infusions of study animals with CM alone resulted in even more profound improvement than observed with MSC. These findings suggest that MSC act in a paracrine fashion through the release of immunomodulatory factors, which based on initial characterization by mass spectroscopic analysis, may include macrophage stimulating factor-1 and osteopontin as potential candidates. Whether late treatment can have further effects on lung regeneration or reversing the arrest in lung development that characterizes established BPD needs to be tested. Several recent studies have shown that xenotransplantation of MSCs in immunocompetent animals attenuates hyperoxia-induced lung injury, such as impaired alveolarization, inflammatory response, increased apoptosis, and fibrosis (Chang et al. 2011; Aslam et al. 2009; Abman and Matthay 2009). Overall, these findings suggest that intratracheal transplantation of allogeneic human UCB-derived MSCs in very preterm infants at the highest risk for developing BPD is safe and feasible. A longterm followup safety study (NCT01632475) on MSC-treated preterm infants and a phase II doubleblind randomized controlled trial to assess the therapeutic efficacy (NCT 01828957) are currently underway (Ahn et al. 2015). However, presently no effective treatments beyond supportive therapies are available for BPD. Finally, the rationale for translating these results to the clinical setting is strong, but how to approach the application of MSC therapy for the prevention of BPD remains a considerable but worthwhile challenge. Therefore, development of new therapeutic modalities to improve the prognosis of BPD in preterm infants is an urgent priority (Table 6.5).

Table 6.5 Clinical trial status of mesenchymal stem cell therapy in BPD as per the information reported in ClinicalTrials.gov

Title	Identifier	Status	Link
Mesenchymal Stem Cell Therapy for Bronchopulmonary Dysplasia in Preterm Babies	NCT02443961	Not yet recruiting patients	https://ClinicalTrials.gov/show/NCT02443961
Intratracheal Umbilical Cord-derived Mesenchymal Stem Cells for Severe Bronchopulmonary Dysplasia	NCT01207869	Unknown[a]	https://ClinicalTrials.gov/show/NCT01207869
Safety and Efficacy of PNEUMOSTEM® in Premature Infants at High Risk for Bronchopulmonary Dysplasia (BPD)—A US Study	NCT02381366	Recruiting patients	https://ClinicalTrials.gov/show/NCT02381366
Safety and Efficacy Evaluation of PNEUMOSTEM® Treatment in Premature Infants with Bronchopulmonary Dysplasia	NCT01297205	Completed	https://ClinicalTrials.gov/show/NCT01297205
Follow-up Study of Safety and Efficacy of PNEUMOSTEM® in Premature Infants with Bronchopulmonary Dysplasia	NCT01632475	Active, not recruiting patients	https://ClinicalTrials.gov/show/NCT01632475
Efficacy and Safety Evaluation of PNEUMOSTEM® Versus a Control Group for Treatment of BPD in Premature Infants	NCT01828957	Active, not recruiting patients	https://ClinicalTrials.gov/show/NCT01828957
Long-term Safety and Efficacy Follow-up Study of PNEUMOSTEM® in Patients Who Completed PNEUMOSTEM® Phase-I Study	NCT02023788	Active, not recruiting patients	https://ClinicalTrials.gov/show/NCT02023788
Follow-up Safety and Efficacy Evaluation on Subjects Who Completed PNEUMOSTEM® Phase-II Clinical Trial	NCT01897987	Recruiting patients	https://ClinicalTrials.gov/show/NCT01897987

[a]Status has not been verified in more than 2 years

6.18 Pulmonary Arterial Hypertension

PAH is a devastating and refractory disease which is defined by a resting mean pulmonary artery pressure at or above 25 mmHg. Even though PAH is classified as a single PH subgroup, it is a heterogeneous disease with several clinical classifications based on pathology, hemodynamic characteristics, and/or treatment strategies. According to the most recent guidelines established at the fifth World Symposium on PH, PAH can be subdivided into the following classifications: (1) PAH of unknown etiology or idiopathic (IPAH); (2) heritable PAH (HPAH); (3) drug- and toxin-induced PAH; and (4) PAH associated with connective tissue disease, HIV infection, portal hyper tension, congenital heart diseases, or schistosomiasis (Simonneau et al. 2013). The disease is characterized by the decreased blood flow through the pulmonary arteries (Fig. 6.8) due to various pathological processes leading to vasoconstriction, vascular remodelling, and obliteration, which contribute to increased pulmonary vascular resistance, right ventricle hypertrophy, and in many cases, death due to right heart failure (Archer et al. 2010). Although the exact mechanisms leading to structural remodelling of the pulmonary vasculature remain yet to be fully elucidated, endothelial dysfunction is widely considered to be an initiating event (Budhiraja et al. 2004; Sakao et al. 2009) that results in the reduced production of vasodilators such as prostacyclin (Tuder et al. 1999), nitric oxide (Giaid and Saleh 1995), and increase the production of vasoconstrictors such as thromboxane A2 (Humbert et al. 2004) and endothelin-1 (Giaid et al. 1993). PAH is further characterized by several histological and functional abnormalities such as intimal hyperplasia and fibrosis, medial smooth muscle cell hypertrophy, increased deposition of ECM inflammatory cell infiltrates, and/or pulmonary endothelial cell injury and dysfunction at the level of the intra-alveolar (precapillary) arteriole (Archer et al. 2010). Although currently there is no cure for this disease, treatment has been improved during the past decade, offering both reliefs from symptoms and prolonged survival.

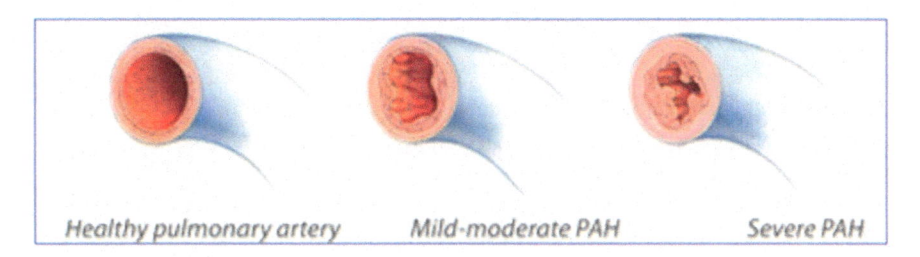

Fig. 6.8 Narrowing of the pulmonary arteries associated with PAH

6.19 Attenuation of PAH with MSC Therapy

Recently, the regenerative method such as transplantation of BM-derived MSCs and gene therapy has been introduced to break the vicious cycle of PAH (Jiang et al. 2002; Zhao et al. 2005). However, efficacy of MSCs transplantation was found to be unsatisfactory, due to the poor viability and massive death of the engrafted MSCs in the injured tissue. Recently, animal studies taking use of hypoxia induced animal model in pulmonary medicine have demonstrated that naive- or gene-modified MSCs from bone marrow can ameliorate some of the pulmonary hypertension symptoms. Interestingly, it has been shown that both intratracheal and intravenous administration of MSCs can attenuate pulmonary hypertension by modulating processes such as endothelial dysfunction (Baber et al. 2007), alveolar loss, lung inflammation (Aslam et al. 2009), and ventricle remodelling (Spees et al. 2008; Umar et al. 2009). Also, engineering of the MSCs to overexpress transgenes, or MSC-based gene therapy, has been shown to enhance the regenerative activity of MSCs in vivo. Recent studies have found that endothelial nitric oxide synthase (eNOS) (Kanki-Horimoto et al. 2006) or prostacyclin synthase (Takemiya et al. 2010) or lung-specific heme oxygenase (HO)-1 (Liang et al. 2011) modified MSCs are able to show ameliorating effects on PH-related right ventricle impairment and improve the prognosis and survival in PH animals. Overexpression of transgenes after ex vivo transfection is one of the important strategies to study specific molecular pathways in MSCs; however, genetic manipulation is rather costly, difficult to scale-up, and introduces potential toxicity of the nonviral or viral transfection vectors. Interestingly, these studies might form the basis for novel, less disruptive strategies, such as pharmacological or physicochemical preconditioning. Although a robust protection against lung injury on MSCs treatment was observed in most of the studies, only a small fraction of administered MSCs were detected in the wall of the pulmonary vessels. Thus, it may be speculated that engraftment and direct tissue repair might not be the sole mechanisms for MSC therapeutic function; paracrine mechanisms may have a major role in the whole process to play (Anversa et al. 2012). It is known that MSCs possess the property to get mobilized from the pool of bone marrow stromal cells (BMSCs) when influenced by hypoxia or other injury factors (Spees et al. 2008) and after mobilization, it can localize into the injured tissue, and even few MSCs can fuse with cells from the host (Rochefort et al. 2006). Under stress by tumor necrosis factor (TNF) or hypoxia, MSCs have also been shown to increase production of growth factors, such as VEGF, insulin-like growth factor (IGF), and HGF (Wang et al. 2006; Rehman et al. 2004). Possibilities from the studies arise that transplanted MSCs may repair injured vascular endothelium by an action involving the release of factors that improve endothelial function or stimulate vascular growth in the injured lung, which can be partly confirmed by the inhibition of lung inflammation after systemic delivery of MSCs-conditioned media (Prockop 1997; Kinnaird et al. 2004). Therefore, it may be considered that MSCs paracrine signalling is the primary mechanism accounting for beneficial effects of MSCs in response to injury (Jurasz et al. 2010; Favre et al. 2013). Among all the paracrine types of MSCs, exosomes, as mediators of cell–cell communication, provide a novel insight into the efficient role of MSCs in PH. Studies suggest that MSCs-derived exosomes can exert a pleiotropic protective effect on the lung and inhibit pulmonary hypertension through suppression of hyperproliferative pathways, such as STAT3-mediated signalling induced by hypoxia (Zhu et al. 2015). Interestingly, syngeneic transfer of bone marrow derived MSCs has been shown to effectively reduce pulmonary pressures, right ventricular hypertrophy, and pulmonary vascular remodelling in the multi-cellular tumor spheroid rat model of PAH (Baber et al. 2007; Umar et al. 2009; Kanki-Horimoto et al. 2006; Luan et al. 2012). In a study that examined cell engraftment, intratracheally administered MSCs integrated mostly in the alveolar epithelium rather than the pulmonary vasculature, yet still provided benefit in the therapy of PAH. These studies support the paracrine hypothesis, in which secreted soluble factors support tissue regeneration and repair. Along these lines, Lee et al. recently demonstrated that enrichment of MSC-derived exosomes ameliorated PAH in the mouse hypoxia model of PAH (Lee et al. 2012). Lim et al. demonstrated that ceramide-1 phosphate priming increased the effects of MSC therapy by enhancing the migratory, self-renewal, and anti-inflammatory activity of MSCs. This therapy can be optimized with priming protocols as a promising option for the treatment of PAH patients (Lim et al. 2016). For example, Poly (I:C) has been used to stimulate toll-like receptor-3 pathway in pathogen recognition and innate immunity (Alexopoulou et al. 2001). There are studies where MSCs preconditioned with poly (I:C), a toll-like receptor-3 ligand, have increased immunosuppressive effects by potentiating key anti-inflammatory mediators such as indoleamine 2,3-dioxygenase and prostaglandin E2 (PGE2) (Waterman et al. 2012). Currently, this preconditioning strategy has shown favorable outcomes as a treatment for diseases of severe inflammation such as experimental sepsis and therefore might have some utility in modulating chronic inflammation in PAH (Waterman et al. 2012; Zhao et al. 2014). Preconditioning methods such as hypoxia might also prime MSCs toward favorable therapeutic phenotypes. These properties and findings make MSC treatment a novel and promising approach for protection from and repair of PAH. As we gain a better understanding of the regulation of this fascinating cell type, we will be able to develop effective and minimally invasive methods of MSC enhancement.

6.20 Future Perspectives and Concerns

Stem cell therapy appears to be a promising strategy to attenuate or even reverse chronic lung diseases. Over the past decade, numerous preclinical studies have demonstrated the capability of EpiSPCs, MSCs, and EPCs from adult lung to facilitate tissue repair and regeneration in a number of pulmonary disease models. Many completed or ongoing clinical trials have supported their safety in the treatment of lung diseases. Research into therapies using MSCs is still in its infancy and a great deal of work is needed before such therapies can be used routinely in patients. These may in principle harness both the ability of these cells to generate skeletal tissues and the ability of these cells to direct the function of hematopoietic and vascular cells. Questions remain about how the cells can be controlled, how they will behave when transplanted into the body, how they can be delivered to the right place so that they work effectively, and so forth. By studying how these cells work and interact within the body, researchers hope to develop safe and effective new treatments in the future. MSCs have generated a great amount of enthusiasm over the past decade as a novel therapeutic strategy for a variety of lung diseases. Although advancements have been made from preclinical studies using MSCs, substantial challenges have yet to be overcome before MSC therapy can be used in clinical practice. Clinical studies published to date have reported that MSC administration is safe, with few adverse effects concerning infusion reactions and late effects. However, due to the relatively small number of patients that have received MSC therapy to date, further investigations should be performed to further characterize its safety profile. Moreover, to ensure MSC quality control, bacteriological tests (to reduce microbial contamination of MSC cultures), viability and phenotype tests, oncogenicity tests, and endotoxin assays should be carefully performed. In addition, the optimal timing and duration of administration, the cell dose (per kilogram of body weight, escalating doses), the source of MSCs, the best delivery route, and the optimal schedule of administration (e.g., single versus repeated doses) all need to be evaluated. The use of MSCs in the clinical setting requires a large number of cells; however, continuous in vitro passaging of MSCs may result in genetic abnormalities (chromosomal abnormalities, increased c-myc levels, and telomerase activity), raising the possibility of cell transformation (Rubio et al. 2005). MSCs are relatively immune-privileged, as they express low levels of major histocompatibility complex I (MHC-I) molecules and do not express MHC-II molecules or costimulatory molecules such as CD80, CD86, or CD40 (Guo et al. 2008), suggesting that both allogeneic and autologous MSCs can be used in the clinical setting. Nevertheless, further studies with larger sample sizes are required to evaluate which source of cells would result in superior beneficial effects. Recent clinical studies have used MSCs manufactured by different companies and by non-commercial cell repositories; thus, regulations and standards for production of clinical-grade MSCs need to be very well defined and include methods and criteria for MSC culture, storage, shipping, and administration. This is a very important issue, as differences in cell production (e.g., MSC passage) may result in different effects. Answering these emerging questions will enable a more rapid, reliable, and effective translation of MSC therapy from bench to bedside for patients with lung diseases. Scientists and physicians alike are excited by the therapeutic potential of MSCs for several lung diseases, as suggested by their anti-inflammatory, antifibrotic, antiapoptotic, antibacterial, and pro-reparative features. The early phase clinical trials conducted to date or currently in progress offer reassurance regarding the safety of MSC therapy for these diseases. Determination of the efficacy of MSC therapies for COPD, ARDS, IPF, BPD, asthma, and pulmonary hypertension will require large scale, international, multicenter phase III clinical trials. These studies are eagerly awaited. In conclusion, studies are needed to fill gaps in our understanding of how resident lung MSCs are impaired during disease and their role during the development of disease which will be useful for designing further intervention provided so far.

References

Abdollahi A et al (2005) Inhibition of platelet-derived growth factor signaling attenuates pulmonary fibrosis. J Exp Med 201:925–935

Abman SH, Matthay MA (2009) Mesenchymal stem cells for the prevention of bronchopulmonary dysplasia: delivering the secretome. Am J Respir Crit Care Med 180:1039–1041

Abreu SC et al (2013) Bone marrow-derived mononuclear cells vs. mesenchymal stromal cells in experimental allergic asthma. Respir Physiol Neurobiol 187:190–198

Admyre C et al (2008) Exosomes - nanovesicles with possible roles in allergic inflammation. Allergy 63:404–408

Aggarwal S, Pittenger MF (2005) Human mesenchymal stem cells modulate allogeneic immune cell responses. Blood 105:1815–1822

Ahn SY, Chang YS, Park WS (2015) Stem cell therapy for bronchopulmonary dysplasia: bench to bedside translation. J Korean Med Sci 30:509–513

Akram KM, Samad S, Spiteri MA, Forsyth NR (2013) Mesenchymal stem cells promote alveolar epithelial cell wound repair in vitro through distinct migratory and paracrine mechanisms. Respir Res 14:9

Alexopoulou L, Holt AC, Medzhitov R, Flavell RA (2001) Recognition of double-stranded RNA and activation of NF-kappaB by Toll-like receptor 3. Nature 413:732–738

Antoniou KM et al (2010) Investigation of bone marrow mesenchymal stem cells (BM MSCs) involvement in idiopathic pulmonary fibrosis (IPF). Respir Med 104:1535–1542

Antunes MA, Laffey JG, Pelosi P, Rocco PRM (2014) Mesenchymal stem cell trials for pulmonary diseases. J Cell Biochem 115:1023–1032

Anversa P, Kajstura J, Leri A, Loscalzo J (2011) Tissue-specific adult stem cells in the human lung. Nat Med 17:1038–1039

Anversa P, Perrella MA, Kourembanas S, Choi AMK, Loscalzo J (2012) Regenerative pulmonary medicine: potential and promise, pitfalls and challenges. Eur J Clin Invest 42:900–913

Archer SL, Weir EK, Wilkins MR (2010) Basic science of pulmonary arterial hypertension for clinicians: new concepts and experimental therapies. Circulation 121:2045–2066

Aslam M et al (2009) Bone marrow stromal cells attenuate lung injury in a murine model of neonatal chronic lung disease. Am J Respir Crit Care Med 180:1122–1130

Augello A et al (2005) Bone marrow mesenchymal progenitor cells inhibit lymphocyte proliferation by activation of the programmed death 1 pathway. Eur J Immunol 35:1482–1490

Baber SR et al (2007) Intratracheal mesenchymal stem cell administration attenuates monocrotaline-induced pulmonary hypertension and endothelial dysfunction. Am J Physiol Heart Circ Physiol 292:H1120–H1128

Badylak SF, Weiss DJ, Caplan A, Macchiarini P (2012) Engineered whole organs and complex tissues. Lancet (London, England) 379:43–52

Baksh D, Boland GM, Tuan RS (2007) Cross-talk between Wnt signaling pathways in human mesenchymal stem cells leads to functional antagonism during osteogenic differentiation. J Cell Biochem 101:1109–1124

Barst RJ (2005) PDGF signaling in pulmonary arterial hypertension. J Clin Invest 115:2691–2694

Belema-Bedada F, Uchida S, Martire A, Kostin S, Braun T (2008) Efficient homing of multipotent adult mesenchymal stem cells depends on FROUNT-mediated clustering of CCR2. Cell Stem Cell 2:566–575

Berg T et al (2010) β-catenin regulates mesenchymal progenitor cell differentiation during hepatogenesis. J Surg Res 164:276–285

Bernardo ME, Locatelli F, Fibbe WE (2009) Mesenchymal stromal cells. Ann N Y Acad Sci 1176:101–117

Beyth S et al (2010) and induce T-cell unresponsiveness human mesenchymal stem cells alter antigen-presenting cell maturation and induce T-cell unresponsiveness. Blood 105:2214–2219

Bianco P, Robey PG, Simmons PJ (2008) Mesenchymal stem cells: revisiting history, concepts, and assays. Cell Stem Cell 2:313–319

Bieback K, Klüter H (2007) Mesenchymal stromal cells from umbilical cord blood. Curr Stem Cell Res Ther 2:310–323

Bitencourt CS et al (2011) Hyaluronidase recruits mesenchymal-like cells to the lung and ameliorates fibrosis. Fibrogenesis Tissue Repair 4:3

Braza F et al (2016) Mesenchymal stem cells induce suppressive macrophages through phagocytosis in a mouse model of asthma. Stem Cells. doi:10.1002/stem.2344

Brewster CE et al (1990) Myofibroblasts and subepithelial fibrosis in bronchial asthma. Am J Respir Cell Mol Biol 3:507–511

Budhiraja R, Tuder RM, Hassoun PM (2004) Endothelial dysfunction in pulmonary hypertension. Circulation 109:159–165

Calió ML et al (2014) Transplantation of bone marrow mesenchymal stem cells decreases oxidative stress, apoptosis, and hippocampal damage in brain of a spontaneous stroke model. Free Radic Biol Med 70:141–154

Caplan AI (2007) Adult mesenchymal stem cells for tissue engineering versus regenerative medicine. J Cell Physiol 213:341–347

Cargnoni A et al (2009) Transplantation of allogeneic and xenogeneic placenta-derived cells reduces bleomycin-induced lung fibrosis. Cell Transplant 18:405–422

Chamberlain G, Fox J, Ashton B, Middleton J (2007) Concise review: mesenchymal stem cells: their phenotype, differentiation capacity, immunological features, and potential for homing. Stem Cells 25:2739–2749

Chang YS et al (2011) Intratracheal transplantation of human umbilical cord blood-derived mesenchymal stem cells dose-dependently attenuates hyperoxia-induced lung injury in neonatal rats. Cell Transplant 20:1843–1854

Cho SW et al (2008) Differential effects of secreted frizzled-related proteins (sFRPs) on osteoblastic differentiation of mouse mesenchymal cells and apoptosis of osteoblasts. Biochem Biophys Res Commun 367:399–405

Churg A et al (2003) Macrophage metalloelastase mediates acute cigarette smoke-induced inflammation via tumor necrosis factor-alpha release. Am J Respir Crit Care Med 167:1083–1089

Corcione A et al (2006) Human mesenchymal stem cells modulate B-cell functions. Blood 107:367–372

Crisan M et al (2008) A perivascular origin for mesenchymal stem cells in multiple human organs. Cell Stem Cell 3:301–313

Cruz FF et al (2015) Systemic administration of human bone marrow-derived mesenchymal stromal cell extracellular vesicles ameliorates aspergillus hyphal extract-induced allergic airway inflammation in immunocompetent mice. Stem Cells Transl Med 4:1302–1316

Cubillo I, Mirones I, Mari L (2014) Mesenchymal stem cells regulate airway contractile tissue remodeling in murine experimental asthma. Allergy 69:730–740

Dazzi F, Marelli-Berg FM (2008) Mesenchymal stem cells for graft-versus-host disease: close encounters with T cells. Eur J Immunol 38:1479–1482

De Bari C, Dell'Accio F, Tylzanowski P, Luyten FP (2001) Multipotent mesenchymal stem cells from adult human synovial membrane. Arthritis Rheum 44:1928–1942

De Boer J, Wang HJ, Van Blitterswijk C (2014) Effects of Wnt signaling on proliferation and differentiation of human mesenchymal stem cells. Tissue Eng 10:393–401

ARDS Definition Task Force et al (2012) Acute respiratory distress syndrome: the Berlin definition. JAMA 307:2526–2533

Di Ianni M et al (2008) Mesenchymal cells recruit and regulate T regulatory cells. Exp Hematol 36:309–318

Di Nicola M et al (2002) Human bone marrow stromal cells suppress T-lymphocyte proliferation induced by cellular or nonspecific mitogenic stimuli. Blood 99:3838–3843

Dimmeler S, Losordo D, Williams AR, Hare JM (2011) Implications for Cardiac Disease. 016960:910–922

Doeing DC, Solway J (2013) Airway smooth muscle in the pathophysiology and treatment of asthma. J Appl Physiol 114:834–843

Dolgachev VA, Ullenbruch MR, Lukacs NW, Phan SH (2009) Role of stem cell factor and bone marrow-derived fibroblasts in airway remodeling. Am J Pathol 174:390–400

Dominici M et al (2006) Minimal criteria for defining multipotent mesenchymal stromal cells. The international society for cellular therapy position statement. Cytotherapy 8:315–317

Etheridge SL, Spencer GJ, Heath DJ, Genever PG (2004) Expression profiling and functional analysis of wnt signaling mechanisms in mesenchymal stem cells. Stem Cells 22:849–860

Favre J, Terborg N, Horrevoets AJG (2013) The diverse identity of angiogenic monocytes. Eur J Clin Invest 43:100–107

Foronjy RF, Majka SM (2012) The potential for resident lung mesenchymal stem cells to promote functional tissue regeneration: understanding microenvironmental cues. Cells 1:874–885

Fredenburgh LE, Perrella MA, Mitsialis SA (2007) The role of heme oxygenase-1 in pulmonary disease. Am J Respir Cell Mol Biol 36:158–165

Friedenstein AJ, Petrakova KV, Kurolesova AI, Frolova GP (1968) Heterotopic of bone marrow. Analysis of precursor cells for osteogenic and hematopoietic tissues. Transplantation 6:230–247

García-Castro J et al (2008) Mesenchymal stem cells and their use as cell replacement therapy and disease modelling tool. J Cell Mol Med 12:2552–2565

Gazdhar A et al (2013) HGF expressing stem cells in usual interstitial pneumonia originate from the bone marrow and are antifibrotic. PLoS One 8:e65453

Ghofrani HA, Seeger W, Grimminger F (2005) Imatinib for the treatment of pulmonary arterial hypertension. N Engl J Med 353:1412–1413

Giaid A, Saleh D (1995) Reduced expression of endothelial nitric oxide synthase in the lungs of patients with pulmonary hypertension. N Engl J Med 333:214–221

Giaid A et al (1993) Expression of endothelin-1 in the lungs of patients with pulmonary hypertension. N Engl J Med 328:1732–1739

Gizycki MJ, Adelroth E, Rogers AV, O'Byrne PM, Jeffery PK (1997) Myofibroblast involvement in the allergen-induced late response in mild atopic asthma. Am J Respir Cell Mol Biol 16:664–673

Glennie S, Soeiro I, Dyson PJ, Lam EW-F, Dazzi F (2005) Bone marrow mesenchymal stem cells induce division arrest anergy of activated T cells. Blood 105:2821–2827

Goodwin M et al (2011) Bone marrow-derived mesenchymal stromal cells inhibit Th2-mediated allergic airways inflammation in mice. Stem Cells 29:1137–1148

Gronthos S et al (2003) Molecular and cellular characterisation of highly purified stromal stem cells derived from human bone marrow. J Cell Sci 116:1827–1835

Guan X-J et al (2013) Mesenchymal stem cells protect cigarette smoke-damaged lung and pulmonary function partly via VEGF-VEGF receptors. J Cell Biochem 114:323–335

Guo F et al (2008) CD28 controls differentiation of regulatory T cells from naive CD4 T cells. J Immunol 181(4):2285–2291

Gupta N et al (2007) Intrapulmonary delivery of bone marrow-derived mesenchymal stem cells improves survival and attenuates endotoxin-induced acute lung injury in mice. J Immunol 179:1855–1863

Gupta N et al (2012) Mesenchymal stem cells enhance survival and bacterial clearance in murine Escherichia coli pneumonia. Thorax 67:533–539

Hanania NA (2008) Targeting airway inflammation in asthma: current and future therapies. Chest 133:989–998

Hansmann G et al (2012) Mesenchymal stem cell-mediated reversal of bronchopulmonary dysplasia and associated pulmonary hypertension. Pulm Circ 2:170–181

Heldin CH, Westermark B (1999) Mechanism of action and in vivo role of platelet-derived growth factor. Physiol Rev 79:1283–1316

Huh JW et al (2011) Bone marrow cells repair cigarette smoke-induced emphysema in rats. Am J Physiol Lung Cell Mol Physiol 301:L255–L266

Humbert M et al (2004) Cellular and molecular pathobiology of pulmonary arterial hypertension. J Am Coll Cardiol 43:13S–24S

Hung C et al (2013) Role of lung pericytes and resident fibroblasts in the pathogenesis of pulmonary fibrosis. Am J Respir Crit Care Med 188:820–830

Igura K et al (2004) Isolation and characterization of mesenchymal progenitor cells from chorionic villi of human placenta. Cytotherapy 6:543–553

Islam MN et al (2012) Mitochondrial transfer from bone-marrow-derived stromal cells to pulmonary alveoli protects against acute lung injury. Nat Med 18:759–765

Jiang Y et al (2002) Pluripotency of mesenchymal stem cells derived from adult marrow. Nature 418:41–49

Jin Z et al (2015) Biological effects and mechanisms of action of mesenchymal stem cell therapy in chronic obstructive pulmonary disease. J Int Med Res 43:303–310

Jones EA et al (2002) Isolation and characterization of bone marrow multipotential mesenchymal progenitor cells. Arthritis Rheum 46:3349–3360

Jones EA et al (2004) Enumeration and phenotypic characterization of synovial fluid multipotential mesenchymal progenitor cells in inflammatory and degenerative arthritis. Arthritis Rheum 50:817–827

Joo S-Y et al (2010) Mesenchymal stromal cells inhibit graft-versus-host disease of mice in a dose-dependent manner. Cytotherapy 12:361–370

Jun D et al (2011) The pathology of bleomycin-induced fibrosis is associated with loss of resident lung mesenchymal stem cells that regulate effector T-cell proliferation. Stem Cells 29:725–735

Jurasz P, Courtman D, Babaie S, Stewart DJ (2010) Role of apoptosis in pulmonary hypertension: from experimental models to clinical trials. Pharmacol Ther 126:1–8

Kajstura J et al (2011) Evidence for human lung stem cells. N Engl J Med 364:1795–1806

Kanazawa H, Yoshikawa J (2005) Elevated oxidative stress and reciprocal reduction of vascular endothelial growth factor levels with severity of COPD. Chest 128:3191–3197

Kanki-Horimoto S et al (2006) Implantation of mesenchymal stem cells overexpressing endothelial nitric oxide synthase improves right ventricular impairments caused by pulmonary hypertension. Circulation 114:I181–I185

Keating A (2008) How do mesenchymal stromal cells suppress T cells? Cell Stem Cell 2:106–108

Kim N, Cho S-G (2013) Clinical applications of mesenchymal stem cells. Korean J Intern Med 28:387–402

Kim S-Y et al (2012) Mesenchymal stem cell-conditioned media recovers lung fibroblasts from cigarette smoke-induced damage. Am J Physiol Lung Cell Mol Physiol 302:L891–L908

Kinnaird T et al (2004) Marrow-derived stromal cells express genes encoding a broad spectrum of arteriogenic cytokines and promote in vitro and in vivo arteriogenesis through paracrine mechanisms. Circ Res 94:678–685

Kirton JP, Crofts NJ, George SJ, Brennan K, Canfield AE (2007) Wnt/beta-catenin signaling stimulates chondrogenic and inhibits adipogenic differentiation of pericytes: potential relevance to vascular disease? Circ Res 101:581–589

Kolf CM, Cho E, Tuan RS (2007) Mesenchymal stromal cells. Biology of adult mesenchymal stem cells: regulation of niche, self-renewal and differentiation. Arthritis Res Ther 9:204

Kolios G, Moodley Y (2013) Introduction to stem cells and regenerative medicine. Respiration 85:3–10

Königshoff M et al (2009) WNT1-inducible signaling protein-1 mediates pulmonary fibrosis in mice and is upregulated in humans with idiopathic pulmonary fibrosis. J Clin Invest 119:772–787

Kotton DN et al (2001) Bone marrow-derived cells as progenitors of lung alveolar epithelium. Development 128:5181–5188

Krampera M et al (2006) Role for interferon-gamma in the immunomodulatory activity of human bone marrow mesenchymal stem cells. Stem Cells 24:386–398

Kulterer B et al (2007) Gene expression profiling of human mesenchymal stem cells derived from bone marrow during expansion and osteoblast differentiation. BMC Genomics 8:70

Kunter U et al (2006) Transplanted mesenchymal stem cells accelerate glomerular healing in experimental glomerulonephritis. J Am Soc Nephrol 17:2202–2212

Lama VN et al (2007) Evidence for tissue-resident mesenchymal stem cells in human adult lung from studies of transplanted allografts. J Clin Invest 117:989–996

Lathrop MJ et al (2014) Mesenchymal stromal cells mediate Aspergillus hyphal extract-induced allergic airway inflammation by inhibition of the Th17 signaling pathway. Stem Cells Transl Med 3:194–205

Le Blanc K et al (2008) Mesenchymal stem cells for treatment of steroid-resistant, severe, acute graft-versus-host disease: a phase II study. Lancet (Lond) 371:1579–1586

Lee C et al (2012) Exosomes mediate the cytoprotective action of mesenchymal stromal cells on hypoxia-induced pulmonary hypertension. Circulation 126:2601–2611

Leyns L, Bouwmeester T, Kim SH, Piccolo S, De Robertis EM (1997) Frzb-1 is a secreted antagonist of Wnt signaling expressed in the Spemann organizer. Cell 88:747–756

Li L et al (2009) Paracrine action mediate the antifibrotic effect of transplanted mesenchymal stem cells in a rat model of global heart failure. Mol Biol Rep 36:725–731

Li J, Li D, Liu X, Tang S, Wei F (2012) Human umbilical cord mesenchymal stem cells reduce systemic inflammation and attenuate LPS-induced acute lung injury in rats. J Inflamm (Lond) 9:33

Liang J et al (2010) Allogenic mesenchymal stem cells transplantation in refractory systemic lupus erythematosus: a pilot clinical study. Ann Rheum Dis 69:1423–1429

Liang OD et al (2011) Mesenchymal stromal cells expressing heme oxygenase-1 reverse pulmonary hypertension. Stem Cells 29:99–107

Lim J et al (2016) Priming with ceramide 1 phosphate promotes the therapeutic effect of mesenchymal stem / stromal cells on pulmonary artery hypertension. PubMed Commons 473:3–4

Ling L, Nurcombe V, Cool SM (2009) Wnt signaling controls the fate of mesenchymal stem cells. Gene 433:1–7

Liu A-R et al (2013) Activation of canonical wnt pathway promotes differentiation of mouse bone marrow-derived MSCs into type II alveolar epithelial cells, confers resistance to oxidative stress, and promotes their migration to injured lung tissue in vitro. J Cell Physiol 228:1270–1283

Luan Y et al (2012) Implantation of mesenchymal stem cells improves right ventricular impairments caused by experimental pulmonary hypertension. Am J Med Sci 343:402–406

Macchiarini P et al (2008) Clinical transplantation of a tissue-engineered airway. Lancet (London, England) 372:2023–2030

Madec AM et al (2009) Mesenchymal stem cells protect NOD mice from diabetes by inducing regulatory T cells. Diabetologia 52:1391–1399

Marigo I, Dazzi F (2011) The immunomodulatory properties of mesenchymal stem cells. Semin Immunopathol 33:593–602

Mariñas-Pardo L et al (2014) Mesenchymal stem cells regulate airway contractile tissue remodeling in murine experimental asthma. Allergy Eur J Allergy Clin Immunol 69:730–740

Martin J, Helm K, Ruegg P, Burnham E, Majka S (2008) Adult lung side population cells have mesenchymal stem cell potential. Cytotherapy 10:140–151

Martínez-González I et al (2013) Human mesenchymal stem cells overexpressing the IL-33 antagonist soluble IL-1 receptor-like-1 attenuate endotoxin-induced acute lung injury. Am J Respir Cell Mol Biol 49:552–562

McQualter JL, Yuen K, Williams B, Bertoncello I (2010) Evidence of an epithelial stem/progenitor cell hierarchy in the adult mouse lung. Proc Natl Acad Sci U S A 107:1414–1419

Mei SHJ et al (2010) Mesenchymal stem cells reduce inflammation while enhancing bacterial clearance and improving survival in sepsis. Am J Respir Crit Care Med 182:1047–1057

Meisel R et al (2004) Human bone marrow stromal cells inhibit allogeneic T-cell responses by indoleamine 2,3-dioxygenase-mediated tryptophan degradation. Blood 103:4619–4621

Minguell JJ, Erices A (2006) Mesenchymal stem cells and the treatment of cardiac disease. Exp Biol Med (Maywood) 231:39–49

Moeller A et al (2009) Circulating fibrocytes are an indicator of poor prognosis in idiopathic pulmonary fibrosis. Am J Respir Crit Care Med 179:588–594

Moodley Y et al (2009) Human umbilical cord mesenchymal stem cells reduce fibrosis of bleomycin-induced lung injury. Am J Pathol 175:303–313

Moodley Y et al (2010) Human amnion epithelial cell transplantation abrogates lung fibrosis and augments repair. Am J Respir Crit Care Med 182:643–651

Moorman JE et al (2007) National surveillance for asthma--United States, 1980–2004. MMWR Surveill Summ 56:1–54

Murphy DM, O'Byrne PM (2010) Recent advances in the pathophysiology of asthma. Chest 137:1417–1426

Nauta AJ, Kruisselbrink AB, Lurvink E, Willemze R, Fibbe WE (2006) Mesenchymal stem cells inhibit generation and function of both CD34+-derived and monocyte-derived dendritic cells. J Immunol 177:2080–2087

Németh K et al (2009) Bone marrow stromal cells attenuate sepsis via prostaglandin E(2)-dependent reprogramming of host macrophages to increase their interleukin-10 production. Nat Med 15:42–49

Nichols JE, Niles JA, Cortiella J (2012) Production and utilization of acellular lung scaffolds in tissue engineering. J Cell Biochem 113:2185–2192

Nihlberg K et al (2006) Tissue fibrocytes in patients with mild asthma: a possible link to thickness of reticular basement membrane? Respir Res 7:50

Noble PW et al (2011) Pirfenidone in patients with idiopathic pulmonary fibrosis (CAPACITY): two randomised trials. Lancet (London, England) 377:1760–1769

Ogulur I et al (2014) Suppressive effect of compact bone-derived mesenchymal stem cells on chronic airway remodeling in murine model of asthma. Int Immunopharmacol 20:101–109

Ohnishi S, Sumiyoshi H, Kitamura S, Nagaya N (2007) Mesenchymal stem cells attenuate cardiac fibroblast proliferation and collagen synthesis through paracrine actions. FEBS Lett 581:3961–3966

Ortiz LA et al (2003) Mesenchymal stem cell engraftment in lung is enhanced in response to bleomycin exposure and ameliorates its fibrotic effects. Proc Natl Acad Sci U S A 100:8407–8411

Ortiz LA et al (2007) Interleukin 1 receptor antagonist mediates the antiinflammatory and antifibrotic effect of mesenchymal stem cells during lung injury. Proc Natl Acad Sci U S A 104:11002–11007

Parekkadan B, Milwid JM (2010) Mesenchymal stem cells as therapeutics. Annu Rev Biomed Eng 12:87–117

Park J-A et al (2012) Tissue factor-bearing exosome secretion from human mechanically stimulated bronchial epithelial cells in vitro and in vivo. J Allergy Clin Immunol 130:1375–1383

Parsons PE et al (2005) Lower tidal volume ventilation and plasma cytokine markers of inflammation in patients with acute lung injury. Crit Care Med 33:1–6; discussion 230–232

Patel SA et al (2010) Mesenchymal stem cells protect breast cancer cells through regulatory T cells: role of mesenchymal stem cell-derived TGF-beta. J Immunol 184:5885–5894

Petersen TH et al (2010) Tissue-engineered lungs for in vivo implantation. Science 329:538–541

Phinney DG, Isakova I (2005) Plasticity and therapeutic potential of mesenchymal stem cells in the nervous system. Curr Pharm Des 11:1255–1265

Pittenger MF et al (1999) Multilineage potential of adult human mesenchymal stem cells. Science 284:143–147

Pochampally RR, Smith JR, Ylostalo J, Prockop DJ (2004) Serum deprivation of human marrow stromal cells (hMSCs) selects for a subpopulation of early progenitor cells with enhanced expression of OCT-4 and other embryonic genes. Blood 103:1647–1652

Prockop DJ (1997) Marrow stromal cells as stem cells for nonhematopoietic tissues. Science 276:71–74

Prockop DJ, Oh JY (2009) Mesenchymal stem/stromal cells (MSCs): role as guardians of inflammation. Mol Ther 20:14–20

Qayyum AA et al (2012) Adipose-derived mesenchymal stromal cells for chronic myocardial ischemia (MyStromalCell Trial): study design. Regen Med 7:421–428

Raghu G, Weycker D, Edelsberg J, Bradford WZ, Oster G (2006) Incidence and prevalence of idiopathic pulmonary fibrosis. Am J Respir Crit Care Med 174:810–816

Raghu G et al (2011) An official ATS/ERS/JRS/ALAT statement: idiopathic pulmonary fibrosis: evidence-based guidelines for diagnosis and management. Am J Respir Crit Care Med 183:788–824

Ramos-Barbón D et al (2010) T Cells localize with proliferating smooth muscle alpha-actin+cell compartments in asthma. Am J Respir Crit Care Med 182:317–324

Ratajczak MZ, Zuba-Surma EK, Machalinski B, Kucia M (2007) Bone-marrow-derived stem cells--our key to longevity? J Appl Genet 48:307–319

Rehman J et al (2004) Secretion of angiogenic and antiapoptotic factors by human adipose stromal cells. Circulation 109:1292–1298

Ren G et al (2008) Mesenchymal stem cell-mediated immunosuppression occurs via concerted action of chemokines and nitric oxide. Cell Stem Cell 2:141–150

Reya T, Clevers H (2005) Wnt signalling in stem cells and cancer. Nature 434:843–850

Ribeiro-Paes JT et al (2011) Unicentric study of cell therapy in chronic obstructive pulmonary disease/pulmonary emphysema. Int J Chron Obstruct Pulmon Dis 6:63–71

Robinson DS (2010) The role of the T cell in asthma. J Allergy Clin Immunol 126:1081–91; quiz 1092–1093

Rochefort GY et al (2006) Multipotential mesenchymal stem cells are mobilized into peripheral blood by hypoxia. Stem Cells 24:2202–2208

Rogers I, Casper RF (2004) Umbilical cord blood stem cells. Best Pract Res Clin Obstet Gynaecol 18:893–908

Rojas M et al (2005) Bone marrow-derived mesenchymal stem cells in repair of the injured lung. Am J Respir Cell Mol Biol 33:145–152

Rubio L, Vera-Sempere FJ, Lopez-Guerrero JA, Padilla J, Moreno-Baylach MJ (2005) A risk model for non-small cell lung cancer using clinicopathological variables, angiogenesis and oncoprotein expression. Anticancer Res 25(1B):497–504

Sakao S, Tatsumi K, Voelkel NF (2009) Endothelial cells and pulmonary arterial hypertension: apoptosis, proliferation, interaction and transdifferentiation. Respir Res 10:95

Saunders R et al (2009) Fibrocyte localization to the airway smooth muscle is a feature of asthma. J Allergy Clin Immunol 123:376–384

Schermuly RT et al (2005) Reversal of experimental pulmonary hypertension by PDGF inhibition. J Clin Invest 115:2811–2821

Schipani E, Kronenberg HM (2008) Adult mesenchymal stem cells. In: StemBook. The Stem Cell Research Community. doi:10.3824/stembook.1.38.1, http://www.stembook.org

Schmidt M, Sun G, Stacey MA, Mori L, Mattoli S (2003) Identification of circulating fibrocytes as precursors of bronchial myofibroblasts in asthma. J Immunol 171:380–389

Semënov MV et al (2001) Head inducer Dickkopf-1 is a ligand for Wnt coreceptor LRP6. Curr Biol 11:951–961

Shigemura N et al (2006) Autologous transplantation of adipose tissue-derived stromal cells ameliorates pulmonary emphysema. Am J Transplant 6:2592–2600

Simonneau G et al (2013) Updated clinical classification of pulmonary hypertension. J Am Coll Cardiol 62:D34–D41

Smith VC et al (2005) Trends in severe bronchopulmonary dysplasia rates between 1994 and 2002. J Pediatr 146:469–473

Spees JL, Olson SD, Whitney MJ, Prockop DJ (2006) Mitochondrial transfer between cells can rescue aerobic respiration. Proc Natl Acad Sci U S A 103:1283–1288

Spees JL et al (2008) Bone marrow progenitor cells contribute to repair and remodeling of the lung and heart in a rat model of progressive pulmonary hypertension. FASEB J 22:1226–1236

Stagg J, Pommey S, Eliopoulos N, Galipeau J (2006) Interferon-gamma-stimulated marrow stromal cells: a new type of nonhematopoietic antigen-presenting cell. Blood 107:2570–2577

Stenderup K, Justesen J, Eriksen EF, Rattan SI, Kassem M (2001) Number and proliferative capacity of osteogenic stem cells are maintained during aging and in patients with osteoporosis. J Bone Miner Res 16:1120–1129

Summer R, Fitzsimmons K, Dwyer D, Murphy J, Fine A (2007) Isolation of an adult mouse lung mesenchymal progenitor cell population. Am J Respir Cell Mol Biol 37:152–159

Takemiya K et al (2010) Mesenchymal stem cell-based prostacyclin synthase gene therapy for pulmonary hypertension rats. Basic Res Cardiol 105:409–417

Taniguchi H et al (2010) Pirfenidone in idiopathic pulmonary fibrosis. Eur Respir J 35:821–829

Thibeault DW, Mabry SM, Ekekezie II, Zhang X, Truog WE (2003) Collagen scaffolding during development and its deformation with chronic lung disease. Pediatrics 111:766–776

Tropea KA et al (2012) The promise of stem cells in bronchopulmonary dysplasia. Pediatrics 2:6

Trounson A, Thakar RG, Lomax G, Gibbons D (2011) Clinical trials for stem cell therapies. BMC Med 9:52

Tsai M-S, Lee J-L, Chang Y-J, Hwang S-M (2004) Isolation of human multipotent mesenchymal stem cells from second-trimester amniotic fluid using a novel two-stage culture protocol. Hum Reprod 19:1450–1456

Tsutsumi S et al (2001) Retention of multilineage differentiation potential of mesenchymal cells during proliferation in response to FGF. Biochem Biophys Res Commun 288:413–419

Tuder RM et al (1999) Prostacyclin synthase expression is decreased in lungs from patients with severe pulmonary hypertension. Am J Respir Crit Care Med 159:1925–1932

Umar S et al (2009) Allogenic stem cell therapy improves right ventricular function by improving lung pathology in rats with pulmonary hypertension. Am J Physiol Heart Circ Physiol 297:H1606–H1616

Uren A et al (2000) Secreted frizzled-related protein-1 binds directly to Wingless and is a biphasic modulator of Wnt signaling. J Biol Chem 275:4374–4382

van Haaften T et al (2009) Airway delivery of mesenchymal stem cells prevents arrested alveolar growth in neonatal lung injury in rats. Am J Respir Crit Care Med 180:1131–1142

Veevers-Lowe J, Ball SG, Shuttleworth A, Kielty CM (2011) Mesenchymal stem cell migration is regulated by fibronectin through α5β1-integrin-mediated activation of PDGFR-β and potentiation of growth factor signals. J Cell Sci 124:1288–1300

Vosdoganes P, Lim R, Moss TJM, Wallace EM (2012) Cell therapy: a novel treatment approach for bronchopulmonary dysplasia. Pediatrics 130:727–737

Walker N et al (2011) Resident tissue-specific mesenchymal progenitor cells contribute to fibrogenesis in human lung allografts. Am J Pathol 178:2461–2469

Walsh MC et al (2006) Summary proceedings from the bronchopulmonary dysplasia group. Pediatrics 117:S52–S56

Wang S, Wilkes MC, Leof EB, Hirschberg R (2005) Imatinib mesylate blocks a non-Smad TGF-beta pathway and reduces renal fibrogenesis in vivo. FASEB J 19:1–11

Wang M, Crisostomo PR, Herring C, Meldrum KK, Meldrum DR (2006) Human progenitor cells from bone marrow or adipose tissue produce VEGF, HGF, and IGF-I in response to TNF by a $p38$ MAPK-dependent mechanism. Am J Physiol Regul Integr Comp Physiol 291:R880–R884

Ware LB et al (2007) Pathogenetic and prognostic significance of altered coagulation and fibrinolysis in acute lung injury/acute respiratory distress syndrome. Crit Care Med 35:1821–1828

Waterman RS, Henkle SL, Betancourt AM (2012) Mesenchymal stem cell 1 (MSC1)-based therapy attenuates tumor growth whereas MSC2-treatment promotes tumor growth and metastasis. PLoS One 7:e45590

Weiss DJ, Casaburi R, Flannery R, LeRoux-Williams M, Tashkin DP (2013) A placebo-controlled, randomized trial of mesenchymal stem cells in COPD. Chest 143:1590–1598

Weiss DJ et al (2014) An official American thoracic society workshop report : stem cells and cell therapies in lung biology and diseases. doi:10.1513/AnnalsATS.201502-086ST

Williams AR, Hare JM (2011) Mesenchymal stem cells: Biology, pathophysiology, translational findings, and therapeutic implications for cardiac disease. Circ Res 109:923–940

Wong AP et al (2007) Targeted cell replacement with bone marrow cells for airway epithelial regeneration. Am J Physiol Lung Cell Mol Physiol 293:L740–L752

Wu P-S, Egger B, Brand AH (2008) Asymmetric stem cell division: lessons from Drosophila. Semin Cell Dev Biol 19:283–293

Xu Y, Malladi P, Wagner DR, Longaker MT (2005) Adipose-derived mesenchymal cells as a potential cell source for skeletal regeneration. Curr Opin Mol Ther 7:300–305

Yang J, Jia Z (2014) Cell-based therapy in lung regenerative medicine. Regen Med Res 2:1–7

Yang J et al (2013) Activated alveolar epithelial cells initiate fibrosis through secretion of mesenchymal proteins. Am J Pathol 183:1559–1570

Zeng SLIN, Wang LIHUI, Li P, Wang WEI, Yang J (2015) Mesenchymal stem cells abrogate experimental asthma by altering dendritic cell function. 2511–2520. doi:10.3892/mmr.2015.3706

Zhao YD et al (2005) Rescue of monocrotaline-induced pulmonary arterial hypertension using bone marrow-derived endothelial-like progenitor cells: efficacy of combined cell and eNOS gene therapy in established disease. Circ Res 96:442–450

Zhao X et al (2014) The toll-like receptor 3 ligand, poly(I:C), improves immunosuppressive function and therapeutic effect of mesenchymal stem cells on sepsis via inhibiting MiR-143. Stem Cells 32:521–533

Zhen G et al (2008) Mesenchymal stem cells transplantation protects against rat pulmonary emphysema. Front Biosci 13:3415–3422

Zhen G et al (2010) Mesenchymal stem cell transplantation increases expression of vascular endothelial growth factor in papain-induced emphysematous lungs and inhibits apoptosis of lung cells. Cytotherapy 12:605–614

Zheng G et al (2014) Treatment of acute respiratory distress syndrome with allogeneic adipose-derived mesenchymal stem cells : a randomized, placebo-controlled pilot study. Respir Res 15:39

Zhu Y-G et al (2014) Human mesenchymal stem cell microvesicles for treatment of Escherichia coli endotoxin-induced acute lung injury in mice. Stem Cells 32:116–125

Zhu Z, Fang Z, Hu X, Zhou S (2015) MicroRNAs and mesenchymal stem cells : hope for pulmonary hypertension. 30:380–385

Zvaifler NJ et al (2000) Mesenchymal precursor cells in the blood of normal individuals. Arthritis Res 2:477–488

Chapter 7
Recent Advances in Lung Regeneration

Kanwal Rehman and Muhammad Sajid Hamid Akash

7.1 Introduction

Among various fatal ailments, lung related diseases including idiopathic pulmonary fibrosis (IPF), acute respiratory damage syndrome (ARDS), and particularly chronic obstructive pulmonary diseases (COPDs) are known to affect approximately 600 million individuals worldwide including about 24 million Americans (Balkissoon et al. 2011; Tansey et al. 2007) and are expected to become the third leading cause of disease mortality by 2020 globally (Orens and Garrity 2009). Diseases like asthma, emphysema, pulmonary and cystic fibrosis are known to have no typical cure and instead are known to progress for the development of COPD and IPF. Though, lung transplantation was considered to be a choice for decreasing the mortality rate resulting from fatal lung diseases, factors like organ rejection by host immune system and/or unavailability and shortage of the organ donor decisively increasing the death rate of patients suffering from lung diseases. Beyond lung transplantation, there are many other slants that have been opted by researchers to be focused for lung repair and regeneration, however, to be desperately serious in moving towards advancement and cure for lung related diseases, it is essential to have a clear understanding of the underlying pathogenesis of distinct chronic lung diseases.

This chapter will particularly cover the major causes and underlying pathophysiology of various lung diseases followed by a critical review of the recent advances done for the development of novel targeted therapeutic approaches to treat lung diseases.

K. Rehman (✉)
Institute of Pharmacy, Physiology and Pharmacology, University of Agriculture,
Faisalabad, Pakistan
e-mail: Kanwal.akash@uaf.edu.pk; kanwalakash@gmail.com

M.S.H. Akash (✉)
Department of Pharmaceutical Chemistry, Government College University,
Faisalabad, Pakistan
e-mail: sajidakash@gcuf.edu.pk; sajidakash@gmail.com

7.2 Lung

To exchange oxygen with carbon dioxide is the major function of lungs which is influenced by various factors like environmental temperature, particles entering the airway and/or allergens (Herriges and Morrisey 2014). Normally, the foregut endoderm that is considered to be the origin of important body organs like liver, esophagus, and thyroid also generates respiratory organs like lung and trachea (Fig. 7.1). As complex respiratory system consists of a multitude of cell lineages, these lineages generate at later endodermal development when the lung mesoderm also known as mesenchyme is established to interact with the lung endoderm (Herriges and Morrisey 2014). The stages of lung development involve multiple and complex process of specification, dividing and imitating that are controlled by a number of signaling pathways like Wnt (Goss et al. 2009), bone morphogenetic protein (Bmp) (Domyan et al. 2011), and fibroblast growth factor (Fgf) (Ohuchi et al. 2000).

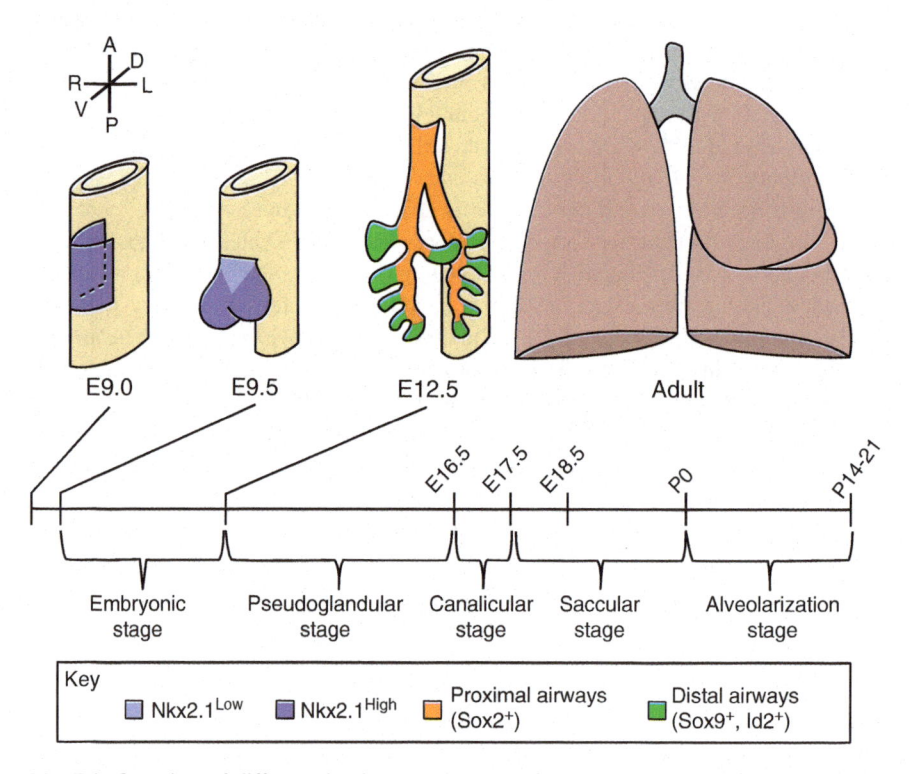

Fig. 7.1 Overview of different developmental stages of lung: Specification of lung endoderm begins at ~E9.0 on the ventral side of anterior foregut endoderm (represented in *yellow color*) where the expression of Nkx2.1 takes place. Formation of trachea is observed at E9.5–E10.0, whereas the embryonic stage of lung development begins at E12.5 and also ends at this point. Other stages of lung development and their corresponding time period have also been displayed in figure. *A* anterior, *D* dorsal, *L* left, *P* posterior, *R* right, *V* ventral. Reproduced with permission from Ref. Herriges and Morrisey (2014)

As far as the role of progenitor cells during the developmental stages of lung diseases are concerned, recent progress has been made in recognizing progenitor cell populations in the embryonic lung identified by lineage-tracing studies. Where a progenitor population at the region of foregut is known to give rise to both trachea and lung (Cardoso and Lu 2006). Similarly, embryonic stem cells are considered really important for lung therapy as the inner cell mass of blastocyst during the embryonic development stage is responsible for the generation of fetal tissue lineages (Loebel et al. 2003).

In short, to maintain the structure of an adult lung and even to efficiently interact during the process of morphogenesis, cells of multiple germinal lineages contribute critically where a single layer is responsible for the generation of other cell lineages ultimately forming airway smooth muscle, fibroblasts, and the vasculature (Neuringer and Randell 2004).

7.3 Lung Disorders

However, just like the complex architecture of the lung and airway, the diseases and disorders associated with lungs are also considered to be multifaceted. Some of them include IPF, ARDS, and particularly COPDs which are briefly discussed in the proceeding sections.

7.3.1 Idiopathic Pulmonary Fibrosis

In IPF, the lung volume is considered to become obstructive having impaired diffusion capacity characterized by interstitial fibrosis and honeycombing (American Thoracic Society 2000). IPF is recognized as a progressive and chronic that predominantly occurs in elderly males. The major risk factor that is considered accountable for lung damage and the prevalence of IPF is inhalation of exogenous noxious compounds (Rabe et al. 2007; Rennard et al. 2007), along with a progressive damage to the alveolar parenchyma (Cottin et al. 2005). According to the definition of American Thoracic Society (ATS)/European Respiratory Society (ERS) classification, IPF is "a particular process of long-lasting fibrosing interstitial pneumonia of unidentified etiology that is specifically restricted to the lung" (American Thoracic Society 2002). IPF has gained great attention of many researchers as median survival for IPF patients has been reported to be around 3 years only (Kim et al. 2006a, b). Moreover, patients suffering from IPF have also been identified further undergoing respiratory disorders leading towards death (Song et al. 2011). Recently, genetic abnormalities are recognized as one of the important pathogenic factors that play a vital role in progression of IPF like telomere dysfunction (Armanios 2009). However, other risk factors that contribute in the progression of IPF may include augmented senescence, abnormalities of surfactant, and/or endoplasmic reticulum stress that can ultimately lead towards exhaustion of precursor cell along with re-epithelialization of abnormal alveolar (Lawson et al. 2008). According to various

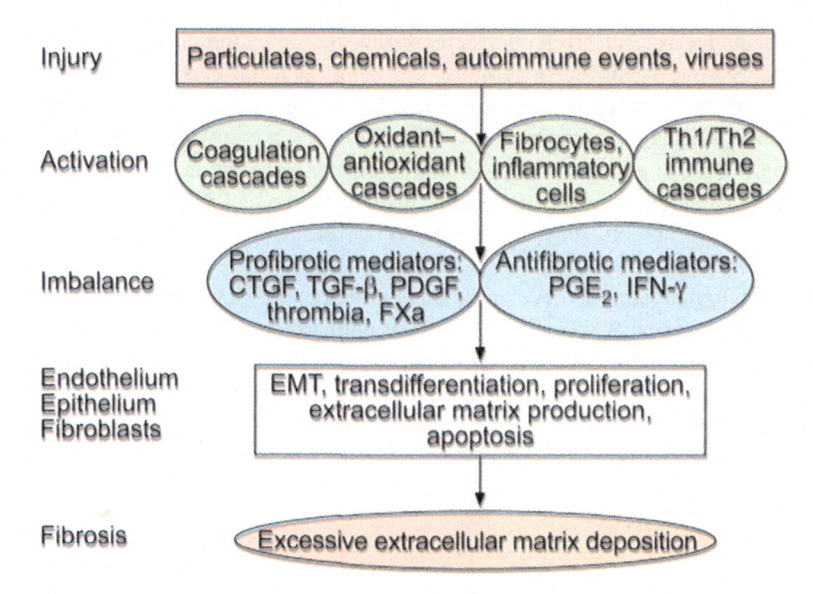

Fig. 7.2 Pathogenesis of idiopathic pulmonary fibrosis: Multi-steps are involved for the pathogenesis of idiopathic fibrosis. Among them, the first step is injury which activates various inflammatory and cell signaling pathways. Activation of these pathways causes an imbalance between the pro-fibrotic and anti-fibrotic mediators which ultimately in return causes the changes in cellular functioning and cell–cell interactions that finally leads to the development of progressive fibrosis. *Th* T-helper cell, *CTGF* connective tissue growth factor, *TGF-β* transforming growth factor-β, *PDGF* platelet-derived growth factor, *FXa* factor Xa, *PG* prostaglandin, *IFN-c* interferon-c, *EMT* epithelial–mesenchymal transition. Reproduced with permission from Ref. Maher et al. (2007)

researchers, the typical pathogenesis of IPF includes altered functioning of epithelial precursors of alveolar tissue by senescent phenotype which may result in mechanical stress and alveolar damage because of inappropriate epithelial renewal (Chilosi et al. 2010; Leslie 2012). Stimuli like injury can provoke inflammatory processes along with signal transduction pathway and repair pathway activation which may ultimately activate further fibrotic mediators causing fibrosis as shown in Fig. 7.2 (Maher et al. 2007). Moreover, biopsy is the most appropriate way to identify the exacerbation of IPF as it reveals histological lesion of acute interstitial pneumonia and alveolar damage (Kim et al. 2006a, b).

Though there is a large literature available on the pathogenesis of IPF, there is still a need of proper manifestation of IPF based not only on histological results but also on including results of clinical evaluation and radiological indications of the disease. Moreover complete understanding of IPF will facilitate better treatment strategies.

7.3.2 Acute Respiratory Distress Syndrome

About 40 to 60 % deaths clinically have been recognized because of ARDS (Rubenfeld and Herridge 2007). Timely identification of ARDS is very essential. SSSSSSSARDS comes as one of the most critical complications of acute lung

Table 7.1 Acute respiratory distress syndrome (ARDS)

Conditions	ARDS (%)	Mortality (%)
Sepsis syndrome	29	32
Pneumonia	38	36
Extra pulmonary source	15	29
Septic shock	37	55
Pulmonary source	48	56
Extra pulmonary source	25	54
Trauma	12–18	10
Blood transfusion	29	57
Gastrointestinal aspiration	22–38	52

Reproduced with permission from Ref. Leaver and Evans (2007)

injury that may further obscure other medical conditions. There are various factors that act as risk factors for the progression of ARDS including alcohol consumption and old age (Wind et al. 2007). The available literature has flawlessly discussed the factors influencing ARDS as depicted in Table 7.1 (Leaver and Evans 2007). Progressive pulmonary fibrosis and pulmonary hypertension are considered as other rare complications associated with ARDS (Leaver and Evans 2007).

Furthermore, there are many complications associated with ARDS observed in 50–70 % of patients. These complications may include pulmonary, neuromuscular, and psychological morbidity (Herridge et al. 2003). The histopathological manifestations of ARDS include injury and inflammation of alveoli which result in increasing the permeability of pulmonary capillaries. Furthermore, phases of exudation, inflammation, and/or fibro-proliferation are considered accountable for the progression of ARDS (Leaver and Evans 2007). For the pathogenesis, the exact mechanism still needs to be elucidated; however, any lung injury or toxicants can provoke body response resulting in activation and release of pulmonary and/or circulatory pro-inflammatory cytokines (Martin 1997) such as interleukin 1 (IL1), IL6, IL8, and tumor necrosis factor (TNF). This results in infiltration of activated neutrophils into interstitial and alveolar spaces (Fulkerson et al. 1996). Further, these neutrophils contribute for a critical role in the pathogenesis of ARDS as their activation can further release cytokines and can also contribute in generating reactive oxygen species leading to further damage. As far as the stages/phases of ARDS are concerned, the primary phase of ARDS increases the vascular permeability that results in accumulation of protein rich fluid in alveoli damaging the alveolar epithelial and causing pulmonary edema (Ware and Matthay 2000). After this, the patients may suffer from different degree of fibrosis, a phase termed as fibro-proliferative phase (Gattinoni et al. 1994).

Various diagnostic parameters and techniques are used for the proper diagnosis of ARDS so that appropriate treatment strategy could be applied. These diagnostic tools may include imaging/CT scanning, hemodynamic evaluation, arterial blood gases which helps in evaluating severity of hypoxemia, and/or assessment of levels of related biomarkers (Dushianthan et al. 2011). However, after appropriate diagnosis of ARDS, suitable treatment and management is required for the patients requiring intensive care, appropriate therapy, and follow-ups.

7.3.3 Chronic Obstructive Pulmonary Disease

COPD has a typical manifestation of limitation in airflow, along with decreased mucociliary clearance, enhanced production of mucus, and increased alveolar dilation; however, it is considered as a preventable and manageable disease (Incalzi et al. 2006). The concurrence of airways inflammation that is usually considered irreversible and emphysema which mainly occurs because of lung parenchyma destruction are known to determine COPD, whereas pulmonary ventilation in COPD is known to face mechanical constraints along other mechanistic pathways disrupting the gas exchange phenomenon (Hogg and Timens 2009).

The major causative factor for COPD recognized globally is cigarette smoking that can further cause airflow limitation varying from individual to individual (Fletcher and Peto 1977). It has been reported that circulating neutrophils increase in count due to cigarette smoking resulting in more neutrophils in lung capillaries (MacNee et al. 1989). The specific relationship between increased neutrophils and progression of COPD still remains to be elucidated; however, it has been seen that there exists severity in patients having more neutrophils in their bronchial specimens (Di Stefano et al. 1998).

As far as the pathogenesis and signaling pathways for the progression of COPD is concerned, it has been reported that there occurs altered or decreases extracellular matrix protein production that leads towards dilation of alveolar spaces along with remodeling of parenchyma. These proteins might be elastin which is known for its important role of maintaining connective tissues and repairing epithelial tissues (Muller et al. 2006). Along with affecting the production of elastin, it has also been reported that during COPD, an altered deregulated role between Wnt and Notch signaling pathway has been observed that in turn causes deregulation of alveologenesis which ultimately results in weak pulmonary parenchyma and alveolar dilation (Chilosi et al. 2012). Moreover the neutrophils that are abundantly found in airways of COPD patients are known to be actually stimulated by few chemotactics like IL-8, growth-related oncogene-α, and/or leukotriene B_4 (Traves et al. 2002).

Though the pathogenesis and progression of COPD involves complex mechanistic detail, we look forward to further experimental investigations that will contribute to unravel the new pathways involve in COPD pathogenesis and hence can open the gateway for more appropriate and targeted therapy for COPD increasing the survival rates of COPD patients in future.

7.4 Lung Regeneration

In above sections, we have briefly discussed the understanding of normal lung development along with some of the major lung disorders that are known to have high prevalence globally. This discussion will help in making a better prospect of strategies that are being used and/or required in future for rebuilding or regeneration of lung for advanced treatment approaches of various lung diseases.

7.4.1 Stem and/or Progenitor Cells for Lung Regeneration

The progenitor cells are primarily present in bone marrow that further help in the generation of various cell lines, these progenitor cells include hematopoietic stem cells, fibroblasts, and/or mesenchymal stem cells (Jones et al. 2009). These progenitor cells have been reported to help in modifying various organ disorders like lung diseases and liver disorders. Interestingly, other than bone marrow, lungs themselves have been reported to possess progenitor cells, including conducting airway stem cells and distil lung epithelial progenitor cells (Rock and Konigshoff 2012). Nevertheless, various studies have varying suggestions regarding the role of progenitor cells in lung regeneration for diseased lung. The details of different experimental studies have been identified in Table 7.2 (Neuringer and Randell 2004) for delivery of lung progenitor cells either having positive or negative results. Moreover, circulating endothelial progenitor cells (EPCs) have been used experimentally to observe their repairing effects on damaged lung vasculature. Further EPCs are known to be helpful in future for pulmonary hypertension. Similarly, when investigated, bone marrow-derived mesenchymal stromal cells (MSCs) have also shown their immunomodulatory effects in lung disease suggesting their future role for the treatment of particular lung disorders like COPD (Lau et al. 2012). As in COPD there occurs loss of proteolytic lung parenchyma and elastic recoil, so use of stem cells rather than drug therapy which can only provide symptomatic relief might be more beneficial. Alveolar type II pneumocytes are well known to have the ability to repair any damage to alveoli, it has been reported that human embryonic stem cells can be transformed into alveolar type II pneumocytes (Wang et al. 2007).

Similarly, the growing literature has provided an evidence for the potential of cell-based therapy for ARDS in particularly MSCs (Matthay et al. 2010). It has been reported that bleomycin-induced liver damage in mice was significantly decreased by the utilization of MSCs (Ortiz et al. 2003). Moreover, the response of host towards bacterial sepsis has also been observed to be improved after the use of MSCs by the reduction of septic lung damage (Gupta et al. 2007; Danchuk et al. 2011).

Fibroblast growth factor (FGF)-10, which is known to regulate early lung epithelial progenitor cells, is known to be secreted by MSCs that are at the distal tip of the branching epithelium (Ramasamy et al. 2007). FGF-10 is also important for signal pathways (Bmp, Wnt, and Shh) involved controlling progenitor cells for lung development (Morrisey and Hogan 2010). Similarly, bronchiolar smooth muscles are reported to be generated through MSCs adjacent to the trachea and extrapulmonary bronchi (Shan et al. 2008). Moreover, the expression of genes of induced pluripotent stem cells (iPSCs) are comparable to ESCs and there is a growing evidence of utilization of iPSCs against ARDS as shown in Fig. 7.3 (Hayes et al. 2012). Interestingly, recently c-kit-positive cells found in distal lung of human have been tried for injured mouse cells showing significant lung regeneration (Kajstura et al. 2011). However, more work is required to elucidate the precise mechanism of these c-kit-positive cells and their role in lung regeneration.

Table 7.2 Acute respiratory distress syndrome (ARDS)

Study	Disease model	Tissue of origin	Lung cell type formed	Method of detection
Animal	BMT	MSC	Undefined mesenchymal cells	PCR for collagen gene marker
Animal	Bleomycin fibrosis	MSC	Type I pneumocytes	B-galactosidase protein
Animal	BMT	HSC enrichment	Type II pneumocytes/up to 20%, bronchial epithelium/4%	Y chromosome FISH, surfactant B mRNA
Animal	Radiation pneumonitis	Whole bone marrow	Type II pneumocytes, bronchial epithelium/up to 20% of type II cells	Y chromosome FISH, surfactant B mRNA
Animal	BMT	Whole bone marrow/ EGFP retrovirus	Type II pneumocytes/1–7%	EGFP, keratin immunostain, surfactant protein B FISH
Animal	BMT and parabiotic Animals	HSC	Hematopoietic chimerism but exceedingly rare lung cell types	EGFP
Animal	Bleomycin fibrosis	MSC	Type II pneumocytes/~1%	Y chromosome FISH
Animal	Radiation fibrosis	MSC or whole bone marrow	Fibroblasts	EGFP, Y chromosome FISH, vimentin immunostain
Animal	BMT	Bone marrow, EGFP labeled	Fibroblasts, type I pneumocyte	Flow cytometry
Animal	Hypoxia-induced pulmonary hypertension	Circulating BM-derived c-kit positive	c-kit positive cells in pulmonary artery vessel wall; In hypoxia, circulating cells generate endothelial and smooth muscle cells in-vitro	Flow cytometry and immunohistochemistry
Animal	Ablative radiation and elastase induced emphysema	GFP+ fetal liver	Alveolar epithelium and endothelium; frequency not reported but increased by G-CSF and retinoic acid	Immunohistochemistry for CD45-, GFP+ cells
Animal	Bleomycin fibrosis	Whole marrow GFP+	GFP+ type I collagen expressing	Flow cytometry and immunohistochemistry, RT-PCR

		MSC and SAEC	Cell fusion	Immunostaining, microarray
Human	Heat shock in cell culture			
Animal and human	OVA-sensitized mouse model, allergen-sensitized asthmatics	CD34 positive, collagen I expressing fibrocytes CD34 positive, collagen I expressing fibrocytes	Myofibroblasts	CD34-positive, collagen I, α-smooth muscle actin CD34-positive, collagen I, α-smooth muscle actin
Human	Human heart and lung transplant	Sex-mismatched donor lung or heart	No lung cell types of recipient origin	X and Y chromosome FISH, antibody stain for hematopoietic cells
Human	Human lung transplant, human BMT	Sex-mismatched donor lung and sex-mismatched donor bone marrow	Bronchial epithelium, type II pneumocytes, glands of recipient origin No lung cell types of donor origin	Y chromosome FISH, short tandem repeat PCR Y chromosome FISH, short tandem repeat PCR
Human	Human BMT	Sex-mismatched donor bone marrow	Lung epithelium and endothelium of donor origin	X and Y chromosome FISH, keratin, and PECAM immunostain
Human	Human BMT	Sex-mismatched donor bone marrow	No nasal epithelium of donor origin	Y chromosome FISH, cytokeratin immunostain

Reproduced with permission from Ref. Neuringer and Randell (2004)
BMT bone marrow transplant (with prior ablation), *MSC* mesenchymal stem cells (bone marrow stromal cells, adherent bone marrow cells), *EGFP* enhanced green fluorescent protein, *HSC* hematopoietic stem cells, *FISH* fluorescence in situ hybridization, *SAEC* small airway epithelial cells

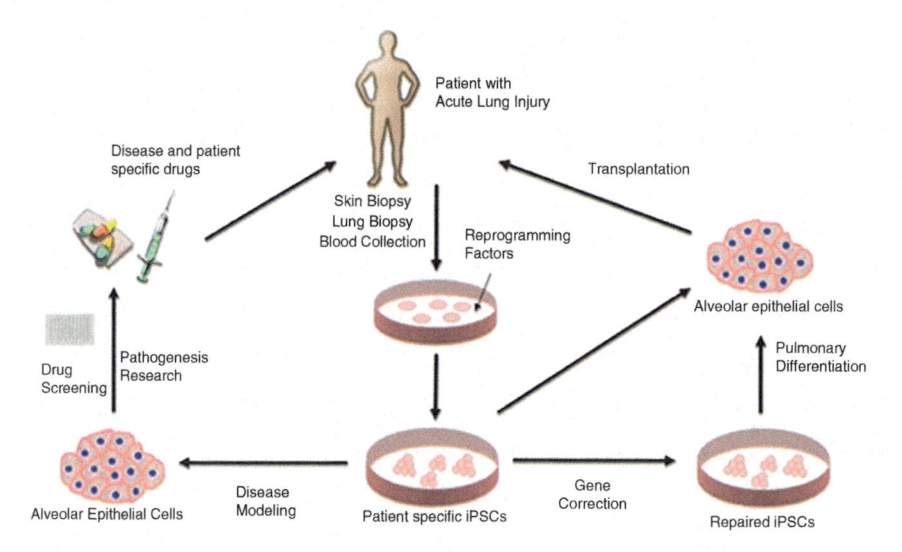

Fig. 7.3 Potential applications of induced pluripotent stem cells for liver diseases: Reproduced with permission from Ref. Hayes et al. (2012)

Various types of pathogenic mechanisms are involved in different types of lung diseases that have been briefly described in the above sections, but for the utilization of safe and effective cell-based treatments for these diseases, use of exogenous stem/progenitor cells is very essential to find out the probable chances for the development of further complications.

7.4.2 Other Treatment Approaches for IPF

Various drug classes including anti-inflammatory, anti-fibrotic agents, and immune modulators have been reported to be investigated and beneficial for the management of IPF. On getting a CT scan of IPF patients having corticosteroids, it was observed that there was a reduction in inflammation in IPF patients that also helped improved the function of pulmonary system (Lee et al. 1992). Though, these agents can provide symptomatic treatment, rather than curing the disease or providing any permanent treatment to IPF. Besides an overuse of these drugs can be challenging as high or accidental dosage can lead towards adverse events like hyperglycemia, hypertension, and/or osteoporosis. Similarly, different anti-fibrotic agents having capability to block fibrogenesis have demonstrated significant results against IPF. For instance, it has been observed that relaxin, an inhibitor of fibrogenesis in patients with progressive systemic sclerosis, has shown to improve pulmonary function test (Seibold et al. 2000). Similarly, in a mouse model of lung injury, pirfenidone, that is known to block fibroblast proliferation, has shown to ameliorate cyclophosphamide-induced pulmonary fibrosis (Kehrer and Margolin 1997). Besides drugs, lung transplant has also provided a gateway for the treatment of life-threatening lung diseases (Meyers et al. 2000); however, it has certain limitations but still is a viable option for IPF patients. This approach of tissue engineering can be promising as recently it has been reported that ex vivo engineered rat lung proved to be functional for a period of time when it was re-implanted into animals (Petersen et al. 2010). Though such studies provide a hope for future, there are some limitations that still need to be addressed like immune system rejection and/or proper functioning of organ for a long period of time. Similar limitations are for lung transplantation that will require use of immunosuppressant drugs and there are serious issues regarding the availability of the donors.

7.4.3 Other Treatment Approaches for ARDS

It may require supportive measures like improved nutrition and diet containing eicosapentaenoic acid, γ-linolenic acid, and anti-oxidants (Pontes-Arruda et al. 2006). Other supportive measures, such as mechanical ventilation strategies for lung protection and positive end expiratory pressure which helps in increasing functional residual capacity and thereby improving ventilation (Leaver and Evans 2007). The restriction of fluid intake can somehow increase ventilation by lowering the alveolar lung edema (Wiedemann et al. 2006).

However, as far as the research work regarding the treatment of ARDS is concerned, various suggestions have been proposed but none have so far provided any satisfactory clinical outcome. These studies have utilized different drugs and agents against ARDS including anti-oxidants (Bernard et al. 1999; Ortolani et al. 2000), immunomodulating agents (Bernard et al. 1999), and corticosteroids (Thompson 2003) which could not show any satisfactory results. However, it has also been reported that administration of corticosteroids during early ARDS showed significant improvement in mortality (Meduri et al. 1998). Similarly, other measures for supportive care that should be adopted for ARDS patients to avoid further complications include proper monitoring of infections, GIT bleeding, and/or thromboembolism (Mortelliti and Manning 2002). General principles of ARDS management are summarized in Table 7.3.

7.4.4 Other Treatment Approaches for COPD

Disappointingly, there have been limited therapeutic improvements in drug therapy for COPD.

However, the pharmacotherapy for COPD includes long acting bronchodilators like β2-agonists; indacaterol and carmoterol are considered effective against COPD (Matera and Cazzola 2007). Similarly, as stated above, Leukotriene B4 is found in

Table 7.3 General principles of acute respiratory distress syndrome (ARDS) management

Sr.#	Non-ventilatory supportive care	Ventilation management
1.	Identify early and aggressively treat underlying cause	Set initial tidal volume at 6 mL per kg (based on predicted body weight)
2.	Minimize potential complications such as nosocomial infections, gastrointestinal bleeding, and thromboembolism	Aim to maintain P_{plat} at <30 mmHg (may require subsequent adjustments of tidal volume)
3.	Initiate appropriate nutrition	Ensure adequate oxygenation: P_{aO2} of 55–80 mmHg or S_{pO2} of 88–95 %
4.	Maintain appropriate level of sedation to facilitate patient-ventilator synchrony and limit patient discomfort	Ensure adequate carbon dioxide removal
5.	Limit excessive and prolonged sedation and use of paralytic agents	Ensure maximal recruitment of alveoli through appropriate levels of PEEP

Reproduced with permission from Ref. Mortelliti and Manning (2002)
P_{plat} plateau pressure, P_{aO2} partial pressure of arterial oxygen, S_{pO2} oxyhemoglobin saturation, *PEEP* positive end expiratory pressure

the sputum of COPD, thereby antagonist of Leukotriene B4 might be a suitable intervention for the treatment of COPD, however, till now no significant result has been obtained (Hicks et al. 2007). Likewise the TNF-α has also been identified in the sputum of COPD patients, it is accounted responsible for the progression of inflammation during COPD. However, the typical anti-inflammatory drugs could not provide significant clinical outcomes when used in COPD (Rennard et al. 2007). As far as the use of corticosteroid is concerned for COPD, high doses of corticosteroids have shown negligible effects on the progression of COPD and mortality (Yang et al. 2007). Thereby, there is a great need particularly for the development of the novel therapeutic approaches for the treatment of COPD.

7.5 Conclusion

In this chapter, we have discussed the basics of different chronic lung diseases requiring critical attention as they have been considered the major cause of population death globally. No doubt, it is a great challenge to take regenerative potential of stem/progenitor cells for the cure/management of lung diseases. However, here, we have briefly addressed the major outcomes of studies that have focused using these stem/progenitor cells for the treatment of lung diseases. The recent findings highlighted in the current chapter have shown that a significant work has been carried out and has also provided future hope for lung repair and regeneration by using appropriate strategies.

7.6 Future Prospects

As far as the future prospects, limitations, and challenges are concerned, the validation of source and lineage of progenitor cells should be considered first along with the signaling pathways that are involved for the proliferation and development. These points need critical evaluation as every chronic lung disease has its own risk factors, pathogenesis, and consequences, thereby, improper understanding and utilization of stem/progenitor cells can further lead towards unwanted complications. These questions need further studies to open novel therapeutic gateway for bioengineering of lung through appropriate identification of progenitor cells and signaling pathways.

Conflict of Interest Nothing to be declared.

References

American Thoracic Society (2000) Idiopathic pulmonary fibrosis: diagnosis and treatment. International consensus statement. American Thoracic Society (ATS), and the European Respiratory Society (ERS). Am J Respir Crit Care Med 161:646–664

American Thoracic Society/European Respiratory Society International Multidisciplinary Consensus Classification of the Idiopathic Interstitial Pneumonias (2002) This joint statement of the American Thoracic Society (ATS), and the European Respiratory Society (ERS) was adopted by the ATS board of directors, June 2001 and by the ERS Executive Committee, June 2001. Am J Respir Crit Care Med 165:277–304

Armanios M (2009) Syndromes of telomere shortening. Annu Rev Genomics Hum Genet 10:45–61

Balkissoon R, Lommatzsch S, Carolan B, Make B (2011) Chronic obstructive pulmonary disease: a concise review. Med Clin North Am 95:1125–1141

Bernard GR, Wheeler AP, Naum CC, Morris PE, Nelson L, Schein R, Wright P, Brigham KL, Russell J, Paz H (1999) A placebo controlled, randomized trial of IL-10 in acute lung injury (ALI). Chest 116:260S

Cardoso WV, Lu J (2006) Regulation of early lung morphogenesis: questions, facts and controversies. Development 133:1611–1624

Chilosi M, Doglioni C, Murer B, Poletti V (2010) Epithelial stem cell exhaustion in the pathogenesis of idiopathic pulmonary fibrosis. Sarcoidosis Vasc Diffuse Lung Dis 27:7–18

Chilosi M, Poletti V, Rossi A (2012) The pathogenesis of COPD and IPF: distinct horns of the same devil? Respir Res 13:3

Cottin V, Nunes H, Brillet PY, Delaval P, Devouassoux G, Tillie-Leblond I, Israel-Biet D, Court-Fortune I, Valeyre D, Cordier JF (2005) Combined pulmonary fibrosis and emphysema: a distinct underrecognised entity. Eur Respir J 26:586–593

Danchuk S, Ylostalo JH, Hossain F, Sorge R, Ramsey A, Bonvillain RW, Lasky JA, Bunnell BA, Welsh DA, Prockop DJ, Sullivan DE (2011) Human multipotent stromal cells attenuate lipopolysaccharide-induced acute lung injury in mice via secretion of tumor necrosis factor-alpha-induced protein 6. Stem Cell Res Ther 2:27

Di Stefano A, Capelli A, Lusuardi M, Balbo P, Vecchio C, Maestrelli P, Mapp CE, Fabbri LM, Donner CF, Saetta M (1998) Severity of airflow limitation is associated with severity of airway inflammation in smokers. Am J Respir Crit Care Med 158:1277–1285

Domyan ET, Ferretti E, Throckmorton K, Mishina Y, Nicolis SK, Sun X (2011) Signaling through BMP receptors promotes respiratory identity in the foregut via repression of Sox2. Development 138:971–981

Dushianthan A, Grocott MP, Postle AD, Cusack R (2011) Acute respiratory distress syndrome and acute lung injury. Postgrad Med J 87:612–622

Fletcher C, Peto R (1977) The natural history of chronic airflow obstruction. Br Med J 1:1645–1648

Fulkerson WJ, MacIntyre N, Stamler J, Crapo JD (1996) Pathogenesis and treatment of the adult respiratory distress syndrome. Arch Intern Med 156:29–38

Gattinoni L, Bombino M, Pelosi P, Lissoni A, Pesenti A, Fumagalli R, Tagliabue M (1994) Lung structure and function in different stages of severe adult respiratory distress syndrome. JAMA 271:1772–1779

Goss AM, Tian Y, Tsukiyama T, Cohen ED, Zhou D, Lu MM, Yamaguchi TP, Morrisey EE (2009) Wnt2/2b and beta-catenin signaling are necessary and sufficient to specify lung progenitors in the foregut. Dev Cell 17:290–298

Gupta N, Su X, Popov B, Lee JW, Serikov V, Matthay MA (2007) Intrapulmonary delivery of bone marrow-derived mesenchymal stem cells improves survival and attenuates endotoxin-induced acute lung injury in mice. J Immunol 179:1855–1863

Hayes M, Curley G, Ansari B, Laffey JG (2012) Clinical review: Stem cell therapies for acute lung injury/acute respiratory distress syndrome - hope or hype? Crit Care 16:205

Herridge MS, Cheung AM, Tansey CM, Matte-Martyn A, Diaz-Granados N, Al-Saidi F, Cooper AB, Guest CB, Mazer CD, Mehta S, Stewart TE, Barr A, Cook D, Slutsky AS (2003) One-year outcomes in survivors of the acute respiratory distress syndrome. N Engl J Med 348:683–693

Herriges M, Morrisey EE (2014) Lung development: orchestrating the generation and regeneration of a complex organ. Development 141:502–513

Hicks A, Monkarsh SP, Hoffman AF, Goodnow R Jr (2007) Leukotriene B4 receptor antagonists as therapeutics for inflammatory disease: preclinical and clinical developments. Expert Opin Investig Drugs 16:1909–1920

Hogg JC, Timens W (2009) The pathology of chronic obstructive pulmonary disease. Annu Rev Pathol 4:435–459

Incalzi RA, Corsonello A, Pedone C, Masotti G, Bellia V, Grassi V, Rengo F (2006) From global initiative for chronic obstructive lung disease (GOLD) guidelines to current clinical practice. Drugs Aging 23:411–420

Jones CP, Pitchford SC, Lloyd CM, Rankin SM (2009) CXCR2 mediates the recruitment of endothelial progenitor cells during allergic airways remodeling. Stem Cells 27:3074–3081

Kajstura J, Rota M, Hall SR, Hosoda T, D'Amario D, Sanada F, Zheng H, Ogorek B, Rondon-Clavo C, Ferreira-Martins J, Matsuda A, Arranto C, Goichberg P, Giordano G, Haley KJ, Bardelli S, Rayatzadeh H, Liu X, Quaini F, Liao R, Leri A, Perrella MA, Loscalzo J, Anversa P (2011) Evidence for human lung stem cells. N Engl J Med 364:1795–1806

Kehrer JP, Margolin SB (1997) Pirfenidone diminishes cyclophosphamide-induced lung fibrosis in mice. Toxicol Lett 90:125–132

Kim DS, Collard HR, King TE Jr (2006a) Classification and natural history of the idiopathic interstitial pneumonias. Proc Am Thorac Soc 3:285–292

Kim DS, Park JH, Park BK, Lee JS, Nicholson AG, Colby T (2006b) Acute exacerbation of idiopathic pulmonary fibrosis: frequency and clinical features. Eur Respir J 27:143–150

Lau AN, Goodwin M, Kim CF, Weiss DJ (2012) Stem cells and regenerative medicine in lung biology and diseases. Mol Ther 20:1116–1130

Lawson WE, Crossno PF, Polosukhin VV, Roldan J, Cheng DS, Lane KB, Blackwell TR, Xu C, Markin C, Ware LB, Miller GG, Loyd JE, Blackwell TS (2008) Endoplasmic reticulum stress in alveolar epithelial cells is prominent in IPF: association with altered surfactant protein processing and herpesvirus infection. Am J Physiol Lung Cell Mol Physiol 294:L1119–L1126

Leaver SK, Evans TW (2007) Acute respiratory distress syndrome. BMJ 335:389–394

Lee JS, Im JG, Ahn JM, Kim YM, Han MC (1992) Fibrosing alveolitis: prognostic implication of ground-glass attenuation at high-resolution CT. Radiology 184:451–454

Leslie KO (2012) Idiopathic pulmonary fibrosis may be a disease of recurrent, tractional injury to the periphery of the aging lung: a unifying hypothesis regarding etiology and pathogenesis. Arch Pathol Lab Med 136:591–600

Loebel DA, Watson CM, De Young RA, Tam PP (2003) Lineage choice and differentiation in mouse embryos and embryonic stem cells. Dev Biol 264:1–14

MacNee W, Wiggs B, Belzberg AS, Hogg JC (1989) The effect of cigarette smoking on neutrophil kinetics in human lungs. N Engl J Med 321:924–928

Maher TM, Wells AU, Laurent GJ (2007) Idiopathic pulmonary fibrosis: multiple causes and multiple mechanisms? Eur Respir J 30:835–839

Martin TR (1997) Cytokines and the acute respiratory distress syndrome (ARDS): a question of balance. Nat Med 3:272–273

Matera MG, Cazzola M (2007) Ultra-long-acting beta2-adrenoceptor agonists: an emerging therapeutic option for asthma and COPD? Drugs 67:503–515

Matthay MA, Thompson BT, Read EJ, McKenna DH Jr, Liu KD, Calfee CS, Lee JW (2010) Therapeutic potential of mesenchymal stem cells for severe acute lung injury. Chest 138:965–972

Meduri GU, Headley AS, Golden E, Carson SJ, Umberger RA, Kelso T, Tolley EA (1998) Effect of prolonged methylprednisolone therapy in unresolving acute respiratory distress syndrome: a randomized controlled trial. JAMA 280:159–165

Meyers BF, Lynch JP, Trulock EP, Guthrie T, Cooper JD, Patterson GA (2000) Single versus bilateral lung transplantation for idiopathic pulmonary fibrosis: a ten-year institutional experience. J Thorac Cardiovasc Surg 120:99–107

Morrisey EE, Hogan BL (2010) Preparing for the first breath: genetic and cellular mechanisms in lung development. Dev Cell 18:8–23

Mortelliti MP, Manning HL (2002) Acute respiratory distress syndrome. Am Fam Physician 65:1823–1830

Muller KC, Welker L, Paasch K, Feindt B, Erpenbeck VJ, Hohlfeld JM, Krug N, Nakashima M, Branscheid D, Magnussen H, Jorres RA, Holz O (2006) Lung fibroblasts from patients with emphysema show markers of senescence in vitro. Respir Res 7:32

Neuringer IP, Randell SH (2004) Stem cells and repair of lung injuries. Respir Res 5:6

Ohuchi H, Hori Y, Yamasaki M, Harada H, Sekine K, Kato S, Itoh N (2000) FGF10 acts as a major ligand for FGF receptor 2 IIIb in mouse multi-organ development. Biochem Biophys Res Commun 277:643–649

Orens JB, Garrity ER Jr (2009) General overview of lung transplantation and review of organ allocation. Proc Am Thorac Soc 6:13–19

Ortiz LA, Gambelli F, McBride C, Gaupp D, Baddoo M, Kaminski N, Phinney DG (2003) Mesenchymal stem cell engraftment in lung is enhanced in response to bleomycin exposure and ameliorates its fibrotic effects. Proc Natl Acad Sci U S A 100:8407–8411

Ortolani O, Conti A, De Gaudio AR, Masoni M, Novelli G (2000) Protective effects of N-acetylcysteine and rutin on the lipid peroxidation of the lung epithelium during the adult respiratory distress syndrome. Shock 13:14–18

Petersen TH, Calle EA, Zhao L, Lee EJ, Gui L, Raredon MB, Gavrilov K, Yi T, Zhuang ZW, Breuer C, Herzog E, Niklason LE (2010) Tissue-engineered lungs for in vivo implantation. Science 329:538–541

Pontes-Arruda A, Aragao AM, Albuquerque JD (2006) Effects of enteral feeding with eicosapentaenoic acid, gamma-linolenic acid, and antioxidants in mechanically ventilated patients with severe sepsis and septic shock. Crit Care Med 34:2325–2333

Rabe KF, Hurd S, Anzueto A, Barnes PJ, Buist SA, Calverley P, Fukuchi Y, Jenkins C, Rodriguez-Roisin R, van Weel C, Zielinski J (2007) Global strategy for the diagnosis, management, and prevention of chronic obstructive pulmonary disease: GOLD executive summary. Am J Respir Crit Care Med 176:532–555

Ramasamy SK, Mailleux AA, Gupte VV, Mata F, Sala FG, Veltmaat JM, Del Moral PM, De Langhe S, Parsa S, Kelly LK, Kelly R, Shia W, Keshet E, Minoo P, Warburton D, Bellusci S (2007) Fgf10 dosage is critical for the amplification of epithelial cell progenitors and for the formation of multiple mesenchymal lineages during lung development. Dev Biol 307:237–247

Rennard SI, Fogarty C, Kelsen S, Long W, Ramsdell J, Allison J, Mahler D, Saadeh C, Siler T, Snell P, Korenblat P, Smith W, Kaye M, Mandel M, Andrews C, Prabhu R, Donohue JF, Watt R, Lo KH, Schlenker-Herceg R, Barnathan ES, Murray J (2007) The safety and efficacy of infliximab in moderate to severe chronic obstructive pulmonary disease. Am J Respir Crit Care Med 175:926–934

Rock J, Konigshoff M (2012) Endogenous lung regeneration: potential and limitations. Am J Respir Crit Care Med 186:1213–1219

Rubenfeld GD, Herridge MS (2007) Epidemiology and outcomes of acute lung injury. Chest 131:554–562

Seibold JR, Korn JH, Simms R, Clements PJ, Moreland LW, Mayes MD, Furst DE, Rothfield N, Steen V, Weisman M, Collier D, Wigley FM, Merkel PA, Csuka ME, Hsu V, Rocco S, Erikson M, Hannigan J, Harkonen WS, Sanders ME (2000) Recombinant human relaxin in the treatment of scleroderma. A randomized, double-blind, placebo-controlled trial. Ann Intern Med 132:871–879

Shan L, Subramaniam M, Emanuel RL, Degan S, Johnston P, Tefft D, Warburton D, Sunday ME (2008) Centrifugal migration of mesenchymal cells in embryonic lung. Dev Dyn 237:750–757

Song JW, Hong SB, Lim CM, Koh Y, Kim DS (2011) Acute exacerbation of idiopathic pulmonary fibrosis: incidence, risk factors and outcome. Eur Respir J 37:356–363

Tansey CM, Louie M, Loeb M, Gold WL, Muller MP, de Jager J, Cameron JI, Tomlinson G, Mazzulli T, Walmsley SL, Rachlis AR, Mederski BD, Silverman M, Shainhouse Z, Ephtimios IE, Avendano M, Downey J, Styra R, Yamamura D, Gerson M, Stanbrook MB, Marras TK, Phillips EJ, Zamel N, Richardson SE, Slutsky AS, Herridge MS (2007) One-year outcomes and health care utilization in survivors of severe acute respiratory syndrome. Arch Intern Med 167:1312–1320

Thompson BT (2003) Glucocorticoids and acute lung injury. Crit Care Med 31:S253–S257

Traves SL, Culpitt SV, Russell REK, Barnes PJ, Donnelly LE (2002) Increased levels of the chemokines GROα and MCP-1 in sputum samples from patients with COPD. Thorax 57:590–595

Wang D, Haviland DL, Burns AR, Zsigmond E, Wetsel RA (2007) A pure population of lung alveolar epithelial type II cells derived from human embryonic stem cells. Proc Natl Acad Sci U S A 104:4449–4454

Ware LB, Matthay MA (2000) The acute respiratory distress syndrome. N Engl J Med 342:1334–1349

Wiedemann HP, Wheeler AP, Bernard GR, Thompson BT, Hayden D, deBoisblanc B, Connors Jr AF, Hite RD, Harabin AL (2006) Comparison of two fluid-management strategies in acute lung injury. N Engl J Med 354:2564–2575

Wind J, Versteegt J, Twisk J, van der Werf TS, Bindels AJ, Spijkstra JJ, Girbes AR, Groeneveld AB (2007) Epidemiology of acute lung injury and acute respiratory distress syndrome in The Netherlands: a survey. Respir Med 101:2091–2098

Yang IA, Fong KM, Sim EHA, Black PN, Lasserson TJ (2007) Inhaled corticosteroids for stable chronic obstructive pulmonary disease (Review). Cochrane Database Syst Rev 2:CD002991

Part III
Heart Regeneration

Chapter 8
Road to Heart Regeneration with Induced Pluripotent Stem Cells

Jun Fujita, Shugo Tohyama, Kazuaki Nakajima, Tomohisa Seki, Hideaki Kanazawa, and Keiichi Fukuda

8.1 Introduction

Heart failure (HF) is one of the leading causes of morbidity and mortality in developed countries. The current radical treatment of severe HF is cardiac transplantation. However, the shortage of donor hearts has hampered the prevalence of cardiac transplantation as a general therapy (Lund et al. 2015). Cell transplantation therapy is the primary candidate as an alternative to cardiac transplantation. The main cell sources for cardiac regenerative therapy were bone marrow stem cells, monocytes, myoblasts, and cardiac progenitor cells (Bolli et al. 2011; Menasché et al. 2008; Strauer and Steinhoff 2011). In the last decade, many clinical trials have been performed to treat HF with these cells. These treatments were considered to be innovative and hopeful therapies, but they have not yet reached the level of standard therapies. Although clinical trials were approved on the basis of the positive results from basic studies with animal's models, the optimal results could not be reproduced in humans. The main issue appeared to be enhancing the differentiation capacity of the cells to cardiomyocytes in order to achieve effective cardiac function to move to the next stage of application of cardiac regenerative therapies.

Pluripotent stem cells (PSCs), including embryonic stem cells (ESCs) and induced pluripotent stem cells (iPSCs), are more recently discovered candidates for cardiac regenerative medicine, because they can potentially differentiate into all cell types, including cardiomyocytes, in vitro. However, the ethical issue of the clinical use of human ESCs has been debated, because they are established from human blastocysts

---

J. Fujita (✉) • S. Tohyama • K. Nakajima • T. Seki • H. Kanazawa • K. Fukuda
Department of Cardiology, Keio University School of Medicine,
35 Shinanomachi Shinjuku-ku, Tokyo 160-8582, Japan
e-mail: jfujita@a6.keio.jp; shugotohyama@gmail.com; kznkjm777@yahoo.co.jp; tomohisaseki@gmail.com; kanazawa@a5.keio.jp; kfukuda@a2.keio.jp

© Springer International Publishing AG 2017
P.V. Pham (ed.), *Liver, Lung and Heart Regeneration*,
Stem Cells in Clinical Applications, DOI 10.1007/978-3-319-46693-4_8

(Thomson et al. 1998). In 2007, human iPSCs emerged as a new form of PSCs (Takahashi et al. 2007). In contrast to human ESCs, iPSCs can be reprogrammed from the patient's own cell source. This avoids the ethical issues associated with ESCs and enables autologous transplantation, which can reduce the amount of immunosuppressive therapies required after cell transplantation. In 2014, human iPSCs were first applied to a patient with macular degeneration as an autologous transplantation as an autologous transplantation (Cyranoski 2014). Therefore, application of iPSCs to more complex organs such as the heart is anticipated as the next step.

The differentiation efficiency of PSCs has been improved over the last decade. Although the establishment of cardiac cell therapies with human iPSCs appears to be straightforward in theory, there are several issues to overcome before these therapies can be applied in standard practice (Fig. 8.1). The quality of iPSCs, achieving efficient and mass cultures of cardiomyocytes, and development of an effective transplantation strategy of cardiomyocytes must be established for clinical application. In particular, the tumorigenicity of PSCs remains the most critical issue for

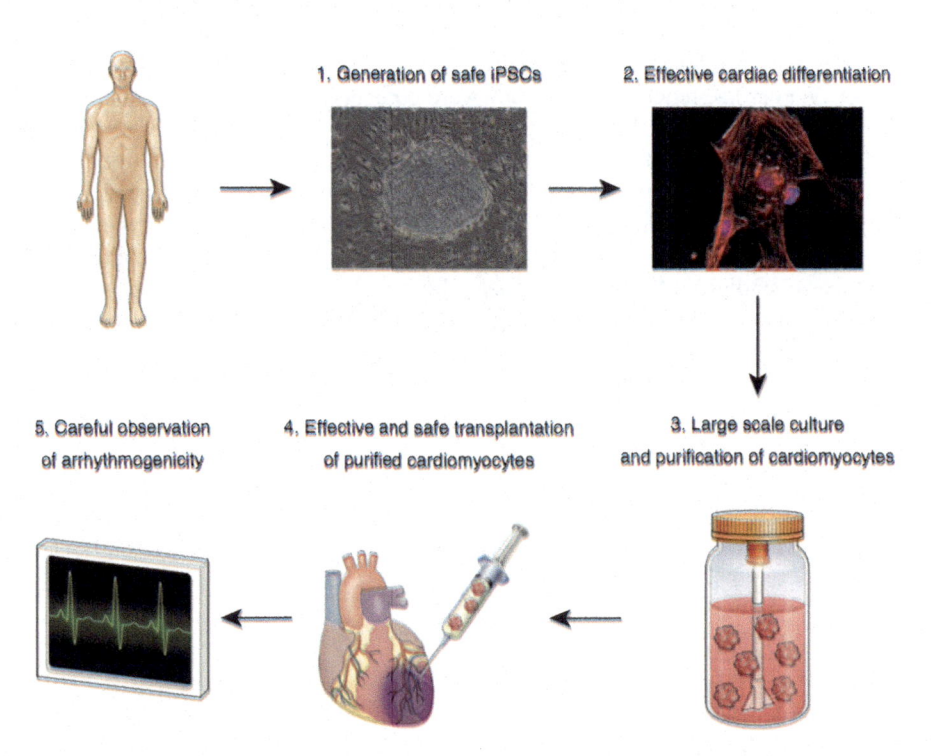

Fig. 8.1 Multiple steps toward to induced pluripotent stem cells (iPSCs) therapy for heart failure. In order to realize the clinical application of human iPSC-derived cardiomyocytes to the treatment of heart failure, each critical step must be carefully validated. Figures are from Fujita and Fukuda (2014)

their clinical application. The follow-up protocol after cell transplantation is also very important for preventing the severe side effects resulting from cell transplantation therapies. In this chapter, we discuss the possibility of cardiac cell therapy as a future treatment for HF.

8.2 Clinical Application of Human iPSCs

The first step toward clinical application is achieving effective quality control of iPSCs. Because iPSCs are artificially produced, their quality must be carefully examined. In addition to the pluripotency and karyotype, the differentiation potential of iPSCs to cardiomyocytes must be evaluated. In order to prevent tumor formation after cell transplantation, whole-genome and exome sequencing may be necessary.

8.2.1 Reprogramming Vectors and Somatic Cell Source for iPSCs

Human iPSCs were initially generated from dermal fibroblasts using retrovirus vectors (Takahashi et al. 2007). As a somatic cell source, blood cells are now the mainstream instead of dermal fibroblasts, because sampling from patients is easy. The episomal vector and the Sendai virus vector were reported to efficiently reprogram peripheral blood T cells (Okita et al. 2013; Seki et al. 2010). Because the episomal vector has potential to integrate into the host genome with low frequency, it is necessary to verify genome integration using genomic polymerase chain reaction, Southern blotting analysis, and genome sequencing. By contrast, the Sendai virus vector is an RNA vector, and will therefore never integrate into the host genome (Fusaki et al. 2009). Because of the integration of a virus can damage the host genome, non-integrating vectors can be preferable for clinical application.

8.2.2 iPSCs Bank

As mentioned above, the main advantage of the clinical use of iPSCs is that they can be established from the patients themselves, which enables autologous transplantation. Therefore, if an institute has a good manufacturing practice (GMP) facility to reprogram patient cells to iPSCs and the capacity to perform the quality control, it is best to establish the clinical-grade iPSCs from the patients themselves. However, in practice, it may be more costly to establish clinically qualified iPSCs. Furthermore, the differentiation capacity may vary among different iPSCs lines; therefore, the most appropriate line must be selected for cardiac differentiation. The establishment of an iPSCs bank is another idea for providing a stable supply of clinical-grade iPSCs. However, the disadvantage of this system is that this removes the main advantage of iPSCs, which is that they are produced from the recipients themselves in contrast to the use of human ESCs for transplantation. Establishment of a human iPSCs bank with a homozygous human leukocyte antigen (HLA) haplotype, which covers the majority of the population, appears to be an effective solution to resolve this problem (Nakatsuji et al. 2008). Transplantation of HLA haplotype-matched human iPSCs can reduce the possibility of immunological rejection, and is expected to support the engraftment of transplanted iPSC-derived cardiomyocytes. Thus, it is expected that completion of an iPSCs bank with a homozygous HLA haplotype will promote regenerative medicine with human iPSCs.

8.3 Regenerative Cardiomyocytes for Stem Cell Therapies

8.3.1 Cardiac Differentiation Protocol

The control of differentiation of human iPSCs is the key factor for clinical application. A stable differentiation protocol must be established for clinical application and industrialization of stem cell therapies. Because the differentiation ratio varies among iPSC lines, the protocol should be adjusted specifically for each cell line. Otherwise, the productivity of cardiomyocytes will be significantly different at each application.

Several cytokines such as bone morphologic protein 4, Wnt3A, activin, and fibroblast growth factor were reported to promote the induction of cardiac differentiation from PSCs (Burridge et al. 2012; Kattman et al. 2011; Takei et al. 2009; Tran et al. 2009; Yang et al. 2008). Vascular endothelial growth factor and the Wnt inhibitor DKK1 are important for the differentiation of mesodermal cells to cardiomyocytes (Yang et al. 2008). However, it is currently too expensive to use these molecules for widespread clinical application. Therefore, an active area of research involves the synthesis of small molecules as alternatives to these cytokines, which are also expected to exhibit a stable effect. Two key molecules, CHIR99021 and IWR-1, which activate and inhibit canonical Wnt signaling, respectively, were shown to be very effective in promoting cardiac differentiation (Lian et al. 2012; Willems et al. 2011). The medium components and nutrient factors are also very important for effective cardiac differentiation. The representative media for cardiac differentiation are StemPro34 and RPMI1640 with B27 components (Kattman et al. 2011; Laflamme et al. 2007). B27 contains 21 components and is also very expensive for use in massive amounts. These media components for cardiac differentiation must be prepared in xeno-free and GMP-grade manners for clinical application (Burridge et al. 2014).

8.3.2 Mass Culture System of iPSC-Derived Cardiomyocytes

In contrast to small tissues, a large number of cardiomyocytes are required for cardiac cell therapies. Therefore, an effective large-scale culture method must be developed. Because cardiomyocytes show relatively low proliferation rates, a large amount of

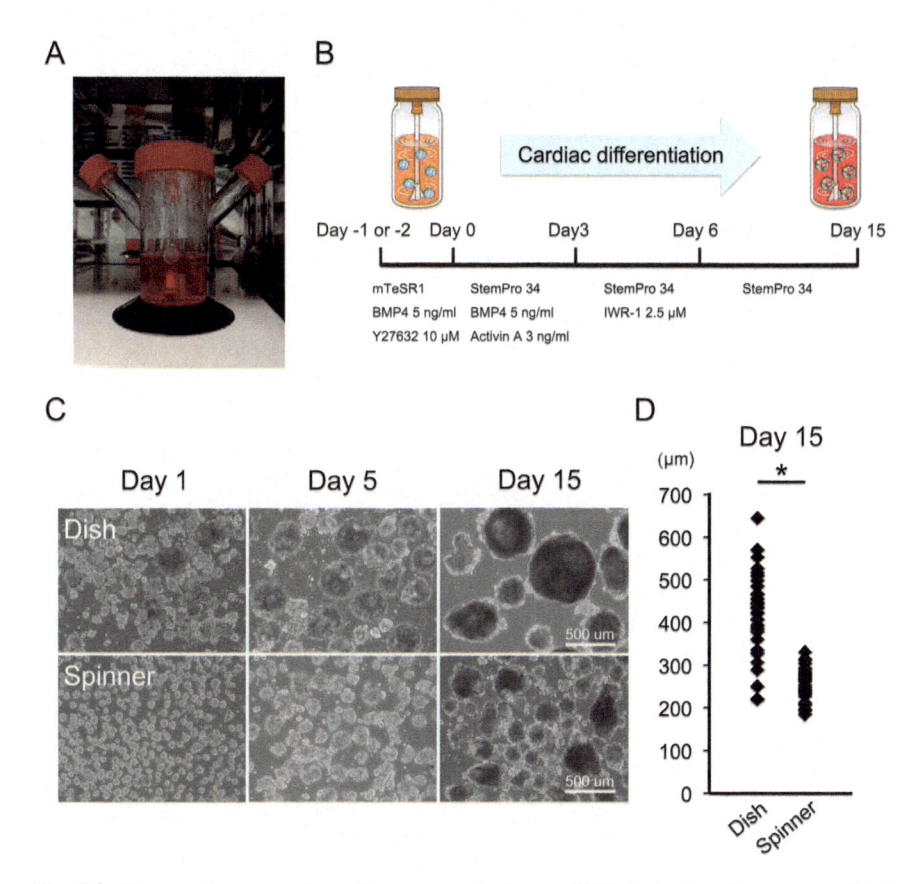

Fig. 8.2 The massive suspension culture system for human iPSC-derived cardiomyocytes. (**a**) A representative figure of the spinner flask used in the massive suspension culture system (MSCS) for human iPSC-derived cardiomyocytes. (**b**) The differentiation protocol for human iPSCs to cardiomyocytes is shown. (**c**) The embryoid bodies (EB), which differentiated into cardiomyocytes in the MSCS, were more homogeneous than the EBs in the dishes. (**d**) The EB size in the dishes varied more than did the EBs in the MSCS at day 15. * $P<0.05$. *BMP4* bone morphogenetic protein 4. Figures are from Hemmi et al. (2014)

undifferentiated iPSCs must be prepared before their differentiation. Undifferentiated iPSCs were reported to be successfully cultured in a bioreactor (Olmer et al. 2012). The problem with suspension culture of undifferentiated iPSCs is that the cells might spontaneously start the differentiation process as embryoid bodies, which would make control and synchronization of their differentiation states difficult.

Nevertheless, use of a suspension culture is more efficient to establish a large-scale culture of cardiomyocytes. A spinner flask and a bioreactor are currently used for a massive suspension culture system (Hemmi et al. 2014; Matsuura et al. 2012); in comparison with suspension culture in dishes, the embryoid bodies tend to be more even and stably differentiated (Fig. 8.2). A mass culture system is very useful for cardiac differentiation, but undifferentiated stem cells remain in the culture, which can lead to teratoma formation (Fig. 8.3). Thus, finding a solution to prevent

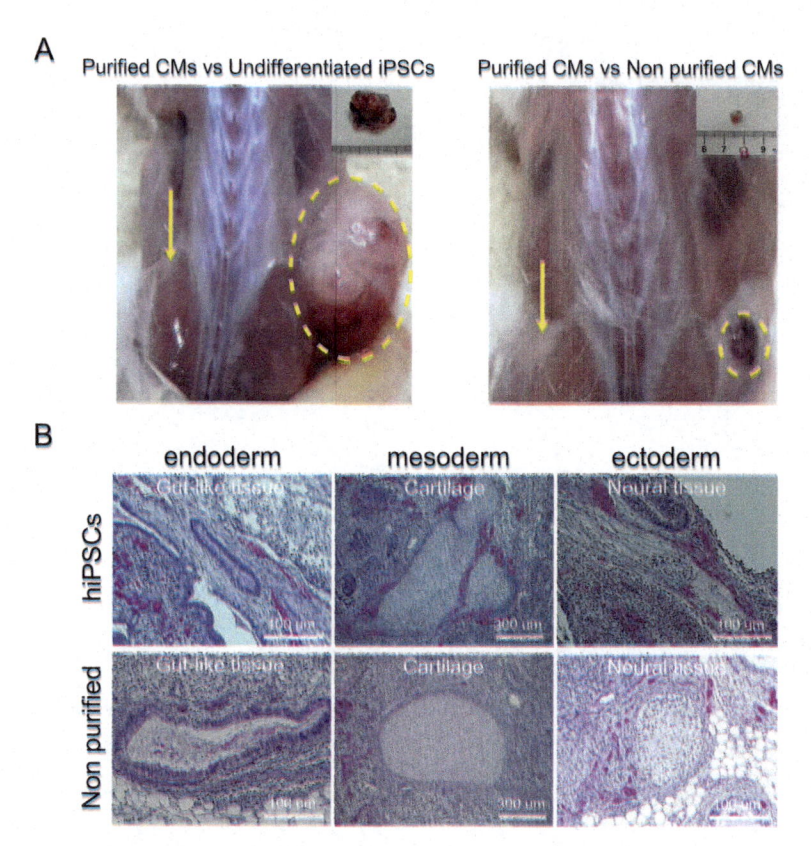

Fig. 8.3 The remained undifferentiated stem cells in human iPSC-derived cardiomyocytes cause teratoma formation. (**a**) In contrast that undifferentiated iPSCs generated a large teratoma, no teratoma was found in purified cardiomyocytes. Non-purified cardiomyocytes still had a potential for tumor formation. (**b**) Teratomas from both human iPSCs and non-purified cardiomyocytes contained the three germ layers. Figures are from Hemmi et al. (2014)

tumorigenicity is the next step for clinical application of a cardiac differentiation system. The cost of the culture medium for massive suspension culture is also an important consideration for clinical application as a standard therapy.

8.3.3 Phenotype of Cardiomyocytes

PSC-derived cardiomyocytes present an infant phenotype after cardiac differentiation, and show fetal gene expression (Cao et al. 2008). In comparison with mature cardiomyocytes, the contraction force of these initially differentiated cardiomyocytes is also very weak. The fetal phenotype may be advantageous for engraftment to the host tissues, but the weak contraction force may not be sufficient to effectively recover the cardiac function. Electrophysiologically, these cells consist of

Fig. 8.4 Electro physiological phenotype of human iPSC-derived cardiomyocytes. Human iPSC-derived cardiomyocytes contained mixture of nodal-, atrial-, and ventricular-like action potentials. Figures are from Hemmi et al. (2014)

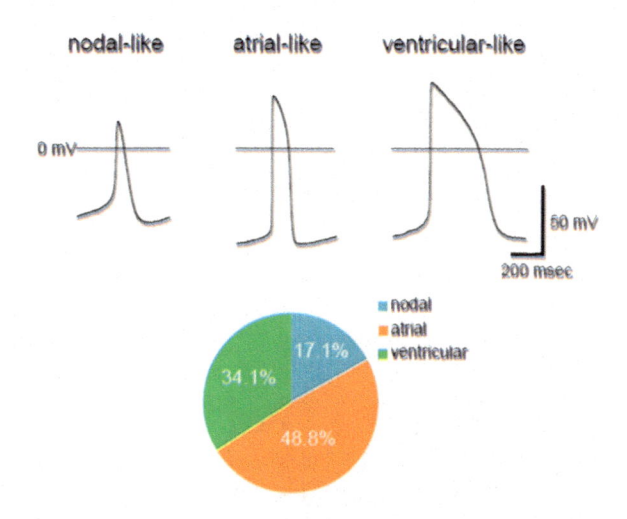

nodular, atrial, and ventricular phenotypes (Fig. 8.4) (Hemmi et al. 2014). To date, preclinical studies have been performed with the mixed phenotypes of cardiomyocytes. The main target phenotype for the treatment of HF is ventricular cardiomyocytes; however, it is currently unknown how they differentiate into the specific phenotype. Therefore, appropriate phenotype selection and maturation in vitro will be critical for future cardiac cell therapies.

8.4 Purification Strategy for PSC-Derived Cardiomyocytes

The most important step in preparation of differentiated human PSCs for transplantation is the elimination of residual undifferentiated stem cells and purification of the differentiated cells. Because large amounts of cardiomyocytes are necessary for clinical application, the non-cardiomyocytes and undifferentiated stem cells could remain in the population of PSC-derived cardiomyocytes. Several strategies have been reported to purify the regenerative cardiomyocytes. A purification method using genetic modification has been the main strategy for purification of the target cells, because there is no surface marker specific for cardiomyocytes (Elliott et al. 2011; Ma et al. 2011). However, this method is not suitable for clinical application because of a lack of safety and stability. To overcome the problem, our group developed a non-genetic purification method using a combination of mitochondrial dye and fluorescence-activated cell sorting (FACS) (Hattori et al. 2010). This method is based on the structural characteristics in cardiomyocytes, which have many well-developed mitochondria. Vascular cell adhesion molecule 1 (VCAM1) or signal-regulatory protein alpha (SIRPA) was reported as a useful surface marker to purify cardiomyocytes (Dubois et al. 2011; Uosaki et al. 2011). Although this FACS-based method is suitable for obtaining small numbers of cardiomyocytes, it is difficult to apply for obtaining large numbers of cardiomyocytes, because it is quite

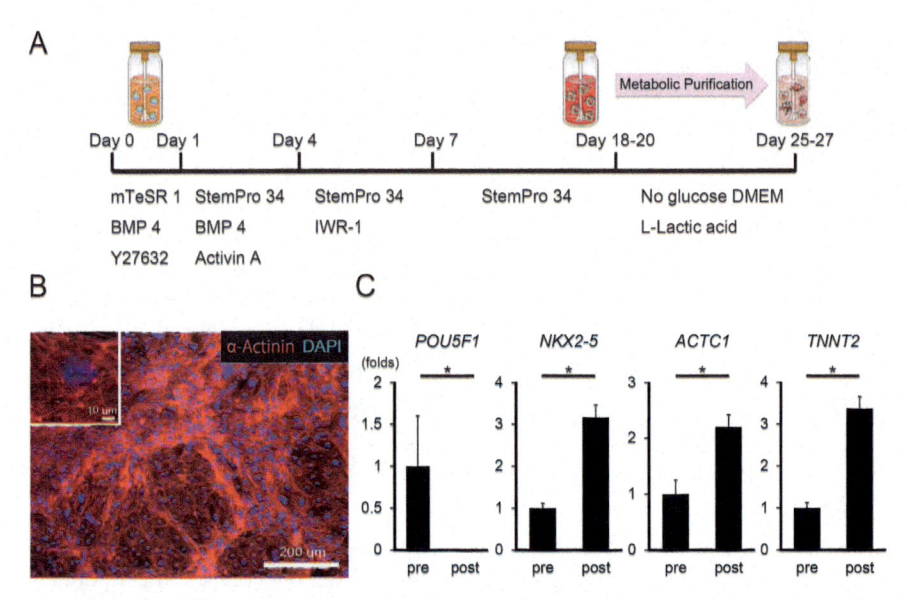

Fig. 8.5 The metabolic purification completely eliminated the undifferentiated stem cells and enriched cardiomyocytes. (a) MSCS was modified with metabolic purification. (b) Condensed cardiomyocytes were positive for cardiac marker, α-actinin, after metabolic purification. (c) The expression of a pluripotent stem cell marker, POU5F1, and cardiac differentiation markers (NKX2-5, ACTC1, and TNNT2) were checked using real-time polymerase chain reaction. After metabolic purification, the expression of POU5F1 was significantly decreased, and the cardiac markers were highly enriched. * $P<0.05$. *BMP4* bone morphogenetic protein 4; *DAPI* 49,6-diamidino-2-phenylindole; *DMEM* Dulbecco's modified Eagle's medium. Figures are from Hemmi et al. (2014)

time-consuming. In addition, FACS requires a single-cell dissociation process, which often causes severe damage to the target cells. Thus, our group has searched for new ideal purification methods that are well suited for obtaining large numbers of cardiomyocytes derived from human PSCs. We first focused on optimization of the culture conditions and evaluated metabolic differences in cardiomyocytes and undifferentiated stem cells using transcriptome and metabolome analyses. As a result, we found that undifferentiated stem cells depend on glycolysis for ATP and biomass production, and cardiomyocytes depend on pyruvate or lactate oxidation in the mitochondria for ATP generation. According to previous reports that fetal and neonatal cardiomyocytes can utilize lactate efficiently (Werner and Sicard 1987), we created glucose-depleted and lactate-supplemented conditions and succeeded in purifying human PSC-derived cardiomyocytes efficiently (Tohyama et al. 2013). This method shows potential for large-scale cultivation to obtain pure cardiomyocytes (Fig. 8.5) (Hemmi et al. 2014). Furthermore, our group evaluated the consumption of amino acids in human PSCs and found that glutamine played a key role in the cell survival of human PSCs under glucose-depleted conditions. In short, human PSCs depend on glycolysis and glutamine oxidation for ATP generation, and culture in glucose- and glutamine-depleted conditions could completely eliminate undifferentiated stem cells (Tohyama et al. 2016). Interestingly, lactate supplementation was essential for the cardiomyocytes to survive under glucose- and glutamine-depleted conditions, because the cells could efficiently utilize lactate to produce ATP and glutamate. These methods have several advantages, including low cost, simplicity, no requirement of a single-cell dissociation process, and suitability for large-scale culture. Thus, it is expected that metabolic approaches will accelerate cardiac regenerative medicine using human PSCs.

8.5 Transplantation Strategies of Regenerative Cardiomyocytes

An effective cell transplantation strategy is the most important component of cell therapies. The most crucial considerations of the method adopted are the safety and feasibility to transplant cells. The engraftment of cells is also one of the most critical factors to achieve better results after cell transplantation. In general, the engraftment success of transplanted cells after cell transplantation with direct injection is very low. In order to improve the efficiency, several transplantation strategies have been established in the last few decades, and biotechnological developments have improved these new strategies.

8.5.1 Direct Cell Injection

Direct cell injection is the most traditional method for transplanting differentiated cells. However, the weakest point of this technique is the low engraftment efficiency of transplanted cells (Hattan et al. 2005). Because the heart is a beating organ, single transplanted cells are easily pumped out from the tissue, which are vulnerable to ischemia and inflammation. The injection of cell aggregates may overcome this critical issue. Indeed, transplantation of aggregated cardiomyocytes drastically improved the engraftment efficiency of cardiomyocytes (Hattori et al. 2010). The advantage of direct cell injection is the noninvasive nature of the transplantation with a catheter or endoscopy (Kimura et al. 2012). These devices can avoid thoracotomy and the need for repeated transplantations.

8.5.2 Biomaterials

The use of supportive materials such as biodegradable materials is expected to promote cell survival and engraftment. The representative biomaterials are fibrin, arginine, and collagen (Rane and Christman 2011). Fibrin and collagen can mimic

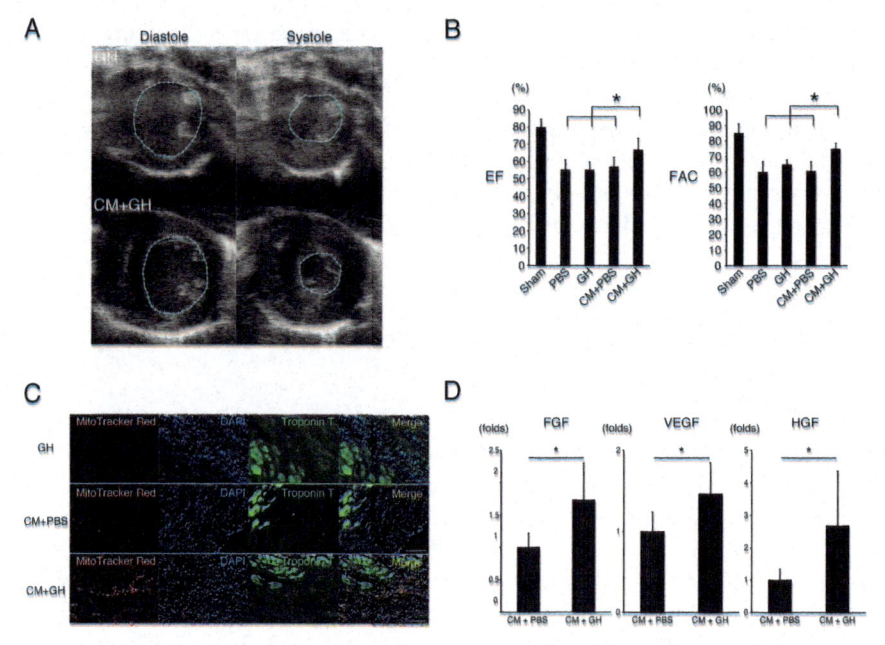

Fig. 8.6 Gelatin hydrogel enhances the engraftment of transplanted rat neonatal cardiomyocytes and angiogenesis to ameliorate cardiac function after myocardial infarction. (**a**) Cardiomyocytes (CM) were transplanted with gelatin hydrogel after myocardial infarction. The representative figures of fractional area change (FAC) in the gelatin hydrogel (GH) and CM + GH groups are shown. (**b**) Left ventricular ejection fraction (EF) and FAC were significantly improved in the CM + GH group. (* $P<0.05$) (**c**) CM were prestained with MitoTracker-Red before transplantation. Sections were co-stained with cardiac troponin T and DAPI. In the CM + GH group, more CM engrafted in the infarcted area. Bars are 100 μm. (**d**) Transplantation of CM with GH increased the release of angiogenic cytokines. Basic fibroblast growth factor (FGF), vascular endothelial growth factor (VEGF), and hepatocyte growth factor (HGF) increased in the CM + GH group. * $P<0.05$. Figures are from Nakajima et al. (2015)

the extracellular matrix surrounding transplanted cardiomyocytes, and these materials can also release growth factors. The most appropriate biomaterials for cell transplantation therapies are also those that are not biologically reactive to the host tissues, such as blood coagulation and inflammation. Although a gelatin hydrogel is made from collagen, unlike collagen, it does not induce blood coagulation. This material has already been clinically used for cytokine-release therapies on ischemic limbs (Marui et al. 2007). We previously provided experimental support that a gelatin hydrogel could promote the engraftment of cardiomyocytes and enhance the cytokine release after myocardial infarction (Fig. 8.6) (Nakajima et al. 2015). These newly developed biomaterials are expected to benefit cell therapies.

8.5.3 Cell Sheet

Transplantation of a cell sheet involves a cell culture method in which the cells are arranged in a monolayer, and the layered sheet is applied to the surface of the organ. The cells are cultured on dishes coated with a temperature-sensitive polymer, poly (*N*-isopropylacrylamide) (PIPAAm) (Okano et al. 1993). This technique is especially useful for the reconstruction of a monolayer tissue such as the retina and skin. In 2014, a human iPSC-derived retinal sheet was first applied to a patient with macular degeneration (Cyranoski 2014). However, application of a cell sheet is more difficult for a complex organ such as the heart. This is because it is a challenge to make a cell sheet multi-layered, since the cells become necrotic when arranged in more than three layers. However, cell sheet transplantation still showed improvement of cardiac function in both small and large animals (Fujita et al. 2012). Nevertheless, the number of surviving cells was decreased in cell sheets, which may have been due to ischemia, apoptosis, inflammation, and/or immunological rejection. The main effect of a cardiac cell sheet is considered to be a paracrine effect to improve the function of a damaged heart (Masumoto et al. 2012). Therefore, long-term engraftment of human iPSC-derived cardiomyocytes must be carefully evaluated.

8.5.4 Future Technology of Cell Therapies: Bioengineering of the Human Myocardium

8.5.4.1 Decellularization of the Heart

The complex heart tissue is difficult to reconstitute with regenerative cardiomyocytes in vitro. In order to reconstitute three-dimensional (3-D) tissues, cells need a scaffold. As a cardiac tissue scaffold, a cytoskeleton was developed from decellularized heart tissue (Ott et al. 2008). The hearts of humans or pigs may be useful for clinical application. As mentioned above, the critical issue is the preparation of a large number of cardiomyocytes, which can fill up the whole or part of the cardiac tissue. In addition, inclusion of components other than cardiomyocytes, such as endothelial cells and smooth muscle cells, will be necessary. If the cardiac cells are stably supplied, human iPSC-derived cardiomyocytes and a decellularized cytoskeleton will enable preparation of a functional human myocardial-like tissue (Guyette et al. 2016). Therefore, it may become possible to transplant constructed heart tissue using a mixture of human iPSC-derived cardiomyocytes and a decellularized cytoskeleton in the near future.

8.5.4.2 Three-dimensional Bioprinting

Recent developments in printing technologies have made it possible to print anything in three dimensions (Nakamura et al. 2010). Organs and tissues in animals can also be printed as a 3-D mass. This technology is also utilized in regenerative medicine. Blood vessels, bones, cartilage, and skeletal muscles have been printed in 3-D using differentiated cells (Kang et al. 2016; Norotte et al. 2009). It is also considered that heart tissues will be able to be printed as a 3-D tissue. However, because there are many types of cells in the heart tissue that are connected to each other in complex manners, the ability to transplant 3-D-printed working heart tissue will likely take more time; however, we expect that this will be possible in the future.

8.6 Adverse Effects of Cell Transplantation

8.6.1 Immunological Rejection

Autologous transplantation of iPSCs can realize the goal of an immunosuppressive therapy-free method of cell transplantation. In the case of the allogeneic transplantation of iPSCs, control of the immune response of a recipient is a key factor for the successful engraftment of transplanted cardiomyocytes. If the HLA of iPSCs in the bank is matched with that of the recipient, the need for immunosuppressive therapies will be minimized (Morizane et al. 2013). However, non-HLA-matched allogeneic cell transplantation requires a full dose of immunosuppressive therapies. The immunosuppressive therapies will be essentially prescribed in the same way as in heart transplantation, because the appropriate dose of immunosuppressive therapies for cardiac cell transplantation remains unknown. Moreover, immunosuppressive therapies can give rise to other side effects such as severe infection and malignancy.

8.6.2 Arrhythmia

Arrhythmogenicity is the most important potential adverse effect of cell therapies. Myoblasts are notorious in their ability to induce arrhythmia after transplantation (Menasché et al. 2008). This is why myoblasts have their own triggered activity to induce arrhythmia (Itabashi et al. 2005). Human ESCs-induced cardiomyocytes can electrically associate with neonatal rat cardiomyocytes, and become synchronized to beat in vitro (Thompson et al. 2012). However, the arrhythmogenicity of human ESC-derived cardiomyocytes in vivo remains controversial. One report showed that after ESC-derived cardiomyocytes were engrafted to guinea pig hearts and they showed electrical coupling with the host heart (Shiba et al. 2012). By contrast, the same group also reported that ventricular arrhythmia was induced within the first 2 weeks after transplantation to a monkey's heart, which disappeared one month after cell transplantation (Chong et al. 2014). These data suggest that it will be necessary to closely monitor the heart for arrhythmia after cell transplantation in humans.

8.7 Conclusion

Cardiac cell transplantation is expected as the future primary strategy for treating HF to replace heart transplantation. Human iPSCs are the most promising cell source as a supply of the regenerative cardiomyocytes for cell transplantation therapy. Although there are still large hurdles to overcome before the first application of this method in humans, achieving the step-by-step solutions of each issue raised herein will undoubtedly lead to the clinical application and industrialization of cardiac regenerative therapies.

Acknowledgements The present work was supported by the Highway Program for Realization of Regenerative Medicine from Japan Science and Technology Agency (to K.F.).

References

Bolli R, Chugh AR, D'Amario D, Loughran JH, Stoddard MF, Ikram S, Beache GM, Wagner SG, Leri A, Hosoda T et al (2011) Cardiac stem cells in patients with ischaemic cardiomyopathy (SCIPIO): initial results of a randomised phase 1 trial. Lancet 378:1847–1857

Burridge PW, Keller G, Gold JD, Wu JC (2012) Production of de novo cardiomyocytes: human pluripotent stem cell differentiation and direct reprogramming. Cell Stem Cell 10:16–28

Burridge PW, Matsa E, Shukla P, Lin ZC, Churko JM, Ebert AD, Lan F, Diecke S, Huber B, Mordwinkin NM et al (2014) Chemically defined generation of human cardiomyocytes. Nat Methods 11:855–860

Cao F, Wagner RA, Wilson KD, Xie X, Fu JD, Drukker M, Lee A, Li RA, Gambhir SS, Weissman IL et al (2008) Transcriptional and functional profiling of human embryonic stem cell-derived cardiomyocytes. PLoS One 3:e3474

Chong JJ, Yang X, Don CW, Minami E, Liu YW, Weyers JJ, Mahoney WM, Van Biber B, Cook SM, Palpant NJ et al (2014) Human embryonic-stem-cell-derived cardiomyocytes regenerate non-human primate hearts. Nature 510:273–277

Cyranoski D (2014) Japanese woman is first recipient of next-generation stem cells. Nature

Dubois NC, Craft AM, Sharma P, Elliott DA, Stanley EG, Elefanty AG, Gramolini A, Keller G (2011) SIRPA is a specific cell-surface marker for isolating cardiomyocytes derived from human pluripotent stem cells. Nat Biotechnol 29:1011–1018

Elliott DA, Braam SR, Koutsis K, Ng ES, Jenny R, Lagerqvist EL, Biben C, Hatzistavrou T, Hirst CE, Yu QC et al (2011) NKX2-5(eGFP/w) hESCs for isolation of human cardiac progenitors and cardiomyocytes. Nat Methods 8:1037–1040

Fujita J, Fukuda K (2014) Future prospects for regenerated heart using induced pluripotent stem cells. J Pharmacol Sci 125:1–5

Fujita J, Itabashi Y, Seki T, Tohyama S, Tamura Y, Sano M, Fukuda K (2012) Myocardial cell sheet therapy and cardiac function. Am J Physiol Heart Circ Physiol 303:H1169–H1182

Fusaki N, Ban H, Nishiyama A, Saeki K, Hasegawa M (2009) Efficient induction of transgene-free human pluripotent stem cells using a vector based on Sendai virus, an RNA virus that does not integrate into the host genome. Proc Jpn Acad Ser B Phys Biol Sci 85:348–362

Guyette JP, Charest JM, Mills RW, Jank BJ, Moser PT, Gilpin SE, Gershlak JR, Okamoto T, Gonzalez G, Milan DJ et al (2016) Bioengineering human myocardium on native extracellular matrix. Circ Res 118:56–72

Hattan N, Kawaguchi H, Ando K, Kuwabara E, Fujita J, Murata M, Suematsu M, Mori H, Fukuda K (2005) Purified cardiomyocytes from bone marrow mesenchymal stem cells produce stable intracardiac grafts in mice. Cardiovasc Res 65:334–344

Hattori F, Chen H, Yamashita H, Tohyama S, Satoh Y-S, Yuasa S, Li W, Yamakawa H, Tanaka T, Onitsuka T et al (2010) Nongenetic method for purifying stem cell-derived cardiomyocytes. Nat Meth 7:61–66

Hemmi N, Tohyama S, Nakajima K, Kanazawa H, Suzuki T, Hattori F, Seki T, Kishino Y, Hirano A, Okada M et al (2014) A massive suspension culture system with metabolic purification for human pluripotent stem cell-derived cardiomyocytes. Stem Cells Transl Med 3:1473–1483

Itabashi Y, Miyoshi S, Yuasa S, Fujita J, Shimizu T, Okano T, Fukuda K, Ogawa S (2005) Analysis of the electrophysiological properties and arrhythmias in directly contacted skeletal and cardiac muscle cell sheets. Cardiovasc Res 67:561–570

Kang HW, Lee SJ, Ko IK, Kengla C, Yoo JJ, Atala A (2016) A 3D bioprinting system to produce human-scale tissue constructs with structural integrity. Nat Biotechnol 34:312–319

Kattman SJ, Witty AD, Gagliardi M, Dubois NC, Niapour M, Hotta A, Ellis J, Keller G (2011) Stage-specific optimization of activin/nodal and BMP signaling promotes cardiac differentiation of mouse and human pluripotent stem cell lines. Cell Stem Cell 8:228–240

Kimura T, Miyoshi S, Okamoto K, Fukumoto K, Tanimoto K, Soejima K, Takatsuki S, Fukuda K (2012) The effectiveness of rigid pericardial endoscopy for minimally invasive minor surgeries: cell transplantation, epicardial pacemaker lead implantation, and epicardial ablation. J Cardiothorac Surg 7:117

Laflamme MA, Chen KY, Naumova AV, Muskheli V, Fugate JA, Dupras SK, Reinecke H, Xu C, Hassanipour M, Police S et al (2007) Cardiomyocytes derived from human embryonic stem cells in pro-survival factors enhance function of infarcted rat hearts. Nat Biotechnol 25:1015–1024

Lian X, Hsiao C, Wilson G, Zhu K, Hazeltine LB, Azarin SM, Raval KK, Zhang J, Kamp TJ, Palecek SP (2012) Robust cardiomyocyte differentiation from human pluripotent stem cells via temporal modulation of canonical Wnt signaling. Proc Natl Acad Sci U S A 109:E1848–E1857

Lund LH, Edwards LB, Kucheryavaya AY, Benden C, Dipchand AI, Goldfarb S, Levvey BJ, Meiser B, Rossano JW, Yusen RD et al (2015) The Registry of the International Society for Heart and Lung Transplantation: thirty-second official adult heart transplantation report--2015; focus theme: early graft failure. J Heart Lung Transplant 34:1244–1254

Ma J, Guo L, Fiene SJ, Anson BD, Thomson JA, Kamp TJ, Kolaja KL, Swanson BJ, January CT (2011) High purity human-induced pluripotent stem cell-derived cardiomyocytes: electrophysiological properties of action potentials and ionic currents. Am J Physiol Heart Circ Physiol 301:H2006–H2017

Marui A, Tabata Y, Kojima S, Yamamoto M, Tambara K, Nishina T, Saji Y, Inui K, Hashida T, Yokoyama S et al (2007) A novel approach to therapeutic angiogenesis for patients with critical limb ischemia by sustained release of basic fibroblast growth factor using biodegradable gelatin hydrogel: an initial report of the phase I-IIa study. Circ J 71:1181–1186

Masumoto H, Matsuo T, Yamamizu K, Uosaki H, Narazaki G, Katayama S, Marui A, Shimizu T, Ikeda T, Okano T et al (2012) Pluripotent stem cell-engineered cell sheets re-assembled with defined cardiovascular populations ameliorate reduction in infarct heart function through cardiomyocyte-mediated neovascularization. Stem Cells 30:1196–1205

Matsuura K, Wada M, Shimizu T, Haraguchi Y, Sato F, Sugiyama K, Konishi K, Shiba Y, Ichikawa H, Tachibana A et al (2012) Creation of human cardiac cell sheets using pluripotent stem cells. Biochem Biophys Res Commun 425:321–327

Menasché P, Alfieri O, Janssens S, McKenna W, Reichenspurner H, Trinquart L, Vilquin J-T, Marolleau J-P, Seymour B, Larghero J et al (2008) The myoblast autologous grafting in ischemic cardiomyopathy (MAGIC) trial. Circulation 117:1189–1200

Morizane A, Doi D, Kikuchi T, Okita K, Hotta A, Kawasaki T, Hayashi T, Onoe H, Shiina T, Yamanaka S et al (2013) Direct comparison of autologous and allogeneic transplantation of iPSC-derived neural cells in the brain of a non-human primate. Stem Cell Reports 1:283–292

Nakajima K, Fujita J, Matsui M, Tohyama S, Tamura N, Kanazawa H, Seki T, Kishino Y, Hirano A, Okada M et al (2015) Gelatin hydrogel enhances the engraftment of transplanted cardiomyocytes and angiogenesis to ameliorate cardiac function after myocardial infarction. PLoS One 10:e0133308

Nakamura M, Iwanaga S, Henmi C, Arai K, Nishiyama Y (2010) Biomatrices and biomaterials for future developments of bioprinting and biofabrication. Biofabrication 2:014110

Nakatsuji N, Nakajima F, Tokunaga K (2008) HLA-haplotype banking and iPS cells. Nat Biotechnol 26:739–740

Norotte C, Marga FS, Niklason LE, Forgacs G (2009) Scaffold-free vascular tissue engineering using bioprinting. Biomaterials 30:5910–5917

Okano T, Yamada N, Sakai H, Sakurai Y (1993) A novel recovery system for cultured cells using plasma-treated polystyrene dishes grafted with poly(N-isopropylacrylamide). J Biomed Mater Res 27:1243–1251

Okita K, Yamakawa T, Matsumura Y, Sato Y, Amano N, Watanabe A, Goshima N, Yamanaka S (2013) An efficient nonviral method to generate integration-free human-induced pluripotent stem cells from cord blood and peripheral blood cells. Stem Cells 31:458–466

Olmer R, Lange A, Selzer S, Kasper C, Haverich A, Martin U, Zweigerdt R (2012) Suspension culture of human pluripotent stem cells in controlled, stirred bioreactors. Tissue Eng Part C Methods 18:772–784

Ott HC, Matthiesen TS, Goh SK, Black LD, Kren SM, Netoff TI, Taylor DA (2008) Perfusion-decellularized matrix: using nature's platform to engineer a bioartificial heart. Nat Med 14:213–221

Rane AA, Christman KL (2011) Biomaterials for the treatment of myocardial infarction a 5-year update. J Am Coll Cardiol 58:2615–2629

Seki T, Yuasa S, Oda M, Egashira T, Yae K, Kusumoto D, Nakata H, Tohyama S, Hashimoto H, Kodaira M et al (2010) Generation of induced pluripotent stem cells from human terminally differentiated circulating T cells. Cell Stem Cell 7:11–14

Shiba Y, Fernandes S, Zhu WZ, Filice D, Muskheli V, Kim J, Palpant NJ, Gantz J, Moyes KW, Reinecke H et al (2012) Human ES-cell-derived cardiomyocytes electrically couple and suppress arrhythmias in injured hearts. Nature 489:322–325

Strauer B-E, Steinhoff G (2011) 10 Years of intracoronary and intramyocardial bone marrow stem cell therapy of the heart: from the methodological origin to clinical practice. J Am Coll Cardiol 58:1095–1104

Takahashi K, Tanabe K, Ohnuki M, Narita M, Ichisaka T, Tomoda K, Yamanaka S (2007) Induction of pluripotent stem cells from adult human fibroblasts by defined factors. Cell 131:861–872

Takei S, Ichikawa H, Johkura K, Mogi A, No H, Yoshie S, Tomotsune D, Sasaki K (2009) Bone morphogenetic protein-4 promotes induction of cardiomyocytes from human embryonic stem cells in serum-based embryoid body development. Am J Physiol Heart Circ Physiol 296:H1793–H1803

Thompson SA, Burridge PW, Lipke EA, Shamblott M, Zambidis ET, Tung L (2012) Engraftment of human embryonic stem cell derived cardiomyocytes improves conduction in an arrhythmogenic in vitro model. J Mol Cell Cardiol 53:15–23

Thomson JA, Itskovitz-Eldor J, Shapiro SS, Waknitz MA, Swiergiel JJ, Marshall VS, Jones JM (1998) Embryonic stem cell lines derived from human blastocysts. Science 282:1145–1147

Tohyama S, Hattori F, Sano M, Hishiki T, Nagahata Y, Matsuura T, Hashimoto H, Suzuki T, Yamashita H, Satoh Y et al (2013) Distinct metabolic flow enables large-scale purification of mouse and human pluripotent stem cell-derived cardiomyocytes. Cell Stem Cell 12:127–137

Tohyama S, Fujita J, Hishiki T, Matsuura T, Hattori F, Ohno R, Kanazawa H, Seki T, Nakajima K, Kishino Y et al (2016) Glutamine oxidation is indispensable for survival of human pluripotent stem cells. Cell Metab 23(4):663–674

Tran TH, Wang X, Browne C, Zhang Y, Schinke M, Izumo S, Burcin M (2009) Wnt3a-induced mesoderm formation and cardiomyogenesis in human embryonic stem cells. Stem Cells 27:1869–1878

Uosaki H, Fukushima H, Takeuchi A, Matsuoka S, Nakatsuji N, Yamanaka S, Yamashita JK (2011) Efficient and scalable purification of cardiomyocytes from human embryonic and induced pluripotent stem cells by VCAM1 surface expression. PLoS One 6:e23657

Werner JC, Sicard RE (1987) Lactate metabolism of isolated, perfused fetal, and newborn pig hearts. Pediatr Res 22:552–556

Willems E, Spiering S, Davidovics H, Lanier M, Xia Z, Dawson M, Cashman J, Mercola M (2011) Small-molecule inhibitors of the Wnt pathway potently promote cardiomyocytes from human embryonic stem cell-derived mesoderm. Circ Res 109:360–364

Yang L, Soonpaa MH, Adler ED, Roepke TK, Kattman SJ, Kennedy M, Henckaerts E, Bonham K, Abbott GW, Linden RM et al (2008) Human cardiovascular progenitor cells develop from a KDR+ embryonic-stem-cell-derived population. Nature 453:524–528

Chapter 9
Myocardial Tissue Engineering for Cardiac Repair

S. Pecha and Y. Yildirim

9.1 Introduction

In our aging population, the prevalence of heart failure is increasing. Heart transplantation still remains the only curative treatment option, however, especially in Europe limited by an increasing organ-donor shortage. Therefore, new organ-independent treatment options are needed. The progress in mechanical circulatory support, achieved in recent years, is especially in the long-term run limited by its several side effects, like bleeding complications, pump thrombosis, stroke, or infections (Kirklin et al. 2014). As the endogenous regeneration capacity of the heart in adult mammals is extremely limited, exogenous repair strategies have become an attractive alternative treatment modality (Soonpaa and Field 1998; Senyo et al. 2013; Laflamme and Murry 2011). Several approaches for repair or replacement therapies of the heart have been published in recent years. The aim of all these approaches is to form a contractile cardiomyocyte formation, which is able to electrically and mechanically couple to the host myocardium and to support the function of the failing heart. The simplest way of the cardiac repair techniques is the direct cell injection. Hereby either multi- or pluripotent stem cells or cardiomyocytes are directly injected intramyocardial. However, recent studies have shown a limited cell survival and retention rate of the intramyocardially injected cells. Therefore, biodistribution of the cells with the risk of subsequent teratoma

---

S. Pecha (✉)
Department of Cardiovascular Surgery, University Heart Center Hamburg,
Martinistr. 52, 20246 Hamburg, Germany

DZHK (German Centre for Cardiovascular Research) Partner Site Hamburg/Kiel/Lübeck,
Lübeck, Germany
e-mail: s.pecha@uke.de

Y. Yildirim
Department of Cardiovascular Surgery, University Heart Center Hamburg,
Martinistr. 52, 20246 Hamburg, Germany

formation might be a problem. Furthermore high cell numbers associated with high costs are needed. A strategy to overcome this problem is the technique of myocardial tissue engineering. Hereby, three-dimensional heart tissue, with properties of native myocardium, is created in-vitro, and afterwards transplanted onto the host myocardium.

9.2 Different Myocardial Tissue Engineering Techniques

The intrinsic capacity of human cardiomyocytes to form spontaneous three-dimensional syncytia is a necessary basic principle of myocardial tissue engineering. Combined with engineering techniques, three-dimensional cardiac tissue constructs of the desired geometry and size can be created. Therefore different methods are used, which are described below.

9.2.1 Decellularized Heart Tissue

This technique, first described by Eschenhagen et al. (1997), aims at reconstituting a native heart by decellularizing a whole heart or some parts of it and to reseed it with cardiac cells. First part of this two-step approach is the decellularization process. The decellularization of the native heart can be achieved by Langendorff-perfusion with Triton X-100 and sodium dodecyl substrate (SDS), a procedure, which eradicates almost all cells from the tissue but leaves connective tissue and blood vessels behind. In a second step, the reseeding process, the remaining matrix is then perfused with cardiac cells (Ott et al. 2008). Thereby, the native heart shape can be reconstituted.

9.2.2 Hydrogel Technique

This technique, developed by Eschenhagen, Elson, and colleagues in 1994 (Zimmermann et al. 2002), was the first described cardiac tissue engineering method. It is based on the use of liquid hydrogels like collagen I (Bian et al. 2009; Eschenhagen et al. 1997; Zimmermann et al. 2002), fibrin (Hansen et al. 2010), or matrigel (Morritt et al. 2007) as matrices. In earlier days, collagen or matrigel was used as matrices, however in recent years, they were replaced by fibrin, to reduce the immunogenicity of the constructs. The liquid hydrogels are coupled with cardiac cells and poured into casting molds, with inserted anchoring constructs. The casting mold affects the geometry of the tissue constructs and thereby helps to build the myocardial tissue of desired form and size. The anchoring constructs are essential as they allow the tissue to fix on two points and to develop forces between the two anchoring points. The liquid hydrogels entrap the cells within the three-dimensional environment and stimulate the native ability of the cardiomyocytes to form intercellular connections to grow electrical and mechanical syncytia. The anchoring points are useful to achieve improved tissue quality, which can be obtained when the constructs grow under continuous mechanical load.

9.2.3 Cell Sheet Technique

The basic principle of the cell sheet technique, first published by Shimizu et al. (2002), is the intrinsic capacity of cardiomyocytes to form two-dimensional monolayers when cultured in cell-dishes. With means of thermosensitive culture dish surfaces, the two-dimensional monolayers can be detached when left at room temperature. Stacking of these thin, spontaneously beating two-dimensional monolayers results in three-dimensional myocardial tissue. The advantage of this approach is that numerous cell layers can be stacked, to achieve tissue of a relevant thickness to replace the lost native myocardium. These scaffold-free cardiac tissues can then be easily transplanted onto the heart without glue or sutures.

9.2.4 Prefabricated Matrices

The prefabricated matrix approach is based on the principle that manufactured solid matrices are seeded with cardiac cells. Several materials including alginate, collagen, gelatin sponges, polyglycolic acid, and poly (glycerol sebatic) acid matrices (Leor et al. 2000; Li et al. 1999) have been used for this technique. One of the advantages of this technique is the possibility to create any desired geometry and size of prefabricated matrices and subsequent engineered heart tissue.

9.3 Requirements in Myocardial Tissue Engineering

9.3.1 Vascularization

Engineered heart tissue (EHT) constructs consisting of native cardiomyocytes, form primitive vascular networks with inner lumina, even without the addition of endothelial cells (Zimmermann et al. 2002; Naito et al. 2006; Stevens et al. 2009a). However it is unlikely that the vascular structures serve more than a paracrine function in the absence of perfusion in-vitro, but they might, however, play a role for in-vivo vascularization after EHT transplantation. Several approaches for in-vitro vascularization have been described, however, all limited by the complexity to form a real three-dimensional capillary network within the tissue construct. One approach is to use dissolvable alginate fibers which are included into the engineered heart tissue construct and which then serve as a primitive vascular network which is perfused with culture media. However, to date, most tissue constructs that have been used in in-vivo small animal models were supplied by diffusion when cultured in-vitro. The impact of an in-vitro vascularization on EHT maturation and furthermore survival after transplantation needs to be determined in future studies.

9.3.2 Hypoxia Resistance

Neonatal rat cardiomyocytes, which have been the cell source of choice for many years, have a high degree of hypoxia resistance. This is clearly an advantage for the generation of EHT from these cells. The immaturity of the cells with their high capacity for anaerobic glycolytic metabolism might be an advantage for cell survival during in-vitro culture and immediately after in vivo transplantation of the EHTs, until vessel in-growth from the host myocardium takes place. This is clearly a crucial period for cell survival, and a higher hypoxia tolerance of immature cells seems to be an advantage here. In a recent publication, measurements of oxygen concentration in large fibrin EHTs have shown 2% and less in the middle of the construct (Vollert et al. 2014). The ideal maturity of hiPSC and hESC derived cardiomyocytes for cardiac repair purposes certainly needs to be further investigated in future experiments.

9.3.3 Non-cardiomyocytes

Besides cardiomyocytes, native cardiac muscle tissue contains several non-myocytes like fibroblasts, endothelial cells, and smooth muscle cells. Theses non-myocytes contribute to the cardiomyocyte growth in in vitro culture, and several investigations have shown that they are also essential for generation of EHT. It has been shown that EHTs made from unpurified heart muscle cell mix (containing cardiac fibroblasts, smooth muscle cells and endothelial cells) developed threefold higher forces compared to EHTs from pure cardiomyocytes. Furthermore the tissue structure was more native heart muscle like (Naito et al. 2006). Additionally, purified human pluripotent stem cell derived cardiomyocytes do not form three-dimensional constructs, but need the addition of cardiac fibroblasts to generate three-dimensional heart tissue (Kensah et al. 2013). Furthermore, the addition of stromal—and endothelial cells to human embryonic stem cell (hESC)—derived cardiomyocytes, improved tissue maturation and its function and structure (Stevens et al. 2009a; Tulloch et al. 2011). On the other hand, fibrin-based EHTs can also be made from almost 100% pure human induced pluripotent stem cell (hiPSC)-derived cardiomyocytes (own unpublished observation), questioning the generality of this conclusion.

9.3.4 Maturation of the Cells

Three-dimensional EHT shows a more mature, native heart muscle-like phenotype than cardiomyocytes in two-dimensional cultures. EHTs show a higher degree of sarcomeric organization have rod-shaped cardiomyocytes. They form the heart muscle tissue specific M-bands and have a rather normal ratio of sarcomeres nuclei and mitochondria. Furthermore, three-dimensional EHTs show physiological responses like Frank–Starling mechanism, force–frequency relationship, and a positive inotropic reaction to calcium and isoprenaline (Eschenhagen et al. 1997; Hansen et al. 2010; de Lange et al. 2011). EHTs also have an action potential shape and duration comparable to native heart muscle tissue (Eschenhagen et al. 1997; Hansen et al. 2010; de Lange et al. 2011).

However, to date EHTs do not achieve a fully mature phenotype of native heart muscle tissue. For example, in 3D EHTs, the ratio of the adult α-myosin heavy chain isoform to the fetal ß-isoform was 7:1, while it was >100:1 in the adult rat heart (Tiburcy et al. 2011). Furthermore, the contractile forces of EHTs are less than those of isolated adult heart preparations. In future, further improvements of cardiac maturation will be necessary. Possible factors influencing the degree of maturation are culture conditions with probably construction of bioreactors (e.g., media composition, nutrient, and oxygen supply), time of in vitro culture, and conditions of contraction during maturation (auxotonic stretch).

9.3.5 Strain

Mechanical strain has been shown to be an essential parameter for cardiomyocyte alignment and tissue maturation in various hydrogel-based EHT studies. The simplest way of mechanical strain is static tension of an EHT between two fixed anchoring points (Eschenhagen et al. 1997; Zimmermann et al. 2002). However, improved cardiac tissue function and structure has been observed in EHTs that were cultured in a manner that they can beat against two flexible anchoring points (Hansen et al. 2010). This auxotonic contraction imitates the physiological conditions of the heart in Myocardial tissue engineering requirements strain vivo.

9.4 Cell Source

Embryonic chicken—or neonatal mouse—or rat cardiomyocytes have been the only available cell sources for myocardial tissue engineering for several years. Due to recent advantages in cell biology the differentiation of cardiomyocytes from human pluripotent stem cells (embryonic and induced pluripotent stem cells, hESC, and hiPSC) at high efficiency is possible (Kattman et al. 2011; Burridge et al. 2011).

Therefore, nowadays there is an ongoing discussion on the ideal cell source for myocardial tissue engineering. Although recent publications demonstrate some limitations of hiPSC-based strategies for clinical applications (Zhao et al. 2011; Koyanagi-Aoi et al. 2013), they also offer advantages over hESC-approaches. Human iPSC lack the ethical concerns associated with hESC and even if some immune responses were noted in a mouse model (Kensah et al. 2013), hiPSC may offer the opportunity for an autologous approach without the need for full long-term immunosuppression. However this autologous approach would carry out serious time constraints and regulatory hurdles. A certain time is necessary to create the patient-specific cell line and furthermore, as every cell line is a new therapeutic substance, each stem cell product will have to undergo extensive safety testing prior to clinical application. Due to these serious concerns, finally an allogeneic approach with well defined, HLA-matched stem cell banks will probably be favorized. In the field of hESCs the technique of somatic cell nucleus transfer (SCNT) has brought several new treatment modalities. As the pluripotent embryonic stem cells, derived from SCNT only have mitochondrial DNA from the donor, and otherwise only consist of recipient DNA, there might be a much lower immune response than in other conventional ESC approaches. However, in a first small animal immunological study by Deuse et al. the mitochondrial DNA in hESC derived cardiomyocytes led to an enormous immune response, when transplanted in immunocompetent mice. These immunological issues need to be addressed in further studies. An advantage of the SNCT derived embryonic stem cells is the reduced ethical concerns with this type of cells.

9.5 Tissue Engineering for Cardiac Repair

Treating failing heart muscles with artificial cardiac tissue is an interesting approach. Intentions of this potentially curative therapy are to supplement or replace malfunctioned myocardium (Kirklin et al. 2014), propose contractile cells (Soonpaa and Field 1998), reconstruct malformations (Senyo et al. 2013), improve electrical coupling (Laflamme and Murry 2011), improve vascularization (Ott et al. 2008), prohibit cell death and stabilize heart structures, e.g., ventricular wall (Bian et al. 2009). Most of the investigations were done with neonatal rat heart cells and demonstrated survival and integration of transplanted cardiomyocytes to recipient heart. It seems to be stunning that grafted tissue containing cardiomyocytes with high oxygen requirement can survive without having initially a connection to recipient circulation. Most likely, the immature phenotype of neonatal rat cardiomyocytes makes them resistant to hypoxia and appears as a survival benefit. The immature cells can survive for a couple of days by dramatically decreasing their metabolism and conducting anaerobic glycolysis. In contrast to native myocardial tissue artificial tissue is loose packed with cells, which allows for enhanced diffusion capacity until the relatively quick process of host vessel ingrowths takes place (Shimizu). Despite encouraging findings engineered heart tissue has to overcome several technical difficulties to reach reliable clinical practice. One of the major problems is to detect a proper cell source. First cells that entered clinical trials were skeletal muscle myoblasts. The therapeutic results of injected myoblasts seem to be questionable and the proarrhythmic observations raised safety concerns. Due to the simplicity of cell therapy divers autologous mesenchymal stem cells were used in clinical trials mainly after myocardial infarction and beneficial effects are discussed controversy (Schachinger et al. 2006; Lunde et al. 2006). However, the simple technique of injecting cells might be limited by low cell retention rate and safety issues in case of potential teratogenic cells like hESC or hiPSC. Therefore, strict quality control of the cells needs to be implemented before this technique is approaching clinical use. Since all mentioned cell sources are not ready or have safety concerns same researchers started focusing to increase endogenic reconstruction by injection of biological scaffolds like alginate. Positive results encouraged to start first prospective randomized human clinical trial (AUGMENT HF). Patients receiving alginate patches had an improvement in peak VO_2 max, NYHA class as well as 6 min walk test after 6 months. But a worse outcome in terms of major adverse events (78 % vs. 45 % of patients) including death (15 % vs. 8 %) was observed in patients receiving alginate patches compared to control group treated with optimal medical therapy only.

9.5.1 Implantation Studies of Artificial Heart Tissue

Myocardial infarction is still a life-threatening condition and a major cause of death. Replacement of approximately one billion cells after myocardial infarction seems to be challenging. Ten billion cells have to be injected for replacement due to maximal engraftment efficiency of 10 % because of prompt wash out and cell death. The beneficial effects of cell injection also appear to be independent of yielding contractile elements and might be explained by paracrine factors. However, it is unlikely to restore the complete defect after myocardial infarction and clinical trials showed low significant functional improvements in patients (Lunde et al. 2006). Implantation of in vitro engineered myocardium might overcome some limitations of cell therapy. Transplanted tissue patches onto the heart have a much higher cell retention rate; although a direct head-to-head comparison with cell injection therapy has not been performed yet. In vitro engineered tissue can be designed demand-oriented in complex geometrical structure and contractile muscle should deliver a direct and rapid support of the damaged ventricle. In the last decades, several in vivo studies of cardiac surrogate tissue pointed out interesting findings. Predominately the groups used neonatal cardiomyocytes in hydrogel-based EHTs, stacked cell sheets (Sekine et al. 2008; Furuta et al. 2006), or scaffold-free cell syncytia (Stevens et al. 2009a). The relevant improvement of failing hearts is the main objective of tissue replacement therapy. In a dainty study with more than 300 rats Zimmermann et al. implanted multiloop-EHTs and revealed improvement of systolic and diastolic function 4 weeks after myocardial infarction. EHTs initially contained 12.5 million cells, of which 4 million cells survived the in-vitro period. An uncertain number of cells survived the transplantation and newly formed myocardium could be observed with a mean thickness of 440 µm. The control group with non-cardiomyocyte constructs did not improve left ventricular function (Zimmermann et al. 2006).

Surprisingly, further studies with much smaller 3D EHT constructs or non-cardiomyocyte patches have reported similar therapeutic effects. This suggests that other effects, like increased mechanical wall stability, angiogenesis, or paracrine effects, might contribute to the observed therapeutic benefits. Furthermore, an alternative explanation is that cardiomyocytes have beneficial paracrine effects but did not support the insufficient heart by adding contractile force. Probably the force created by the multiloop EHTs is too low to contribute to an improvement in left ventricular contractility, or the force conduction in the fibrotic infarction area of the left ventricle is reduced. Definitely, a synchronic systolic contraction of the graft and the recipient heart is essential and requires electrical coupling. However, to facilitate contractile benefits, tissue engineered cardiac grafts have to be thicker, stronger, vascularized, and need to be electrical syncytia with the host heart. In-vitro constructed engineered heart tissue contains loose muscle bundles with maximal thickness of 200 μm due to diffusion limitations. Even if the fact that implanted cardiomyocytes are able to survive in absence of immediate perfusion is surprising, the inserted tissue thickness is limited. Shimizu et al. demonstrated in a methodical implantation study the vascularization of sequential implanted cell sheets. Grafting more than three cell sheets (80 μm) resulted in extensive cell death (Shimizu et al. 2006). Researchers tried to solve this restriction by different approaches. Zimmerman et al. fused four loop EHTS to multiloop-EHT patches and Shimizu et al. implanted sequentially layers of cell sheets periodically within 1–2 days between the implants to increase the thickness of proposed cardiac tissue (Shimizu et al. 2006). Nonetheless, further advanced techniques have to be identified to support significantly failing hearts. One of the major disadvantages of the mentioned methods is the period until true vascularization by host vessel ingrowth. This critical point is one of the major circumstances, contributing to cell death after implantation of artificial cardiac tissue. Therefore, several groups put more effort to improve vascularization of the artificial heart tissue before transplantation. In a very interesting in-vivo vascularization approach by Moritt et al. (2007) arteriovenous rat epigastric arteries were incubated together with matrigel and neonatal rat cardiomyocytes in a casting mold. In a second step this well-vascularized tissue construct, with a defined arterial and venous pedicle, could be anastomosed on a different vasculature position in the same animal. Furthermore investigators improved in-vitro vascularization of engineered myocard surrogates by installation of micro-channels, generated by enzymatically dissolving alginate fibers or collagen-based micro-channels. Those micro-channels were populated by endothelial cells and, when perfused with culture medium, contributed to enhanced tissue maturation (Vollert et al. 2014; Sakaguchi et al. 2013). Nevertheless, further studies with innovative approaches are required to create thicker myocardial tissue. Many tissue engineering studies have focused on the repair of regional myocardial defects, e.g., after myocardial infarction, and not on offering passive (restraint) and active (contractility) support to the entire failing heart. Besides the implantation of tissue patches onto failing hearts, there are some efforts to generate whole biological ventricular assist devices. Although this is a quite challenging approach, it might be a future possibility to overcome biventricular heart failure (Yildirim et al. 2007; Sekine et al. 2006). Electrical conduction of engineered heart tissue is another important item while in-vitro cultivation and after transplantation on to the recipient myocardium. The coupling of the transplanted graft with the host heart is necessary for an effective synchronic systolic support. Continuous electrical stimulation during in-vitro culture leads to qualitative improved myocard surrogates. Dvir et al. could generate stronger, thicker, and better orientated constructs via embedding gold-nano particles in an alginate matrix (Dvir et al. 2011).

Probably some cell–cell connections are enough to provide electrical coupling, or the improved conduction of remote cells within the infarction area due to paracrine effects might play a role. Further explanation would be the electric conduction via cardiac fibroblasts, which has been shown in another context (Kohl and Camelliti 2007). Although a negative induction of rhythm disorders (ventricular tachycardia) should be considered and investigated before clinical application. Even if studies indicated that EHTs and cell sheets electrically coupled to the host myocardium an alternative approach could be a synchronic external pacing of the constructs.

9.5.2 Engineered Heart Tissue from hPSC

First three-dimensional tissue engineered constructs (e.g., hydrogel-based EHTs and scaffold-based 3D constructs, cell sheets) have been produced from hiPSC and hESC derived cardiomyocytes (Tulloch et al. 2011; Schaaf et al. 2011, 2014; Stevens et al. 2009b). Although tissue quality is still not as good as that from neonatal rat cardiomyocyte EHTs, improving protocols with cocultures, addition of growth factors and mechanical and pacing conditioning improves tissue quality (Hirt et al. 2014). First promising animal studies have been performed with cell sheets and hydrogel-based hiPSC EHTs. In a preliminary study, implantation of fibrin-based hiPSC-cardiomyocyte EHTs resulted in a guinea pig myocardial injury model in significant improvement of left ventricular cardiac function (Pecha et al. 2015). Similar results were reported for implantation of hiPSC derived cardiomyocyte cell sheet in a porcine cardiomyopathy model (Kawamura et al. 2012). Further studies are needed to confirm first promising results, and to go further steps towards a clinical use of hPSC-based EHTs.

9.5.3 Hurdles on the Way to Clinical Application

Recent advantages in terms of cell biology, with the possibility to generate unlimited numbers of hiPSCs or hESCs derived cardiomyocytes, have been an important progress regarding cell availability. Nevertheless, refinement of available differentiation protocols is needed, to receive reproducible differentiation efficacy with little intercellular variability of cardiomyocytes and cell lines.

Use of pluripotent stem cells is always associated with safety regarding teratoma formation. Especially the application of iPSC-derived cardiomyocytes additionally carries some risk for chromosome instability and induction of mutations leading to malignancy. Therefore, strict quality control of the cell lines is essential prior to clinical use. Antibody-related selection protocols might help to overcome this limitation (Dubois et al. 2011). Furthermore, graft size and thickness are very important issues in terms of cardiac repair to support failing. Actually, the biological diffusion capacity limits the graft thickness (50–200 µm), and reliable concepts for in-vitro as well as in-vivo vascularization and perfusion are necessary to enlarge the thickness of the three-dimensional grafts. Different experimental approaches for the generation of prevascularized grafts have been described (Tee et al. 2012); however, the ideal method has yet to be defined. In terms of immunological issues, the use of hiPSC derived cardiomyocytes at least theoretically allows for an autologous approach without need for immunosuppression. However, this autologous solution is time-consuming and would at least need 6–9 months for generation of a patient-specific EHT patch. Besides the time concerns, logistic and economic hurdles would have to be handled, as each iPSC cell line is from a regulatory board view, a new product, and needs to undergo extensive toxicity testing prior to clinical application. A more realistic scenario might be a bank of different iPSC or ESC lines, which can be transplanted human leucocyte antigen (HLA) matched, requiring only minimal immunosuppressive regimen.

References

Bian W, Liau B, Badie N, Bursac N (2009) Mesoscopic hydrogel molding to control the 3D geometry of bioartificial muscle tissues. Nat Protoc 4:1522–1534

Burridge PW, Thompson S, Millrod MA et al (2011) A universal system for highly efficient cardiac differentiation of human induced pluripotent stem cells that eliminates interline variability. PLoS One 6:e18293

de Lange WJ, Hegge LF, Grimes AC et al (2011) Neonatal mouse-derived engineered cardiac tissue: a novel model system for studying genetic heart disease. Circ Res 109:8–19

Dubois NC, Craft AM, Sharma P et al (2011) SIRPA is a specific cell-surface marker for isolating cardiomyocytes derived from human pluripotent stem cells. Nat Biotechnol 29:1011–1018

Dvir T, Timko BP, Brigham MD et al (2011) Nanowired three-dimensional cardiac patches. Nat Nanotechnol 6:720–725

Eschenhagen T, Fink C, Remmers U et al (1997) Three-dimensional reconstitution of embryonic cardiomyocytes in a collagen matrix: a new heart muscle model system. FASEB J 11:683–694

Furuta A, Miyoshi S, Itabashi Y et al (2006) Pulsatile cardiac tissue grafts using a novel three-dimensional cell sheet manipulation technique functionally integrates with the host heart, in vivo. Circ Res 98:705–712

Hansen A, Eder A, Bonstrup M et al (2010) Development of a drug screening platform based on engineered heart tissue. Circ Res 107:35–44

Hirt MN, Boeddinghaus J, Mitchell A et al (2014) Functional improvement and maturation of rat and human engineered heart tissue by chronic electrical stimulation. J Mol Cell Cardiol 74:151–161

Kattman SJ, Witty AD, Gagliardi M et al (2011) Stage-specific optimization of activin/nodal and BMP signaling promotes cardiac differentiation of mouse and human pluripotent stem cell lines. Cell Stem Cell 8:228–240

Kawamura M, Miyagawa S, Miki K et al (2012) Feasibility, safety, and therapeutic efficacy of human induced pluripotent stem cell-derived cardiomyocyte sheets in a porcine ischemic cardiomyopathy model. Circulation 126:S29–S37

Kensah G, Roa Lara A, Dahlmann J et al (2013) Murine and human pluripotent stem cell-derived cardiac bodies form contractile myocardial tissue in vitro. Eur Heart J 34:1134–1146

Kirklin JK, Naftel DC, Pagani FD et al (2014) Sixth INTERMACS annual report: a 10,000-patient database. J Heart Lung Transplant 33:555–564

Kohl P, Camelliti P (2007) Cardiac myocyte-nonmyocyte electrotonic coupling: implications for ventricular arrhythmogenesis. Heart Rhythm 4:233–235

Koyanagi-Aoi M, Ohnuki M, Takahashi K et al (2013) Differentiation-defective phenotypes revealed by large-scale analyses of human pluripotent stem cells. Proc Natl Acad Sci U S A 110:20569–20574

Laflamme MA, Murry CE (2011) Heart regeneration. Nature 473:326–335

Leor J, Aboulafia-Etzion S, Dar A et al (2000) Bioengineered cardiac grafts: a new approach to repair the infarcted myocardium? Circulation 102:III56–III61

Li RK, Jia ZQ, Weisel RD, Mickle DA, Choi A, Yau TM (1999) Survival and function of bioengineered cardiac grafts. Circulation 100:II63–II69

Lunde K, Solheim S, Aakhus S et al (2006) Intracoronary injection of mononuclear bone marrow cells in acute myocardial infarction. N Engl J Med 355:1199–1209

Morritt AN, Bortolotto SK, Dilley RJ et al (2007) Cardiac tissue engineering in an in vivo vascularized chamber. Circulation 115:353–360

Naito H, Melnychenko I, Didie M et al (2006) Optimizing engineered heart tissue for therapeutic applications as surrogate heart muscle. Circulation 114:I72–I78

Ott HC, Matthiesen TS, Goh SK et al (2008) Perfusion-decellularized matrix: using nature's platform to engineer a bioartificial heart. Nat Med 14:213–221

Pecha S, Weinberger F, Breckwoldt K et al (2015) Human induced pluripotent stem cells for tissue engineered cardiac repair [e-pub ahead of print]. J Heart Lung Transplant. http://dx.doi.org/10.1016healun.2015.01.093

Sakaguchi K, Shimizu T, Horaguchi S et al (2013) In vitro engineering of vascularized tissue surrogates. Sci Rep 3:1316

Schaaf S, Shibamiya A, Mewe M et al (2011) Human engineered heart tissue as a versatile tool in basic research and preclinical toxicology. PLoS One 6:e26397

Schaaf S, Eder A, Vollert I, Stohr A, Hansen A, Eschenhagen T (2014) Generation of strip-format fibrin-based engineered heart tissue (EHT). Methods Mol Biol 1181:121–129

Schachinger V, Erbs S, Elsasser A et al (2006) Intracoronary bone marrow-derived progenitor cells in acute myocardial infarction. N Engl J Med 355:1210–1221

Sekine H, Shimizu T, Yang J, Kobayashi E, Okano T (2006) Pulsatile myocardial tubes fabricated with cell sheet engineering. Circulation 114:I87–I93

Sekine H, Shimizu T, Hobo K et al (2008) Endothelial cell coculture within tissue-engineered cardiomyocyte sheets enhances neovascularization and improves cardiac function of ischemic hearts. Circulation 118:S145–S152

Senyo SE, Steinhauser ML, Pizzimenti CL et al (2013) Mammalian heart renewal by pre-existing cardiomyocytes. Nature 493:433–436

Shimizu T, Yamato M, Isoi Y et al (2002) Fabrication of pulsatile cardiac tissue grafts using a novel 3-dimensional cell sheet manipulation technique and temperature-responsive cell culture surfaces. Circ Res 90:e40

Shimizu T, Sekine H, Yang J et al (2006) Polysurgery of cell sheet grafts overcomes diffusion limits to produce thick, vascularized myocardial tissues. FASEB J 20:708–710

Soonpaa MH, Field LJ (1998) Survey of studies examining mammalian cardiomyocyte DNA synthesis. Circ Res 83:15–26

Stevens KR, Kreutziger KL, Dupras SK et al (2009a) Physiological function and transplantation of scaffold-free and vascularized human cardiac muscle tissue. Proc Natl Acad Sci U S A 106:16568–16573

Stevens KR, Pabon L, Muskheli V, Murry CE (2009b) Scaffold-free human cardiac tissue patch created from embryonic stem cells. Tissue Eng Part A 15:1211–1222

Tee R, Morrison WA, Dusting GJ et al (2012) Transplantation of engineered cardiac muscle flaps in syngeneic rats. Tissue Eng Part A 18:1992–1999

Tiburcy M, Didie M, Boy O et al (2011) Terminal differentiation, advanced organotypic maturation, and modeling of hypertrophic growth in engineered heart tissue. Circ Res 109:1105–1114

Tulloch NL, Muskheli V, Razumova MV et al (2011) Growth of engineered human myocardium with mechanical loading and vascular coculture. Circ Res 109:47–59

Vollert I, Seiffert M, Bachmair J et al (2014) In vitro perfusion of engineered heart tissue through endothelialized channels. Tissue Eng Part A 20:854–863

Yildirim Y, Naito H, Didie M et al (2007) Development of a biological ventricular assist device: preliminary data from a small animal model. Circulation 116:I16–I23

Zhao T, Zhang ZN, Rong Z, Xu Y (2011) Immunogenicity of induced pluripotent stem cells. Nature 474:212–215

Zimmermann WH, Schneiderbanger K, Schubert P et al (2002) Tissue engineering of a differentiated cardiac muscle construct. Circ Res 90:223–230

Zimmermann WH, Melnychenko I, Wasmeier G et al (2006) Engineered heart tissue grafts improve systolic and diastolic function in infarcted rat hearts. Nat Med 12:452–458

Chapter 10
Stem Cell Therapy for Ischemic Heart Disease

Truc Le-Buu Pham, Ngoc Bich Vu, and Phuc Van Pham

10.1 Overview of Cardiovascular Disease

Cardiovascular disease is a general concept that refers to the disorder of the heart or blood vessels or both. It can be divided into two groups. Group 1 consists of cardiovascular diseases associated with the obstruction of atherosclerotic fragments, including coronary artery disease, cerebrovascular disease, and peripheral vascular disease. Group 2 consists of cardiovascular diseases which cause conditions, including hypertension, congestive heart failure, rheumatic heart disease, cardiomyopathy, and congenital heart disease (Luepker 2011). According to statistical data from WHO, in 2008, there were approximately 17.3 million people deaths from cardiovascular diseases. It is predicted that about 23.6 million people will die by 2030. Of these deaths, coronary heart disease is one of the main leading causes.

Coronary artery disease is caused by the narrowing or blockage of blood vessels that feed the heart. It is divided into two cases: acute myocardial infarction (with or without reperfusion) and myocardial ischemia (Luepker 2011). At present, there are many treatments for coronary artery disease, such as lifestyle changes, medical therapy, coronary intervention, and/or coronary bypass graft surgery.

10.1.1 Lifestyle Changes

Smoking, drinking, unreasonable diet (high-fat, high-sodium, and high-cholesterol foods), and physical inactivity are risk factors for heart disease. In addition, obesity, diabetes, and high lipid levels are also likely to lead to cardiovascular disease.

T.L.-B. Pham • N.B. Vu • P. Van Pham (✉)
Laboratory of Stem Cell Research and Application, University of Science, Vietnam National University, Ho Chi Minh City, Vietnam
e-mail: plbtruc@hcmus.edu.vn; vbngoc@hcmus.edu.vn; pvphuc@hcmuns.edu.vn

Therefore, to prevent the risk of heart disease, patients need to adjust their lifestyles by refraining from smoking, limiting alcohol, and following a reasonable diet and exercise regimen as instructed by their physicians (Becker et al. 2008).

10.1.2 Medical Therapy

When lifestyle changes are ineffective for cardiovascular disease treatment, a combination therapy is necessary. Medicines that are currently used to treat heart disease include anti-platelets, ES 3 antianginals, beta blockers, angiotensin-converting enzyme inhibitors, angiotensin II receptor antagonists, and lipid-lowering drugs. The medical procedures are recommended based on the doctor's guidelines. Moreover, patients must pay attention to drug allergies, such as those to aspirin, beta blockers, and cholesterol-lowering drugs (Boden et al. 2007).

10.1.3 Interventional Procedures

Interventional procedures are used for patients with coronary artery stenosis. A balloon catheter-carrying stent or drug-eluting stent is put into a narrow coronary artery. The balloon is then inflated within the stent to cause the stent to dilate. Consequently, the narrowed segment is expanded. The balloon is deflated and withdrawn with the catheter whereas the stent is left in the narrowed artery segment to help blood flow normally. After stenting, patients must use anticoagulant drugs and follow regular health check-ups (Bravata et al. 2007; Eisenstein et al. 2007). At present, there are several types of balloons and catheters for doctors and patients to choose from, depending on each individual's case.

10.1.4 Coronary Artery Bypass Graft Surgery

In severe cases where heart disease cannot be treated by interventional procedures, the physician will most likely recommend coronary artery bypass surgery. The physician will choose to draw blood from vessels in the chest, arms, or legs of the patient and pair the occlusion segments with these blood vessels. The grafts will create "bridges" which connect blocked areas together and result in restoration of normal blood flow to the heart. This method is invasive and patients must be monitored closely after surgery and undergo regular follow-ups (Bravata et al. 2007; Ott et al. 2008; Serruys et al. 2009).

With the advancement of science, especially in the area related to stem cell discovery and research, new therapeutic methods emerged for heart diseases. To date, the use of stem cells for coronary artery disease curing is widely studied and there have been a number of clinical applications of stem cell therapy for heart diseases (Boyle et al. 2006; Condorelli et al. 2001; Fuentes and Kearns-Jonker 2013; Ichim et al. 2010; Segers and Lee 2008).

10.2 The Ability of Stem Cells to Differentiate Into Angiogenic Cells and Myocardial Cells

Stem cells are defined as cells capable of self-renewal for long periods and with potential to differentiate into many different cell types in the body. Stem cells can be classified according to their origin, e.g., embryonic stem cells (ESCs), adult stem cells, and induced pluripotent stem cells (iPSCs). These three types of cells are capable of differentiation into myocardial cells and/or angiogenic cells. They represent, therefore, a potentially useful cell-based therapy for cardiovascular diseases. In particular, adult stem cells and iPSCs have become valuable candidates for clinical applications of heart disease; their use has advantages but also minimizes controversial moral issues.

10.2.1 Potential for Angiogenesis

The angiogenic potential of cells are measured by differentiation into endothelial cells (ECs), pericytes, vascular smooth muscle cells, and mural cells. ECs establish the inner lining of the vessel wall. Pericytes and vascular smooth muscle cells envelope the surface of the vascular tube. Mural cells produce extracellular matrix. ECs derived from endothelial progenitor cells (EPCs), and pericytes are differentiated from smooth muscle progenitor cells.

During the embryonic stage, mesenchymal stem cells (MSCs), EPCs, and vascular progenitor cells are three main types of cells involved in blood vessel formation. Capillaries are the simplest of the vascular system, and are formed by layers of ECs and surrounded by pericytes. The basal lamina layers are established by secretion of mural cells. These layers serve to stabilize the network. For large vessel such as arteries and veins, smooth muscle cells also participate to build and form functional blood vessels.

In the adult, circulating EPCs, bone marrow cells, and myofibroblasts are mobilized to participate in formation of new blood vessels. Mononuclear cells obtained from bone marrow or umbilical cord blood contain EPCs and the CD34+ fraction (Botta et al. 2004; Isner and Asahara 1999). A study demonstrated the cells can be induced, both in vivo and in vitro, to differentiate into ECs (Isner and Asahara 1999).

Adipose tissue contains a population of EPCs with potential to differentiate into ECs (Madonna and De Caterina 2010). Furthermore, ECs and EPCs are also found in both white and brown fat cells. Capillary sprouts in adipose tissue were found to be a potential source of cells for repairing damaged tissues, in a model for angiogenesis after heart failure (Tran et al. 2012).

MSCs, derived from the mesoderm layer, possess the ability to differentiate into chondrocytes, adipocytes, cartilage, and many other different types of cells, under specific culture conditions. It was found that MSCs were able to differentiate into cardiomyocytes and stimulate the growth of blood vessels during heart injury repair (Hatzistergos et al. 2010; MacKenzie and Flake 2002). On the other hand, when cultured with growth factor, MSCs increased survival rate in vivo, leading to improved vascularity in ischemic cardiac tissue (Rodrigues et al. 2010).

10.2.2 Potential to Differentiate Into Myocardial Cells

Bone marrow contains many types of cells, including immature cells and functional cells. In most pathological processes, bone marrow derived stem cells (BMSCs) were frequently mobilized to repair the damage. Since 2001, Yerebakan and his colleagues have performed transplant of CD133+ bone marrow cells for the treatment of myocardial ischemia. However, use of BMSC populations for myocardial disease did not show high treatment efficacy. In sheep studies, it was found that BMSCs injected into fibrous tissue or damaged heart muscle did not improve function nor blood kinetics (Bel et al. 2003; Menasche 2003). Moreover, immunohistochemistry analyses did not find differentiation or trans-differentiation of transplanted cells into myocardiocytes or vascular cells (Bel et al. 2003). This may be due to the fact that the BMSC cell populations contained only a small amount of stem cells.

Purified MSCs and hematopoietic stem cells (HSCs) can be isolated from many sources, such as bone marrow, placenta, umbilical cord blood, umbilical cord, teeth, and adipose tissue. MSCs can be generated into cardiac muscle cells by treatment with DNA demethylating agent 5-azacytidine (5-aza); following 5-aza treatment, MSCs can be induced in vitro to differentiate into spontaneously beating cells (Davani et al. 2003). Human BMSCs engrafted in the left ventricle of mice were seen to differentiate into cardiomyocytes. Furthermore, after transplant into the ischemic myocardium, the grafted cells could replace damaged heart muscle cells and almost completely restore systolic and diastolic function to normal levels (Makino et al. 1999; Tambara et al. 2003). Moreover, HSCs have been shown to significantly regenerate the infarcted myocardium. In fact, Orlic et al. demonstrated in rat that injection of c-kit+/lin− BMSCs into the contracting wall bordering the infarct led to generation of smooth muscle cells, ECs, and cardiomyocytes; new myocardium covered 68 % of the infarcted zone (Orlic et al. 2001).

Skeletal muscle-derived stem cells (SMSCs) are considered candidate myocardial cells because they possess strong proliferative potential and capability to different into muscle cell lineage. In one study by Tambara et al., SMSCs were transplanted into myocardial infarction rat model; the results showed that these cells could survive in the host and completely replace damaged cardiac muscle cells (Tambara et al. 2003). In other studies, transplantation of skeletal myoblasts in ischemic cardiomyopathies resulted in an increased efficiency of the ejection fraction, which improved cardiac contractility and ventricular wall thickening (Durrani et al. 2010; Ghostine et al. 2002; Haider et al. 2004; Hata et al. 2006; Khan et al. 2007).

Adult cardiomyocytes are considered terminally differentiated cells. They divide rapidly in the body and cannot live under normal conditions (Reinecke et al. 1999). Therefore, the regeneration of cardiac muscle cells after heart injury is quite limited. However, some studies have demonstrated that fetal or neonatal cardiac muscle cells can survive for long periods after injection into mature heart muscle (Müller-Ehmsen et al. 2004; Reinecke et al. 1999). Transplantation of these cells into the infarction also showed effective reduction of left ventricular wall thickness, limited growth of scar tissue, and improvement of cardiac function (Li et al. 1996). However, these cells could not reach the final stage of mature cardiac muscle cells, and thus were not capable of contraction (Müller-Ehmsen et al. 2004).

Resident cardiac stem cells have been described that are capable of long-term self-renewal and differentiation in vitro to cardiomyocytes. This population of cardiac stem cells are positive for c-kit, while being negative for blood lineage markers CD34, CD45, CD20, CD45RO, and Lin. These cells have significantly contributed to the improvement of cardiac function following infarctions (Messina et al. 2004).

In the recent years, the in vitro cardiomyogenic potential of stroma vascular fractions (SVFs) derived from adipose tissues has also been reported. In a study by Planat-Bénard et al., rare beating cells with cardiomyocyte features were identified after culture of SVFs without addition of 5-azacytidine (Planat-Benard et al. 2004).

10.3 Sources of Stem Cells for Use in Therapy

Several different types of cells and some methods of cell transplantation have been evaluated and selected for preclinical studies and clinical trials of heart damages. Initial evaluation showed satisfactory results. However, to optimize clinical trials, it is vital that the transplanted cells survive and exist in high numbers within the transplanted heart tissue. Therefore, ideal cell types as well as ideal transplantation methods are studied frequently (Fig. 10.1).

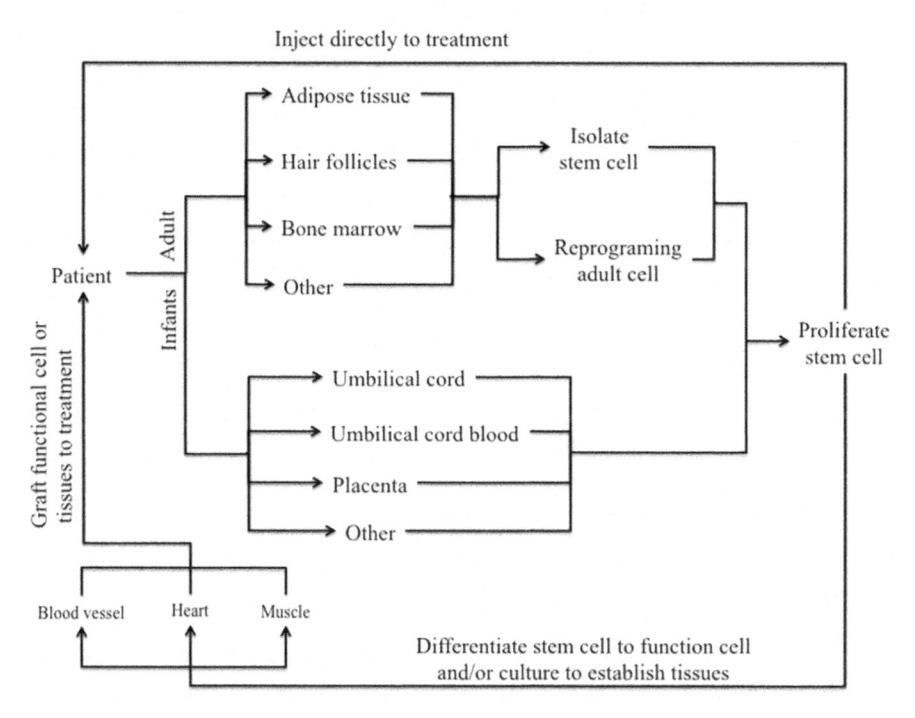

Fig. 10.1 Sources of stem cells for use in therapy. There are various stem cells used in the therapy. They can be derived from adult tissue as adipose tissue, bone marrow, and hair follicle; from infant waste as umbilical cord blood, placenta, and umbilical cord or induced from somatic tissue (reprogramming cells). Then, they can be proliferated to expand or differentiated into functional cells and finally transplanted to patients

To date, there have been several different types of stem cell which have been studied for use in treatment of cardiovascular disease. These cells include ESCs, BMSCs, and MSCs from adipose tissue, umbilical cord blood stem cells, resident cardiac stem cells, and iPSCs. Each type has its own distinct advantages and disadvantages.

10.3.1 Embryonic Stem Cells

ESCs are derived from blastocysts and have the ability to differentiate into any cell type of the body. This means they have the ability to differentiate into heart muscle cells and have been tested for treatment of cardiovascular diseases (Kehat et al. 2002; Ma et al. 2005; Wollert and Drexler 2010). However, its use raises ethical issues, which has been one of the biggest concerns for scientists working with this kind of cell type (Wert and Mummery 2003). In addition, ESCs have the potential to form tumors, such as teratomas; this phenomenon has been observed in animal models (Beitnes et al. 2011; Rosenstrauch et al. 2005).

10.3.2 Bone Marrow Derived Stem Cells

BMSCs have been commonly used to study heart regeneration in animal models of myocardial infarction (Orlic et al. 2001). Preliminary studies showed that when cells were implanted into the heart, they differentiated into myocardium. Moreover, a significant improvement in left ventricular ejection fraction was observed only 9 days after transplantation (Kocher et al. 2001; Orlic et al. 2001). In rats with left anterior coronary artery ligation, BMSCs improved cardiac function and angiogenesis induction (Kocher et al. 2001). There are two important points to consider when choosing stem cells from bone marrow: firstly, the procedure is invasive (Bain 2001; Delling et al. 2012) and secondly, the cells will expire with age (Sethe et al. 2006).

10.3.3 Adipose Derived Stem Cells

Compared to BMSCs, adipose derived stem cells (ADSCs) is an easier source of stem cells to obtain and requires a less invasive method for isolation. A number of clinical studies, nowadays, show that ADSC transplantation improves strength and ejection fraction and induces angiogenesis through paracrine factors (Perin 2006). However, more studies need to be conducted to understand the impact of ADSC transplantation, including the capacity for island formation in the graft and the potential for arrhythmia.

10.3.4 Human Umbilical Cord Blood Stem Cells

Human umbilical cord blood-derived MSCs (hUC-MSCs) are easy to acquire, non-invasive, and a good source of young cells. They rarely express human leukocyte antigen (HLA) class II and have capability of immune modulation, thereby minimizing the possibility of immune rejection after transplant in the body (Lu et al. 2006). Infusion of hUC-MSCs has shown to significantly reduce the infarct size of acute myocardial infarction (AMI) in animal models (Hirata et al. 2005; Kocher et al. 2001; Ma et al. 2005).

10.3.5 Resident Cardiac Stem Cells

Stem cells positive for Sca-1, and negative for CD45, CD34, and c-kit, are detectable in the heart. When appropriate signals are received, these cells will differentiate into various cell lineages, including smooth muscle cells and angiogenesis-associated myocardial cells. Research in animal models have shown that these cells can reduce infarction area size and improve left ventricular function (Oh et al. 2003). However, the number of cells is limited for transplantation after infarction.

10.3.6 Induced Pluripotent Stem Cells

Another type of stem cells is iPSC. They are reprogrammed from somatic cells and carry characteristics found in ESCs. They also have the ability to differentiate into other cell types in the body, including angiogenesis-associated blood vessel cells and cardiac muscle cells. Immune rejection is not a problem since the cell type chosen for therapy is taken from the patient. Similar to the ESCs, iPSCs possess an ability to generate tumors after transplantation (Cantz and Martin 2010; Kuzmenkin et al. 2009). Although these cells have the same characteristics of ESCs, their telomeres are shorter than telomeres of ESCs because they are derived from somatic cells (Lister et al. 2011; Marion et al. 2009).

10.4 Delivery Methods for Cell Transplantation

Presently, there are two main methods to mobilize cells to damaged areas of the heart; these include: (1) direct intra-myocardial injection and (2) trans-vascular infusion. Each method has its own advantages as well as limitations.

10.4.1 Direct Intra-Myocardial Injection

This is the first method for stem cell therapy of heart disease and is still widely utilized in animal models and human clinical studies. In this method, stem cells are injected directly into the left ventricle wall without risk of vascular occlusion (George et al. 2008; Hale et al. 2008). The amount of surviving injected cells, however, is low due to the leakage of injected cells from the injection site and through coronary veins (Anderl et al. 2009; Hale et al. 2008; Yasuda et al. 2005). To circumvent this issue, cells should be injected in scattered places.

There are three ways to inject cells into heart muscle, including: (1) trans-epicardial injection; (2) trans-endocardial injection, and (3) coronary vein injection.

10.4.1.1 Trans-Epicardial Injection

This method has been used for both humans and animals. Cells are suspended in albumin and injected into the region around the infarction area (Dib et al. 2005; Hagege et al. 2006; Menasche et al. 2008). Autologous myoblast transplantation using this method, unexpectedly, did not improve cardiac function but caused arrhythmias. Because myoblasts are differentiated orientation cells, they have their own rhythm in vitro and are unable to fuse with local cells in harmony, leading to arrhythmia (Menasche et al. 2008). On the contrary, when stem cells from bone marrow are transplanted using this method, cardiac function is improved (van Ramshorst et al. 2009).

10.4.1.2 Trans-Endocardial Injection

This technique is done with an electromagnetic mapping system to place the catheter into the femoral artery through the aortic valve into the left ventricular chamber. Mapping systems include Helix, MyoCath, MyoStar, Stiletto, and Noga (Oh et al. 2003). Unlike trans-epicardial injection, this method does not cause bleeding or damage to the tissue surface. Clinical trials with bone marrow cells showed an improvement of ejection fraction and systolic volume depletion (Perin et al. 2008).

10.4.1.3 Coronary Vein Injection

In this method, a catheter system is guided by ultrasound (IVUS system) to the myocardium and surrounding areas (Sherman et al. 2006; Thompson et al. 2003, 2005). This procedure poses no risk of death, does not require access to the left ventricle (Thompson et al. 2005), and improves ejection fraction (Siminiak et al. 2005). However, due to the complexity of the coronary system, it is difficult to correctly identify a particular area of the heart that needs to be injected.

To increase the efficiency of transplantation, this method can be combined and implemented with TMR (trans-myocardial laser re-vascularization) technique. The TMR method creates "channels" in the environment by using laser beams to enhance survival of transplanted cells in the graft for a longer period of time (Sherman et al. 2006).

10.4.2 Trans-Vascular Infusion

10.4.2.1 Intracoronary Infusion

In this method, cells are delivered into the infarct area through a catheter placed in the coronary artery. Cells are infused during the reperfusion period so that coronary blood flow will stop for a few minutes and a short acute infarction is created (Strauer et al. 2002). Compared to the direct injection methods, this procedure is less invasive and damaging. In addition, this approach also allows an increased volumetric quantity of cells to be delivered into the area of interest in the damaged heart. The main downside of this method is that cells are rapidly washed out of the tissue. Therefore, the number of cells that stay in the heart tissue is lower than those for the direct injection methods. Finally, this method limits the cell size and requires that cells must be able to travel through the vessel wall (Schachinger et al. 2004, 2006; Wollert et al. 2004; Zohlnhofer et al. 2006). Some trials using this method are TopCare (Schachinger et al. 2004), REPAIR-AMI (Schächinger et al. 2006; Schachinger et al. 2006), ASTAMI (Patel et al. 2016), and BOOST (Wollert et al. 2004).

10.4.2.2 Intravenous Infusion

Cells are transplanted into recipients via intravenous infusion. The cells then migrate through the circulatory system and settle in the damaged regions of myocardial tissue. This method is easy to implement and non-invasive. However, the number of surviving cells in infused heart after transplantation is less than those in other implant methods (Fukushima et al. 2008; Kocher et al. 2001). Most of the cells are retained in the lung, liver, spleen, and kidney, rather than the heart (Fukushima et al. 2008). Similar to the intracoronary infusion method, intravenous infused cells need to be capable of trans-endothelial migration.

10.5 Mechanism of the Therapy

Stem cells may be able to repair and/or replace damaged heart muscle cells by the following three mechanisms: (1) directly differentiating into functional cells of the heart; (2) secreting growth factors that mobilize stem cells repair and/or replace damaged tissues; and (3) secreting factors that inhibit myocardial necrosis.

For the first mechanism (direct differentiation), stem cells represent an ideal source for differentiation into functional cells of the heart, such as cardiomyocytes. They can replace necrotic or damaged cells, vascular ECs, and vascular smooth muscle cells to form new blood vessels (Suzuki et al. 2015). Following transplant of ADSCs (Nagata et al. 2016) and peripheral blood CD34+ cells (Yeh and Zhang 2006) in ischemic myocardium (of a mouse myocardial infarction model), ADSCs showed expression of markers for ECs, vascular smooth muscle cells, and cardiomyocytes.

Mechanisms for replacing injured tissues may also occur via fusion of transplanted stem cells with tissue-resident cells. Bone marrow transplantation has demonstrated that bone marrow derived cells (i.e., BMSCs) fuse in vivo with cardiac muscle in the heart, resulting in formation of multinucleated cells (Alvarez-Dolado et al. 2003). Some studies have reported that stem cell transplantation leads to reprogramming of the recipient cell, including re-entry of cell cycle. As a result, the fate of the recipient cardiomyocytes may be changed, myocardial cells regenerated, and cardiac function improved (Fig. 10.2) (Yeh and Zhang 2006).

In the indirection mechanism (secretion of factors), transplanted cells produce paracrine factors, which can mobilize stem cells, leading to repair of internal damaged tissues. ADSCs are adult stem cells. They possess the characteristic to produce

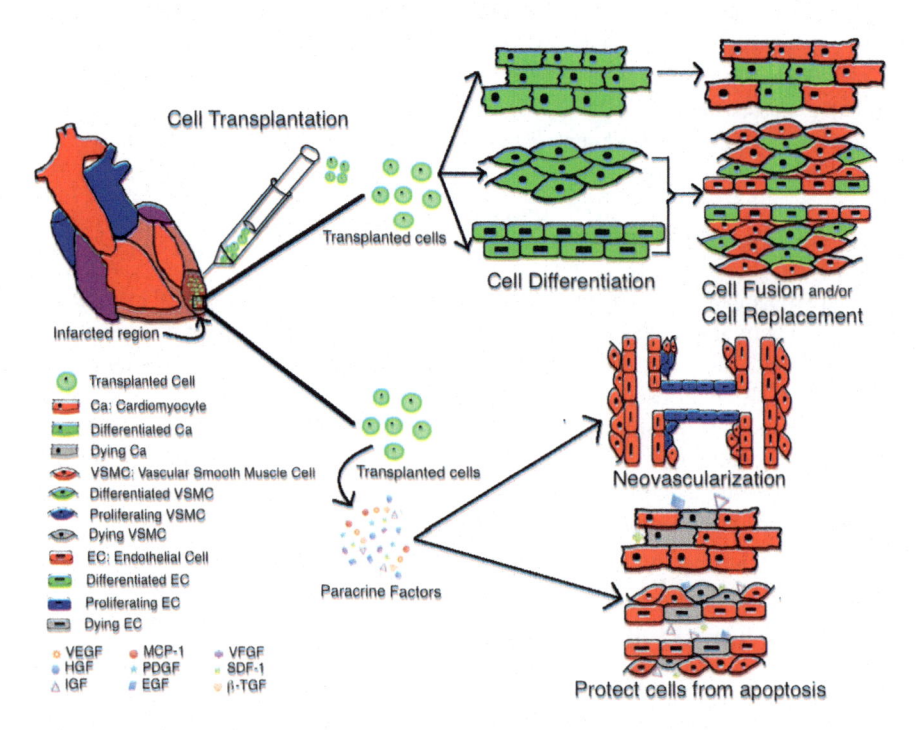

Fig. 10.2 Mechanism of the hear disease therapy. Transplanted cells may be differentiated into cardiomyocytes, smooth muscle cells, or ECs to perform cell fusion or cell replacement. They may also release paracrine factors that mobilize local stem cells to damaged region for neovascularization or protect cells from apoptosis

a variety of paracrine factors such as cytokines and growth factors, which induce cytoprotection and neovascularization. Vascular endothelial growth factor (VEGF), hepatocyte growth factor (HGF), and insulin-like growth factor (IGF) are all growth factors, secreted by ADSCs, which can activate resident cardiac stem cells (Gnecchi et al. 2008; Suzuki et al. 2015). Pelacho et al. demonstrated that transplantation of murine BM-derived multipotent adult progenitor cells (MAPCs) significantly improved left ventricular contractile function and reduced infarct size. Damage repair mechanisms may be indirectly contributed by secreted inflammatory factors, including monocyte chemo-attractant protein (MCP)-1, VEGF, platelet-derived growth factor (PDGF)-BB, transforming growth factor (TGF)-beta 1, and others (Pelacho et al. 2007). In ischemic limbs, recruited human EPCs also have the ability to release pro-angiogenic factors (e.g., VEGF), which exert direct effects on mature endothelial cell migration, thereby improving neovascularization (Burchfield and Dimmeler 2008).

As a third mechanism, stem cells can repair heart muscle damage through secretion of factors, which inhibit cardiac cell necrosis. Stem cells injected into the infarct heart may directly protect against cardiac myocyte cell death. Marrow-derived stromal cells and ADSCs secrete a wide array of arteriogenic cytokines, such as epidermal growth factor (EGF), basic fibroblast growth factor (bFGF), IGF, and stromal cell-derived factor-1 (SDF-1) (Kim et al. 2015a; Xu et al. 2007). Transplantation of these cells led to a decrease in cardiac myocyte apoptosis and an upregulation of BCL-2 (well-known anti-apoptotic protein) in cardiac myocytes (Mirotsou et al. 2007).

10.6 Cell Doses in Transplantation

Many studies have shown that the efficiency of cell therapy significantly depends on cells dosage, frequency, route of administration, and other factors. Cell dose in transplantation for heart disease treatment has been variable; for example, one method for calculating the optimal cell dose for transplantation takes into account the area of injured tissue (Durrani et al. 2010; Ye et al. 2007).

In a study, autologous AC133+ bone marrow cells (1.5 million cells) were injected into the infarct border zone of myocardial infarction patients. The result showed that all patients were alive and well, global left ventricular function was enhanced, and infarct tissue perfusion had improved at 3–9 months after surgery (Stamm et al. 2003).

In another trial, myoblast autologous grafting was used to treat ischemic cardiomyopathy. The high-dose cell group (800 million) showed a significant decrease in left ventricular volumes and an improvement of echocardiographic heart function compared to the low-dose cell group (400 million). Both doses led to a significant improvement of left ventricular function (Menasche et al. 2008).

In addition, there have been studies comparing allogeneic versus autologous BMSCs, delivered by trans-endocardial injection into patients with ischemic cardiomyopathy. The efficacy of treatment was compared using 3 different doses (20, 100, and 200 million) of allogeneic and autologous MSCs in patients with ischemic heart disease. It was demonstrated that all doses significantly improved patient functional capacity, quality of life, as well as ventricular remodeling. However, the high-dose cell group (200 million) had a lower effect than the low-dose cell group (20 million) (Hare et al. 2012). These results, however, differ from those observed in a study of heart disease in a swine model. In the swine model, MSCs at high dose (200 million) and low dose (20 million) trended towards a reduction in scar size and amelioration of cardiac function; only the 200 million dose affected reverse remodeling (Schuleri et al. 2009).

In another animal study (pig model of chronic ischemic heart disease), animals were injected with BM-derived mononuclear cells at 3 different cell doses (50, 100, and 200 million). The results showed that stem cell transplanted pigs showed significant recovery, compared to untreated pigs. No arrhythmias or cardiac inflammatory reactions occurred in the treated pigs. In the pigs injected with 100 million cells, fibrosis was significantly lower compared to the untreated pigs. Capillary density progressively increased with cell dose; the 200 million dose group showed higher capillary-density numbers than did the 100 million dose group (Silva et al. 2011).

10.7 Safety Aspects of Stem Cell Therapy for Ischemic Heart Disease

Preclinical trials using stem cell therapy for treatment of cardiovascular diseases continue to be conducted, with the hope of improving therapeutic efficacy and prolonging patient survival. However, potential safety concerns remain an overriding priority in these clinical trials. Results to date, in regard to safety, have been encouraging. Many studies have provided evidence that autologous stem cells are safe and feasible platforms to treat ischemic myocardial ischemia.

The safety of these clinical trials is evaluated by assessing major cardiac adverse events (MCAE). An MCAE is defined as the composite of cardiovascular- and noncardiovascular-related deaths, myocardial infarctions, congestive heart failures, resuscitated sudden deaths, stroke, and arrhythmias (Menasche et al. 2008). Other events that are evaluated as well are oncogenic transformation, multi-organ seeding, aberrant cell differentiation, and accelerated atherosclerosis (Gersh et al. 2009). Recent studies using BMSCs provide reassurance about the long-term safety of intramyocardial (Leistner et al. 2011; Strauer et al. 2001), epicardial (Stamm et al. 2003), and endocardial (Tse et al. 2003) cell therapy in humans with severe heart failure.

Cell therapy for heart diseases is still in its infancy and requires thorough evaluation of both short- and long-term complications of the therapy. Previous studies have reported that prolonged culture of human embryonic stem cells (hESCs) may lead to structural and chromosomal abnormalities as well as variations in DNA copy number during cultivation (Maitra et al. 2005; Narva et al. 2010). Moreover, it is difficult for iPSCs to maintain genomic stability during reprogramming (Laurent et al. 2011). Therefore, to date, the applications of hESCs and iPSCs in clinical trials have been limited. For MSCs, Wang et al. have demonstrated that majority of cultured hUC-MSCs develop genomic alterations but do not undergo malignant transformation. Despite this, management of genomic stability is recommended before these cells are used for clinical applications (Wang et al. 2013).

Several earlier studies have documented malignant ventricular arrhythmias after skeletal myoblast transplant. Inherent electrophysiological properties of myoblasts differ from stem cells; their ability to couple electromechanically among themselves or with host cardiomyocytes is not the same (Makkar et al. 2003). Clinical data derived from BMSC-based therapy for heart disease has been suggested that stem cell transplantation is less likely to cause arrhythmias than myoblast transplantation (Schachinger et al. 2004; Stamm et al. 2003; Strauer et al. 2001).

10.8 Efficiency of Stem Cell Therapy for Ischemic Heart Disease

It has been more than 20 years since the first trial of cell transplantation for heart disease was published (Marelli et al. 1992). Until now, cell therapy for heart diseases has made remarkable progress. Early on, scientists focused on the use of skeletal muscle satellite cells (Marelli et al. 1992; Murry et al. 2002). However, these cells did not show effective differentiation into heart muscle cells; as well, their ability to connect to heart muscle cells in vivo after transplantation was very low (Table 10.1) (Murry et al. 2005; Reinecke et al. 2002).

Further studies were conducted to find suitable cells and sources. BMSCs emerged as suitable cells (Mazo et al. 2012). ESCs can differentiate into mature cell types and have the ability to regenerate heart cells in animal models. However, bioethics, legal issues (Blum and Benvenisty 2008), and immunity (Zhu et al. 2009) have hampered their use in human trials. ESCs have been replaced by iPSCs (Wernig et al. 2007) or induced myocardial cells (Ieda et al. 2010). Currently, heart cells derived from adult stem cells are being tested and evaluated in clinical trials; these stem cells have included bone marrow monocytes (Perin et al. 2012; Williams and Hare 2011), MSCs (Gong et al. 2014; Tao et al. 2015; Zhou et al. 2014), ADSCs (Mazo et al. 2012), and heart derived stem cells (Malliaras et al. 2014).

10.8.1 Preclinical Trials

In the study by Dan et al., transplanted cardiac stem cells (CSCs) improved left ventricular functions, reduced 29 % of the infarct area, and positively impacted restructuring of the left ventricle in ischemic injury/reperfusion rats (Dawn et al. 2005). In rat models of myocardial infarction, Tang and colleagues demonstrated that exogenous cardiac progenitors (CPs) not only regenerated cardiac muscle through differentiation into cardiac muscle cells but also activated the endogenous CPs (Tang et al. 2010). Cardiosphere derived cells (CDCs) have also been studied a

Table 10.1 The summary of cell transplantation on clinical trials

Trial name	Trial design				Results			References
	Code	Cell delivery	Cell sources	Transplantation timing after injury	Safety	Efficacy		
A Phase II Dose-escalation Study to Assess the Feasibility and Safety of Transendocardial Delivery of Three Different Doses of Allogeneic Mesenchymal Precursor Cells (MPCs) in Subjects With Heart Failure	NCT00721045	Transendothelial injection procedure	Allogeneic mesenchymal precursor cells (MPCs)	36 months	Feasible, safe, and well tolerated	Improvement of left ventricular volumes and ejection fraction		Perin et al. (2015)
Myocardial regeneration strategy using Wharton's jelly MSCs as an off-the-shelf "unlimited" therapeutic agent: results from the Acute Myocardial Infarction First-in-Man Study	K/ZDS/005644 and 265761 "CIRCULATE"	Coronary-non-occlusive method	Wharton's jelly MSCs (WJMSCs)	Up to 12 months	Cardiac arrhythmias might be occurred but trivial	No clinical symptoms or signs of myocardial ischemia, no new ischemia Reduce Troponin level after 24 h of transplantation		Musialek et al. (2015)
Cardiac stem cells in patients with ischemic cardiomyopathy (SCIPIO): initial results of a randomized phase 1 trial	NCT00474461	Intracoronary infusion	Autologous cardiac stem cells (CSCs)	12 months	Safe	Improving LV systolic function Reducing infarct size		Bolli et al. (2011)
CArdiosphere-Derived aUtologous Stem CEllls to Reverse ventricUlar dySfunction (CADUCCEUS)	NCT00893360	Intracoronary infusion	Autologous cardiosphere-derived cells (CDCs)	12 months	Did not raise significant safety concerns Safe No patients had died, developed cardiac tumors, or MACE	Decrease scar size Increase viable myocardium Improve infarcted myocardium function Reductions in scar mass, increases in viable heart mass, regional contractility, and regional systolic wall thickening		Makkar et al. (2012) and Malliaras et al. (2014)

Study	NCT	Delivery	Cell type	Follow-up	Safety	Efficacy	Reference
AutoLogous Human CArdiac-Derived Stem Cell to Treat Ischemic cArdiomyopathy (ALCADIA)	NCT00981006	Intramyocardial injections	Autologous human cardiac-derived stem cells (hCSCs)				Takehara et al. (2008) and Tateishi et al. (2007a, b)
Safety and Efficacy Evaluation of Intracoronary Infusion of Allogeneic Human Cardiac Stem Cells in Patients With AMI (CAREMI)	NCT02439398	Intracoronary Infusion	Allogeneic human cardiac stem cells	12 months	Safety	Attenuate inflammation Prevent adverse cardiac remodeling Have a tolerogenic immune behavior	Boukouaci et al. (2014) and Lauden et al. (2013)
Effectiveness of Stem Cell Treatment for Adults With Ischemic Cardiomyopathy (The FOCUS Study)	NCT00824005	Transendocardial injections	Autologous bone marrow mononuclear cells	6 months	Safety	Did not improve LVESV, maximal oxygen consumption, or reversibility on SPECT	Cogle et al. (2014)
Stem Cell Study for Patients With Heart Disease	NCT00081913	Intramyocardial, transendocardial injections	Autologous CD34+ stem cells	12 months	Serious adverse events were evenly distributed. Safety	Efficacy parameters including angina frequency, nitroglycerine usage, exercise time	Losordo et al. (2007)
Stem Cell Therapy for Vasculogenesis in Patients With Severe Myocardial Ischemia	NCT00260338	Intramyocardial injection	Autologous bone marrow derived mesenchymal stromal cells (BMMSCs)	36 months	Excellent long-term safety	Improve symptoms Slows down disease progression	Mathiasen et al. (2013)
A Randomized Clinical Trial of Adipose-derived Stem Cells in Treatment of Non-Re-vascularizable Ischemic Myocardium	NCT00426868	Trans-endocardial injections	Adipose derived stem and regenerative cells (ADRCs)	36 months	Safe and feasible	No malignant arrhythmias Significant improvements in total left ventricular mass Improve wall motion score index Reduction in inducible ischemia	Perin et al. (2014)

(continued)

Table 10.1 (continued)

Trial name	Trial design				Results		References
	Code	Cell delivery	Cell sources	Transplantation timing after injury	Safety	Efficacy	
Cell Therapy for Coronary Heart Disease	NCT00289822	Intracoronary injection	Autologous ex vivo cultivated EPCs Autologous bone marrow progenitor cells	3 months	Adequate safety	Significant improvement in the left ventricular ejection fraction	De Rosa et al. (2013) and Leistner et al. (2012)
Stem Cell Injection to Treat Heart Damage During Open Heart Surgery	NCT01557543	Intracoronary infusion	Autologous blood-derived progenitor cells Autologous bone marrow-derived progenitor cells	4 months	No signs of an inflammatory response or malignant arrhythmias	Significant increase in myocardial viability in the infarct zone	Assmus et al. (2002) and Britten et al. (2003)
Cell-Wave Study: Combined Extracorporal Shock Wave Therapy and Intracoronary Cell Therapy in Chronic Ischemic Myocardium	NCT00326989	Shock wave-facilitated intracoronary administration of BMCs	Autologous bone marrow-derived mononuclear cells (BMCs)	4 months	Safety	Significant, albeit modest, improvement in LVEF	Leistner et al. (2012)

Safety and Efficacy Study of Intramyocardial Stem Cell Therapy in Patients With Dilated Cardiomyopathy (NOGA-DCM)	NCT01350310	Intramyocardial injection Intracoronary injection	Peripheral blood CD34+ stem cells	12 months	Safety	Improved left ventricular ejection fraction Reduced left ventricular dimensions Improved exercise capacity Reduced levels of NT-proBNP	Lezaic et al. (2015), Poglajen et al. (2014), and Vrtovec et al. (2013a, b)
Intramyocardial Transplantation of Bone Marrow Stem Cells in Addition to Coronary Artery Bypass Graft (CABG) Surgery (PERFECT)	NCT00950274	Intramyocardial injection	Peripheral blood CD34+ stem cells	6 months	Safety	Significant improvement in rest myocardial perfusion scores, LVEF, and 6-min walking distance	Donndorf et al. (2012)

lot with desirable results. These cells can differentiate into heart muscle cells, reduce the rate of programmed cell death, and increase capillary density (Chimenti et al. 2010). CDCs exist for a long time in infarction area and improve cardiac functions in a rat model of myocardial infarction (Carr et al. 2011).

Zhang et al. (2014) inhibited the activity of histone deacetylase 4 (HDAC4) in c-kit positive CSCs by HDAC4 siRNA. Their study involved a left anterior descending artery (LAD) ligation mouse model and showed that inhibition of HDAC4 promoted tissue regeneration and functional recovery (Zhang et al. 2014). Avolio et al. (2015) compared cardiac repair potential between saphenous vein derived pericytes (SVPs) and CSCs in an LAD ligation mouse model. In their study, transplantation of either SVPs or CSCs improved cardiac contractility after 14 days and reduced both the infarct size and fibrosis process. SVPs stimulated angiogenesis while CSCs promoted myocardial differentiation and mobilization of endogenous stem cells. The combination of 2 types of cells reduced infarct size, promoted proliferation, and enhanced angiogenesis, which helped repair the infarcted heart (Avolio et al. 2015).

Kim et al. (2015a) assessed myocardial viability and cell grafts following transplantation of human placental derived iPSCs, human placental derived c-kit positive cells, and human placental derived cells without selective c-kit. They also assessed the impact of each cell type on LAD ligation in a mouse model. Evaluating after transplantation showed that transplantation of human placental cells derived iPSCs transplantation yields the most effective results (Kim et al. 2015b).

Tang et al. (2015) identified an effective dose for CSC injection in a rat model of acute infarction. Their results showed that a dose of 0.3 million cells yielded good efficacy; moreover, doses of 0.75, 1.5, and 3 million cells significantly improved left ventricular function, thickness of the infarcted heart wall, end-systolic volume, end-diastolic volume, and ejection fraction (Tang et al. 2015).

Additionally, EPC transplantation has been shown to reduce inflammation and augment angiogenesis in mouse models of myocardial infarction (Tang et al. 2015). Park et al. (2011) transplanted EPCs in a mouse model of coronary ligation. The authors showed that factors related to lymph angiogenesis such as VEGF-C, VEGF-D, Lyve-1, and high-Prox-1 were expressed by the placebo group at a higher level than the EPC-transplanted group. Transplantation of EPCs reduced the formation of lymphatic vessels, which reduced inflammation and positively impacted the EPC transplanted mice (Park et al. 2011).

Cheng et al. (2012) studied the effect of implanted EPCs and erythropoietin (EPO) in a mouse model of cardiac ischemia. Results showed that EPO improves the efficiency of EPC transplantation. These positive effects were included enhancement of survival of transplanted EPCs and mobilization of endogenous EPCs (Cheng et al. 2012). Furthermore, Cheng et al. (2013) studied the effect of EPC injection directly into the infarct border zone after coronary ligation surgery in mice. Their results showed that EPCs improve cardiac functions, which are mediated via activation of VEGF-PI3K/Akt-eNOS pathway (Cheng et al. 2013).

Chen et al. (2013) overexpressed eNOS in EPCs and transplanted these cells into a mouse model of coronary ligation. Their results demonstrated that EPCs could improve angiogenesis and prevent heart functions from further deterioration

(Chen et al. 2013). Chang et al. (2013) transplanted peripheral blood-derived EPCs in a rat model of myocardial ischemia; at 4 weeks after transplantation, the expression of the proteins involved in the process of blood vessel growth, such as PDGF, VEGF, and FLK-1, was particularly high in the EPC transplanted group compared to the control group (Chang et al. 2013).

Kyu-Tae Kang et al. (2013) injected human endothelial colony forming cells (ECFCs) and mesenchymal progenitor cells (MPCs) into a rat model of coronary ligation. After 14 days, ECFCs were found in the transplanted tissue. After 4 months, the infarct size was significantly improved and blood vessels, created by ECFCs, were found in the heart tissue (Kang et al. 2013).

Moreover, Like et al. showed from their studies in 2005 that surviving transplanted cells in a dog heart were positive for c-kit, MDR1, and SCA-1 (Linke et al. 2005). Then, they demonstrated that IGF-1 and HGF injection into the area around the infarct, following coronary artery ligation, increased the number of progenitor cells (per square centimeter) in the infarct center by 11-fold and in the area around the infarct center by 16-fold. They also demonstrated that the number of cardiac muscle cells also increased, though to a much lesser extent, and that only 17 % dead of myocyte mass was re-established.

Takehara et al. used bFGF to enhance the effectiveness of human CDC implantation in a pig model of myocardial infarction (Kraitchman and Bulte 2008; Nussbaum et al. 2007). The results showed that there was a significant improvement in cardiac function and stability of the capillary system. In a recent study in pigs, CSCs were capable of reducing fibrosis, improving left ventricular function, and promoting revascularization (Bolli et al. 2011). Additionally, CSCs have been transplanted in pigs at 1 week (Yee et al. 2014), 2–3 weeks (Yee et al. 2014), and 8 weeks (Yee et al. 2014), immediately after infarction. Efficiency of treatment in the treated group was analyzed by magnetic resonance imaging (MRI), and compared to the placebo injection group at 2 months of intervention; MRI results showed improvement in heart function, new cardiac tissue formation, and infarct size reduction of 3.6 %.

To improve treatment efficiency, new methods have been implemented, including combination therapy with other cell types (Williams et al. 2013), combination with growth factors (Takehara et al. 2008), or magnetic targeting for stem cell delivery (Cheng et al. 2010). Recently, from studies in a pig model of myocardial infarction, Williams et al. showed that the regenerative potential of heart tissue was enhanced when a combination of BMSC and c-kit positive cells were transplanted (instead of just the individual cell types). Moreover, in the combined transplant group, not only did heart function improve twofold but transplanted cells also increased sevenfold after 4 weeks of treatment (Williams et al. 2013).

In a different approach, Cheng et al. used iron-labeled cells in order to improve transplantation efficiency. The iron-labeled cells were subjected to energy from 1.3T while injected into a rat model of myocardial infarction. After ensuring that the use of ferromagnetic force did not affect the viability of the cells, the authors evaluated the existence and impact of labeled CDCs after 24 h and after 3 weeks. Their results showed that after 3 weeks of implantation, the number of marked CDCs were three times higher than the number of unmarked CDCs. As well, heart structure and function were improved in the target cell transplantation group. There was a strong correlation between left ventricular ejection fraction (LVEF) and maintenance of cells (Cheng et al. 2010).

Chong et al. (2014) grafted hESC-cardiomyocytes in a pig-tailed macaque model of cardiac ischemia. Two weeks after transplantation, an electrical connection between the grafted cells and the local cells was formed. After 3 months of transplantation, heart function was well recovered, and while arrhythmia did occur, it was not fatal (Chong et al. 2014).

Kanazawa et al. (2015) infused CDCs into a pig model of coronary artery ligation/reperfusion after 30 min. After 24 h, the amount of Troponin I levels in the cell transplanted group was lower than that of the placebo's. Transfusion of CDCs was safe, feasible, and effective in protecting the heart; it also reduced the infarct size and prevented the acute remodeling process (Kanazawa et al. 2015).

Zhang et al. (2012) grafted EPCs in a pig model of myocardial infarction. After 14 days, the EPCs had a positive impact; they reduced heart failure processes, preserved heart function and left ventricular size, and inhibited the spread of the infarcted area. When using an Akt inhibitor, Ly294002, the positive impact of EPCs was abrogated in the pig model of myocardial infarction. However, use of angiogenesis-associated factors (Akt, Ang-1, bFGF, and IL-1) increased the efficiency of EPC transplantation. The results demonstrated that the positive impact of EPCs was related to Akt-mediated signaling pathway (Zhang et al. 2012).

Mitchell et al. (2013) transplanted peripheral blood-derived EPCs into a dog model of myocardial infarction, with or without reperfusion. Different time points were assessed: immediately, 1 week, and 4–5 weeks after infarction. Results showed that there were no significant differences between the various injection timelines. Cardiac rehabilitation was related to cell longevity. EPC transplantation was shown to be safe (Mitchell et al. 2013).

Jansen Of Lorkeers et al. (2015) conducted a comprehensive data analysis from 82 studies of cell-based therapy of heart ischemia. In the studies, there were over 1415 large animals treated and assessed—including 1141 pigs (from 67 studies), 64 dogs (from 5 studies), and 210 sheep (from 10 studies). Overall, the results showed that there was a significant improvement in left ventricular ejection fraction in these treated animals compared to placebo-treated animals. This improvement was similar between allograft and autograft therapies (Antanaviciute et al. 2015).

10.8.2 Clinical Trials

Results of recent clinical trials support the notion that stem cell therapy is safe and capable of repairing cardiac structure as well as restoring heart function (Karantalis et al. 2012; Williams and Hare 2011). The results of these stem cell studies have been interesting. The study by Perin et al. (2015) evaluated MSC allograft therapy in 45 patients with chronic heart failure (due to ischemia or non-ischemia). None of the patients showed symptoms of clinical immune responses; heart structure and function were improved at 3 years after transplantation (Perin et al. 2015). Musialek et al. (2015) transplanted MSCs derived from Wharton's Jelly membrane into 10 patients with acute myocardial infarction (Musialek et al. 2015). The results showed that the patients did not have signs of cardiac ischemia. There was no heart failure status and no other adverse events. Troponin level was reduced after 24 h of injection. Some cardiac arrhythmias occurred but were trivial (Musialek et al. 2015). Research on EPC infusion in patients with acute myocardial infarction have shown that EPC transplantation does not affect the density of proteins and leukocytes (Schachinger et al. 2004), and does not induce tumor formation even at 5 years after transplantation (Leistner et al. 2011). Moreover, different studies have demonstrated that CSCs present in the hearts of mammals, including humans, can be used to regenerate myocardial infarction in different ways; studies on human, using autologous CSCs, have shown good results (Makkar et al. 2012).

From the encouraging autologous cell transplantation results, two new clinical trials are underway: ALLSTAR and CAREMI. These trials use allogeneic cells for acute myocardial infarction. The preliminary results from a trial using allogeneic cardiac-derived cell have been discussed at the 63th American College of Cardiology Conference in March 2014. At this conference, preliminary results (followed for at least 3 months after injection) from the ALLSTAR phase I trial were reported. CDCs were infused into the heart through the coronary artery approximately 4 weeks and 12 months after the infarction. This trial was determined to be safe, and the results of the phase I study were encouraging. At 1 month after infusion, there was no acute myocarditis caused by the infusion, no deaths due to ventricular fibrillation or tachycardia, no sudden death, and no major adverse events to the heart. A phase II study was subsequently approved. For the CAREMI trial, dose escalation studies have just recently begun and the results from this trial are impending (Baez-Diaz et al. 2014; Crisostomo et al. 2014; Lauden et al. 2013).

Currently, three clinical trials involving CSC transplantation have been conducted to evaluate the safety and efficacy of transplantation. These trials include SCIPIO, CADUCEUS, and ALCADIA. All three trials showed that CDC implantation was safe. In the SCIPIO trial, there were 16 patients: 14 in the treatment group and 2 in the control group. After 4 months of treatment, the infarct area decreased 24 % in the treatment group and left ventricle ejection fraction was increased by 8.2 % (thus an improvement). The positive impact of SCIPIO remains to be assessed, even after over a year of treatment (Bolli et al. 2011). The CADUCEUS trial was conducted in 31 people, including 25 in the treatment group. Although there was no improvement in the left ventricle ejection fraction, scarring blocks were 12.3 % lower in the 12th month; as well, movement of the heart wall was also markedly improved (Malliaras et al. 2012). The ALCADIA trial was also a phase I clinical trial, but conducted in 6 patients. It was a small test, with no control group. While the results have not been published in full, it was observed at 6 months after transplantation that ejection fraction increased 12.1 %, infarct area decreased 3.3 %, and heart wall contraction improved.

The results of the above preclinical studies and clinical studies suggest that stem/progenitor cell transplantation is safe and effective for myocardial ischemia. They also suggest varied mechanisms of action elicited by stem/progenitor cells, such as direct differentiation into cardiac muscle cells and blood vessel cells, production of growth factors, and mobilization of local CSCs, as well as the proliferation and differentiation of grafted stem cells and endogenous stemiciency of Stem Cell Therapy for Ischemic Heart Dise cells.

10.9 Conclusion

Cardiovascular diseases encompass heart and blood vessel disease, and can be divided into two groups, one related to atherosclerosis and the other caused by other reasons. In particular, coronary artery disease is common worldwide and poses a high mortality risk. Currently, there are a number of treatments for this disease, such as lifestyle adjustments, medical therapy, stenting, and coronary artery bypass surgery. The most novel treatment is stem cell therapy, which has been studied for heart disease and shown to elicit some efficacy. Stem cells can be injected either directly into the heart muscle or transmitted into the body through the vascular system. Stem cells can also be cultured on scaffolds to form special "cell plates" for "pasting" into the damaged heart. Alternatively, stem cells can be packaged and transported to the damaged heart tissue. Researchers are also studying ways to create an artificial heart using stem cells derived from the patient's own heart and scaffolds consisting of animal or corpse hearts. In the near future, advances in cardiovascular disease therapy should gradually increase with the innovation of science and advancement of stem cells in medicine.

References

Alvarez-Dolado M, Pardal R, Garcia-Verdugo JM, Fike JR, Lee HO, Pfeffer K, Lois C, Morrison SJ, Alvarez-Buylla A (2003) Fusion of bone-marrow-derived cells with Purkinje neurons, cardiomyocytes and hepatocytes. Nature 425:968–973

Anderl JN, Robey TE, Stayton PS, Murry CE (2009) Retention and biodistribution of microspheres injected into ischemic myocardium. J Biomed Mater Res A 88:704–710

Antanaviciute I, Ereminiene E, Vysockas V, Rackauskas M, Skipskis V, Rysevaite K, Treinys R, Benetis R, Jurevicius J, Skeberdis VA (2015) Exogenous connexin43-expressing autologous skeletal myoblasts ameliorate mechanical function and electrical activity of the rabbit heart after experimental infarction. Int J Exp Pathol 96:42–53

Assmus B, Schachinger V, Teupe C, Britten M, Lehmann R, Dobert N, Grunwald F, Aicher A, Urbich C, Martin H et al (2002) Transplantation of progenitor cells and regeneration enhancement in acute myocardial infarction (TOPCARE-AMI). Circulation 106:3009–3017

Avolio E, Meloni M, Spencer HL, Riu F, Katare R, Mangialardi G, Oikawa A, Rodriguez-Arabaolaza I, Dang Z, Mitchell K et al (2015) Combined intramyocardial delivery of human pericytes and cardiac stem cells additively improves the healing of mouse infarcted hearts through stimulation of vascular and muscular repair. Circ Res 116:e81–e94

Baez-Diaz C, Crisostomo V, Maestre J, Garcia-Lindo M, Sun F, Casado JG, Palacios I, Nunes V, Sanchez-Margallo FM (2014) Safety and efficacy assessment of intracoronary delivery of porcine cardiac stem cells in a swine model of acute myocardial infarction: comparison of two different cell doses. J Am Coll Cardiol 63

Bain B (2001) Bone marrow aspiration. J Clin Pathol 54:657–663
Becker RC, Meade TW, Berger PB, Ezekowitz M, O'Connor CM, Vorchheimer DA, Guyatt GH, Mark DB, Harrington RA (2008) The primary and secondary prevention of coronary artery disease: American College of Chest Physicians Evidence-Based Clinical Practice Guidelines (8th Edition). Chest 133:776s–814s
Beitnes JO, Lunde K, Brinchmann JE, Aakhus S (2011) Stem cells for cardiac repair in acute myocardial infarction. Expert Rev Cardiovasc Ther 9:1015–1025
Bel A, Messas E, Agbulut O, Richard P, Samuel JL, Bruneval P, Hagege AA, Menasche P (2003) Transplantation of autologous fresh bone marrow into infarcted myocardium: a word of caution. Circulation 108(Suppl 1):Ii247–Ii252
Blum B, Benvenisty N (2008) The tumorigenicity of human embryonic stem cells. Adv Cancer Res 100:133–158
Boden WE, O'Rourke RA, Teo KK, Hartigan PM, Maron DJ, Kostuk WJ, Knudtson M, Dada M, Casperson P, Harris CL et al (2007) Optimal medical therapy with or without PCI for stable coronary disease. N Engl J Med 356:1503–1516
Bolli R, Chugh AR, D'Amario D, Loughran JH, Stoddard MF, Ikram S, Beache GM, Wagner SG, Leri A, Hosoda T et al (2011) Cardiac stem cells in patients with ischaemic cardiomyopathy (SCIPIO): initial results of a randomised phase 1 trial. Lancet (Lond) 378:1847–1857
Botta R, Gao E, Stassi G, Bonci D, Pelosi E, Zwas D, Patti M, Colonna L, Baiocchi M, Coppola S et al (2004) Heart infarct in NOD-SCID mice: therapeutic vasculogenesis by transplantation of human CD34+ cells and low dose CD34+KDR+ cells. FASEB J 18:1392–1394
Boukouaci W, Lauden L, Siewiera J, Dam N, Hocine HR, Khaznadar Z, Tamouza R, Borlado LR, Charron D, Jabrane-Ferrat N et al (2014) Natural killer cell crosstalk with allogeneic human cardiac-derived stem/progenitor cells controls persistence. Cardiovasc Res 104:290–302
Boyle AJ, Schulman SP, Hare JM, Oettgen P (2006) Is stem cell therapy ready for patients? Stem cell therapy for cardiac repair ready for the next step. Circulation 114:339–352
Bravata DM, Gienger AL, McDonald KM, Sundaram V, Perez MV, Varghese R, Kapoor JR, Ardehali R, Owens DK, Hlatky MA (2007) Systematic review: the comparative effectiveness of percutaneous coronary interventions and coronary artery bypass graft surgery. Ann Intern Med 147:703–716
Britten MB, Abolmaali ND, Assmus B, Lehmann R, Honold J, Schmitt J, Vogl TJ, Martin H, Schachinger V, Dimmeler S et al (2003) Infarct remodeling after intracoronary progenitor cell treatment in patients with acute myocardial infarction (TOPCARE-AMI): mechanistic insights from serial contrast-enhanced magnetic resonance imaging. Circulation 108:2212–2218
Burchfield JS, Dimmeler S (2008) Role of paracrine factors in stem and progenitor cell mediated cardiac repair and tissue fibrosis. Fibrogenesis Tissue Repair 1:4
Cantz T, Martin U (2010) Induced pluripotent stem cells: characteristics and perspectives. Adv Biochem Eng Biotechnol 123:107–126
Carr CA, Stuckey DJ, Tan JJ, Tan SC, Gomes RS, Camelliti P, Messina E, Giacomello A, Ellison GM, Clarke K (2011) Cardiosphere-derived cells improve function in the infarcted rat heart for at least 16 weeks--an MRI study. PLoS One 6:e25669
Chang ZT, Hong L, Wang H, Lai HL, Li LF, Yin QL (2013) Application of peripheral-blood-derived endothelial progenitor cell for treating ischemia-reperfusion injury and infarction: a preclinical study in rat models. J Cardiothorac Surg 8:33
Chen X, Gu M, Zhao X, Zheng X, Qin Y, You X (2013) Deterioration of cardiac function after acute myocardial infarction is prevented by transplantation of modified endothelial progenitor cells overexpressing endothelial NO synthases. Cellular Physiol Biochem 31:355–365
Cheng K, Li TS, Malliaras K, Davis DR, Zhang Y, Marban E (2010) Magnetic targeting enhances engraftment and functional benefit of iron-labeled cardiosphere-derived cells in myocardial infarction. Circ Res 106:1570–1581
Cheng Y, Hu R, Lv L, Ling L, Jiang S (2012) Erythropoietin improves the efficiency of endothelial progenitor cell therapy after myocardial infarction in mice: effects on transplanted cell survival and autologous endothelial progenitor cell mobilization. J Surg Res 176:e47–e55

Cheng Y, Jiang S, Hu R, Lv L (2013) Potential mechanism for endothelial progenitor cell therapy in acute myocardial infarction: Activation of VEGF- PI3K/Akte-NOS pathway. Ann Clin Lab Sci 43:395–401

Chimenti I, Smith RR, Li TS, Gerstenblith G, Messina E, Giacomello A, Marban E (2010) Relative roles of direct regeneration versus paracrine effects of human cardiosphere-derived cells transplanted into infarcted mice. Circ Res 106:971–980

Chong JJ, Yang X, Don CW, Minami E, Liu YW, Weyers JJ, Mahoney WM, Van Biber B, Cook SM, Palpant NJ et al (2014) Human embryonic-stem-cell-derived cardiomyocytes regenerate non-human primate hearts. Nature 510:273–277

Cogle CR, Wise E, Meacham AM, Zierold C, Traverse JH, Henry TD, Perin EC, Willerson JT, Ellis SG, Carlson M et al (2014) Detailed analysis of bone marrow from patients with ischemic heart disease and left ventricular dysfunction: BM CD34, CD11b, and clonogenic capacity as biomarkers for clinical outcomes. Circ Res 115:867–874

Condorelli G, Borello U, De Angelis L, Latronico M, Sirabella D, Coletta M, Galli R, Balconi G, Follenzi A, Frati G et al (2001) Cardiomyocytes induce endothelial cells to trans-differentiate into cardiac muscle: implications for myocardium regeneration. Proc Natl Acad Sci U S A 98:10733–10738

Crisostomo V, Baez-Diaz C, Maestre J, Garcia-Lindo M, Sun F, Casado JG, Rodriguez-Borlado L, Abad JL, Sanchez-Margallo FM (2014) Allogeneic cardiac stem cell administration for acute myocardial infarction. A timing experimental study in swine. J Am Coll Cardiol 63

Davani S, Marandin A, Mersin N, Royer B, Kantelip B, Herve P, Etievent JP, Kantelip JP (2003) Mesenchymal progenitor cells differentiate into an endothelial phenotype, enhance vascular density, and improve heart function in a rat cellular cardiomyoplasty model. Circulation 108(Suppl 1):Ii253–Ii258

Dawn B, Stein AB, Urbanek K, Rota M, Whang B, Rastaldo R, Torella D, Tang XL, Rezazadeh A, Kajstura J et al (2005) Cardiac stem cells delivered intravascularly traverse the vessel barrier, regenerate infarcted myocardium, and improve cardiac function. Proc Natl Acad Sci U S A 102:3766–3771

De Rosa S, Seeger FH, Honold J, Fischer-Rasokat U, Lehmann R, Fichtlscherer S, Schachinger V, Dimmeler S, Zeiher AM, Assmus B (2013) Procedural safety and predictors of acute outcome of intracoronary administration of progenitor cells in 775 consecutive procedures performed for acute myocardial infarction or chronic heart failure. Circ Cardiovasc Interv 6:44–51

Delling U, Lindner K, Ribitsch I, Julke H, Brehm W (2012) Comparison of bone marrow aspiration at the sternum and the tuber coxae in middle-aged horses. Can J Vet Res 76:52–56

Dib N, Michler RE, Pagani FD, Wright S, Kereiakes DJ, Lengerich R, Binkley P, Buchele D, Anand I, Swingen C et al (2005) Safety and feasibility of autologous myoblast transplantation in patients with ischemic cardiomyopathy: four-year follow-up. Circulation 112:1748–1755

Donndorf P, Kaminski A, Tiedemann G, Kundt G, Steinhoff G (2012) Validating intramyocardial bone marrow stem cell therapy in combination with coronary artery bypass grafting, the PERFECT Phase III randomized multicenter trial: study protocol for a randomized controlled trial. Trials 13:99

Durrani S, Konoplyannikov M, Ashraf M, Haider KH (2010) Skeletal myoblasts for cardiac repair. Regen Med 5:919–932

Eisenstein EL, Anstrom KJ, Kong DF, Shaw LK, Tuttle RH, Mark DB, Kramer JM, Harrington RA, Matchar DB, Kandzari DE et al (2007) Clopidogrel use and long-term clinical outcomes after drug-eluting stent implantation. JAMA 297:159–168

Fuentes T, Kearns-Jonker M (2013) Endogenous cardiac stem cells for the treatment of heart failure. Stem Cells Clon Adv Appl 6:1–12

Fukushima S, Coppen SR, Lee J, Yamahara K, Felkin LE, Terracciano CMN, Barton PJR, Yacoub MH, Suzuki K (2008) Choice of cell-delivery route for skeletal myoblast transplantation for treating post-infarction chronic heart failure in rat. PLoS One 3:e3071

George JC, Goldberg J, Joseph M, Abdulhameed N, Crist J, Das H, Pompili VJ (2008) Transvenous intramyocardial cellular delivery increases retention in comparison to intracoronary delivery in a porcine model of acute myocardial infarction. J Interv Cardiol 21:424–431

Gersh BJ, Simari RD, Behfar A, Terzic CM, Terzic A (2009) Cardiac cell repair therapy: a clinical perspective. Mayo Clin Proc 84:876–892

Ghostine S, Carrion C, Souza LC, Richard P, Bruneval P, Vilquin JT, Pouzet B, Schwartz K, Menasche P, Hagege AA (2002) Long-term efficacy of myoblast transplantation on regional structure and function after myocardial infarction. Circulation 106:I131–I136

Gnecchi M, Zhang Z, Ni A, Dzau VJ (2008) Paracrine mechanisms in adult stem cell signaling and therapy. Circ Res 103:1204–1219

Gong Y, Zhao Y, Li Y, Fan Y, Hoover-Plow J (2014) Plasminogen regulates cardiac repair after myocardial infarction through its noncanonical function in stem cell homing to the infarcted heart. J Am Coll Cardiol 63:2862–2872

Hagege AA, Marolleau JP, Vilquin JT, Alheritiere A, Peyrard S, Duboc D, Abergel E, Messas E, Mousseaux E, Schwartz K et al (2006) Skeletal myoblast transplantation in ischemic heart failure: long-term follow-up of the first phase I cohort of patients. Circulation 114:I108–I113

Haider H, Ye L, Jiang S, Ge R, Law PK, Chua T, Wong P, Sim EK (2004) Angiomyogenesis for cardiac repair using human myoblasts as carriers of human vascular endothelial growth factor. J. Mol Med (Berl) 82:539–549

Hale SL, Dai W, Dow JS, Kloner RA (2008) Mesenchymal stem cell administration at coronary artery reperfusion in the rat by two delivery routes: a quantitative assessment. Life Sci 83:511–515

Hare JM, Fishman JE, Gerstenblith G et al (2012) Comparison of allogeneic vs autologous bone marrow–derived mesenchymal stem cells delivered by transendocardial injection in patients with ischemic cardiomyopathy: The poseidon randomized trial. JAMA 308:2369–2379

Hata H, Matsumiya G, Miyagawa S, Kondoh H, Kawaguchi N, Matsuura N, Shimizu T, Okano T, Matsuda H, Sawa Y (2006) Grafted skeletal myoblast sheets attenuate myocardial remodeling in pacing-induced canine heart failure model. J Thorac Cardiovasc Surg 132:918–924

Hatzistergos KE, Quevedo H, Oskouei BN, Hu Q, Feigenbaum GS, Margitich IS, Mazhari R, Boyle AJ, Zambrano JP, Rodriguez JE et al (2010) Bone marrow mesenchymal stem cells stimulate cardiac stem cell proliferation and differentiation. Circ Res 107:913–922

Hirata Y, Sata M, Motomura N, Takanashi M, Suematsu Y, Ono M, Takamoto S (2005) Human umbilical cord blood cells improve cardiac function after myocardial infarction. Biochem Biophys Res Commun 327:609–614

Ichim TE, Solano F, Lara F, Rodriguez JP, Cristea O, Minev B, Ramos F, Woods EJ, Murphy MP, Alexandrescu DT et al (2010) Combination stem cell therapy for heart failure. Int Arch Med 3:5

Ieda M, Fu JD, Delgado-Olguin P, Vedantham V, Hayashi Y, Bruneau BG, Srivastava D (2010) Direct reprogramming of fibroblasts into functional cardiomyocytes by defined factors. Cell 142:375–386

Isner JM, Asahara T (1999) Angiogenesis and vasculogenesis as therapeutic strategies for postnatal neovascularization. J Clin Invest 103:1231–1236

Jansen Of Lorkeers SJ, Eding JE, Vesterinen HM, van der Spoel TI, Sena ES, Duckers HJ, Doevendans PA, Macleod MR, Chamuleau SA (2015) Similar effect of autologous and allogeneic cell therapy for ischemic heart disease: systematic review and meta-analysis of large animal studies. Circ Res 116:80–86

Kanazawa H, Tseliou E, Malliaras K, Yee K, Dawkins JF, De Couto G, Smith RR, Kreke M, Seinfeld J, Middleton RC et al (2015) Cellular postconditioning: allogeneic cardiosphere-derived cells reduce infarct size and attenuate microvascular obstruction when administered after reperfusion in pigs with acute myocardial infarction. Circ Heart Fail 8:322–332

Kang KT, Coggins M, Xiao C, Rosenzweig A, Bischoff J (2013) Human vasculogenic cells form functional blood vessels and mitigate adverse remodeling after ischemia reperfusion injury in rats. Angiogenesis 16:773–784

Karantalis V, Balkan W, Schulman IH, Hatzistergos KE, Hare JM (2012) Cell-based therapy for prevention and reversal of myocardial remodeling. Am J Physiol Heart Circ Physiol 303:H256–H270

Kehat I, Gepstein A, Spira A, Itskovitz-Eldor J, Gepstein L (2002) High-resolution electrophysiological assessment of human embryonic stem cell-derived cardiomyocytes: a novel in vitro model for the study of conduction. Circ Res 91:659–661

Khan M, Kutala VK, Vikram DS, Wisel S, Chacko SM, Kuppusamy ML, Mohan IK, Zweier JL, Kwiatkowski P, Kuppusamy P (2007) Skeletal myoblasts transplanted in the ischemic myocardium enhance in situ oxygenation and recovery of contractile function. Am J Physiol Heart Circ Physiol 293:H2129–H2139

Kim J-H, Joo HJ, Kim M, Choi S-C, Park C-Y, Lee JI, Hong SJ, Lim D-S (2015a) Abstract 15623: transplantation of adipose-derived stem cells sheet for accelerated neovascularization and engraftment in acute myocardial infarction rats. Circulation 132:A15623

Kim PJ, Mahmoudi M, Ge X, Matsuura Y, Toma I, Metzler S, Kooreman NG, Ramunas J, Holbrook C, McConnell MV et al (2015b) Direct evaluation of myocardial viability and stem cell engraftment demonstrates salvage of the injured myocardium. Circ Res 116:e40–e50

Kocher AA, Schuster MD, Szabolcs MJ, Takuma S, Burkhoff D, Wang J, Homma S, Edwards NM, Itescu S (2001) Neovascularization of ischemic myocardium by human bone-marrow-derived angioblasts prevents cardiomyocyte apoptosis, reduces remodeling and improves cardiac function. Nat Med 7:430–436

Kraitchman DL, Bulte JW (2008) Imaging of stem cells using MRI. Basic Res Cardiol 103:105–113

Kuzmenkin A, Liang H, Xu G, Pfannkuche K, Eichhorn H, Fatima A, Luo H, Saric T, Wernig M, Jaenisch R et al (2009) Functional characterization of cardiomyocytes derived from murine induced pluripotent stem cells in vitro. FASEB J 23:4168–4180

Lauden L, Boukouaci W, Borlado LR, Lopez IP, Sepulveda P, Tamouza R, Charron D, Al-Daccak R (2013) Allogenicity of human cardiac stem/progenitor cells orchestrated by programmed death ligand 1. Circ Res 112:451–464

Laurent LC, Ulitsky I, Slavin I, Tran H, Schork A, Morey R, Lynch C, Harness JV, Lee S, Barrero MJ et al (2011) Dynamic changes in the copy number of pluripotency and cell proliferation genes in human ESCs and iPSCs during reprogramming and time in culture. Cell Stem Cell 8:106–118

Leistner DM, Fischer-Rasokat U, Honold J, Seeger FH, Schachinger V, Lehmann R, Martin H, Burck I, Urbich C, Dimmeler S et al (2011) Transplantation of progenitor cells and regeneration enhancement in acute myocardial infarction (TOPCARE-AMI): final 5-year results suggest long-term safety and efficacy. Clinical Res Cardiol 100:925–934

Leistner DM, Seeger FH, Fischer A, Roxe T, Klotsche J, Iekushi K, Seeger T, Assmus B, Honold J, Karakas M et al (2012) Elevated levels of the mediator of catabolic bone remodeling RANKL in the bone marrow environment link chronic heart failure with osteoporosis. Circ Heart Fail 5:769–777

Lezaic L, Socan A, Poglajen G, Peitl PK, Sever M, Cukjati M, Cernelc P, Wu JC, Haddad F, Vrtovec B (2015) Intracoronary transplantation of CD34(+) cells is associated with improved myocardial perfusion in patients with nonischemic dilated cardiomyopathy. J Card Fail 21:145–152

Li RK, Jia ZQ, Weisel RD, Mickle DA, Zhang J, Mohabeer MK, Rao V, Ivanov J (1996) Cardiomyocyte transplantation improves heart function. Ann Thorac Surg 62:654–660, discussion 660–651

Linke A, Muller P, Nurzynska D, Casarsa C, Torella D, Nascimbene A, Castaldo C, Cascapera S, Bohm M, Quaini F et al (2005) Stem cells in the dog heart are self-renewing, clonogenic, and multipotent and regenerate infarcted myocardium, improving cardiac function. Proc Natl Acad Sci U S A 102:8966–8971

Lister R, Pelizzola M, Kida YS, Hawkins RD, Nery JR, Hon G, Antosiewicz-Bourget J, O/Malley R, Castanon R, Klugman S et al (2011) Hotspots of aberrant epigenomic reprogramming in human induced pluripotent stem cells. Nature 471:68–73

Losordo DW, Schatz RA, White CJ, Udelson JE, Veereshwarayya V, Durgin M, Poh KK, Weinstein R, Kearney M, Chaudhry M et al (2007) Intramyocardial transplantation of autologous CD34+ stem cells for intractable angina: a phase I/IIa double-blind, randomized controlled trial. Circulation 115:3165–3172

Lu LL, Liu YJ, Yang SG, Zhao QJ, Wang X, Gong W, Han ZB, Xu ZS, Lu YX, Liu D et al (2006) Isolation and characterization of human umbilical cord mesenchymal stem cells with hematopoiesis-supportive function and other potentials. Haematologica 91:1017–1026

Luepker RV (2011) Cardiovascular disease: rise, fall, and future prospects. Annu Rev Public Health 32:1–3

Ma N, Stamm C, Kaminski A, Li W, Kleine HD, Muller-Hilke B, Zhang L, Ladilov Y, Egger D, Steinhoff G (2005) Human cord blood cells induce angiogenesis following myocardial infarction in NOD/scid-mice. Cardiovasc Res 66:45–54

MacKenzie TC, Flake AW (2002) Human mesenchymal stem cells: insights from a surrogate in vivo assay system. Cells Tissues Organs 171:90–95

Madonna R, De Caterina R (2010) Adipose tissue: a new source for cardiovascular repair. J Cardiovasc Med (Hagerstown) 11:71–80

Maitra A, Arking DE, Shivapurkar N, Ikeda M, Stastny V, Kassauei K, Sui G, Cutler DJ, Liu Y, Brimble SN et al (2005) Genomic alterations in cultured human embryonic stem cells. Nat Genet 37:1099–1103

Makino S, Fukuda K, Miyoshi S, Konishi F, Kodama H, Pan J, Sano M, Takahashi T, Hori S, Abe H et al (1999) Cardiomyocytes can be generated from marrow stromal cells in vitro. J Clin Invest 103:697–705

Makkar RR, Lill M, Chen PS (2003) Stem cell therapy for myocardial repair: is it arrhythmogenic? J Am Coll Cardiol 42:2070–2072

Makkar RR, Smith RR, Cheng K, Malliaras K, Thomson LE, Berman D, Czer LS, Marban L, Mendizabal A, Johnston PV et al (2012) Intracoronary cardiosphere-derived cells for heart regeneration after myocardial infarction (CADUCEUS): a prospective, randomised phase 1 trial. Lancet (Lond) 379:895–904

Malliaras K, Li TS, Luthringer D, Terrovitis J, Cheng K, Chakravarty T, Galang G, Zhang Y, Schoenhoff F, Van Eyk J et al (2012) Safety and efficacy of allogeneic cell therapy in infarcted rats transplanted with mismatched cardiosphere-derived cells. Circulation 125:100–112

Malliaras K, Makkar RR, Smith RR, Cheng K, Wu E, Bonow RO, Marban L, Mendizabal A, Cingolani E, Johnston PV et al (2014) Intracoronary cardiosphere-derived cells after myocardial infarction: evidence of therapeutic regeneration in the final 1-year results of the CADUCEUS trial (CArdiosphere-Derived aUtologous stem CElls to reverse ventricUlar dySfunction). J Am Coll Cardiol 63:110–122

Marelli D, Desrosiers C, el-Alfy M, Kao RL, Chiu RC (1992) Cell transplantation for myocardial repair: an experimental approach. Cell Transplant 1:383–390

Marion RM, Strati K, Li H, Tejera A, Schoeftner S, Ortega S, Serrano M, Blasco MA (2009) Telomeres acquire embryonic stem cell characteristics in induced pluripotent stem cells. Cell Stem Cell 4:141–154

Mathiasen AB, Haack-Sorensen M, Jorgensen E, Kastrup J (2013) Autotransplantation of mesenchymal stromal cells from bone-marrow to heart in patients with severe stable coronary artery disease and refractory angina--final 3-year follow-up. Int J Cardiol 170:246–251

Mazo M, Arana M, Pelacho B, Prosper F (2012) Mesenchymal stem cells and cardiovascular disease: a bench to bedside roadmap. Stem Cells Int 2012:175979

Menasche P (2003) Cell transplantation in myocardium. Ann Thorac Surg 75:S20–S28

Menasche P, Alfieri O, Janssens S, McKenna W, Reichenspurner H, Trinquart L, Vilquin JT, Marolleau JP, Seymour B, Larghero J et al (2008) The myoblast autologous grafting in ischemic cardiomyopathy (MAGIC) trial: first randomized placebo-controlled study of myoblast transplantation. Circulation 117:1189–1200

Messina E, De Angelis L, Frati G, Morrone S, Chimenti S, Fiordaliso F, Salio M, Battaglia M, Latronico MV, Coletta M et al (2004) Isolation and expansion of adult cardiac stem cells from human and murine heart. Circ Res 95:911–921

Mirotsou M, Zhang Z, Deb A, Zhang L, Gnecchi M, Noiseux N, Mu H, Pachori A, Dzau V (2007) Secreted frizzled related protein 2 (Sfrp2) is the key Akt-mesenchymal stem cell-released paracrine factor mediating myocardial survival and repair. Proc Natl Acad Sci U S A 104:1643–1648

Mitchell AJ, Sabondjian E, Blackwood KJ, Sykes J, Deans L, Feng Q, Stodilka RZ, Prato FS, Wisenberg G (2013) Comparison of the myocardial clearance of endothelial progenitor cells injected early versus late into reperfused or sustained occlusion myocardial infarction. Int J Cardiovasc Imaging 29:497–504

Müller-Ehmsen J, Leor J, Kedes L, Peterson KL, Kloner RA (2004) Fetal and Neonatal Cardiomyocyte Transplantation for the Treatment of Myocardial Infarction. In: Dhalla NS, Rupp H, Angel A, Pierce GN (eds) Pathophysiology of cardiovascular disease. Springer US, Boston, pp 535–544

Murry CE, Field LJ, Menasche P (2005) Cell-based cardiac repair: reflections at the 10-year point. Circulation 112:3174–3183

Murry CE, Whitney ML, Reinecke H (2002) Muscle cell grafting for the treatment and prevention of heart failure. J Card Fail 8:S532–S541

Musialek P, Mazurek A, Jarocha D, Tekieli L, Szot W, Kostkiewicz M, Banys RP, Urbanczyk M, Kadzielski A, Trystula M et al (2015) Myocardial regeneration strategy using Wharton's jelly mesenchymal stem cells as an off-the-shelf 'unlimited' therapeutic agent: results from the Acute Myocardial Infarction First-in-Man Study. Postepy Kardiol Interwencyjnej 11:100–107.

Nagata H, Ii M, Kohbayashi E, Hoshiga M, Hanafusa T, Asahi M (2016) Cardiac adipose-derived stem cells exhibit high differentiation potential to cardiovascular cells in C57BL/6 mice. Stem Cells Transl Med 5:141–151

Narva E, Autio R, Rahkonen N, Kong L, Harrison N, Kitsberg D, Borghese L, Itskovitz-Eldor J, Rasool O, Dvorak P et al (2010) High-resolution DNA analysis of human embryonic stem cell lines reveals culture-induced copy number changes and loss of heterozygosity. Nat Biotechnol 28:371–377

Nussbaum J, Minami E, Laflamme MA, Virag JA, Ware CB, Masino A, Muskheli V, Pabon L, Reinecke H, Murry CE (2007) Transplantation of undifferentiated murine embryonic stem cells in the heart: teratoma formation and immune response. FASEB J 21:1345–1357

Oh H, Bradfute SB, Gallardo TD, Nakamura T, Gaussin V, Mishina Y, Pocius J, Michael LH, Behringer RR, Garry DJ et al (2003) Cardiac progenitor cells from adult myocardium: homing, differentiation, and fusion after infarction. Proc Natl Acad Sci U S A 100:12313–12318

Orlic D, Kajstura J, Chimenti S, Jakoniuk I, Anderson SM, Li B, Pickel J, McKay R, Nadal-Ginard B, Bodine DM et al (2001) Bone marrow cells regenerate infarcted myocardium. Nature 410:701–705

Ott HC, Matthiesen TS, Goh S-K, Black LD, Kren SM, Netoff TI, Taylor DA (2008) Perfusion-decellularized matrix: using nature's platform to engineer a bioartificial heart. Nat Med 14:213–221

Park JH, Yoon JY, Ko SM, Jin SA, Kim JH, Cho CH, Kim JM, Lee JH, Choi SW, Seong IW et al (2011) Endothelial progenitor cell transplantation decreases lymphangiogenesis and adverse myocardial remodeling in a mouse model of acute myocardial infarction. Exp Mol Med 43:479–485

Patel AN et al (2016) Ixmyelocel-T for patients with ischaemic heart failure: a prospective randomised double-blind trial. Lancet 387(10036):2412–2421

Pelacho B, Nakamura Y, Zhang J, Ross J, Heremans Y, Nelson-Holte M, Lemke B, Hagenbrock J, Jiang Y, Prosper F et al (2007) Multipotent adult progenitor cell transplantation increases vascularity and improves left ventricular function after myocardial infarction. J Tissue Eng Regen Med 1:51–59

Perin EC (2006) Stem cell therapy for cardiovascular disease. Tex Heart Inst J 33:204–208

Perin EC, Borow KM, Silva GV, DeMaria AN, Marroquin OC, Huang PP, Traverse JH, Krum H, Skerrett D, Zheng Y et al (2015) A phase II dose-escalation study of allogeneic mesenchymal precursor cells in patients with ischemic or nonischemic heart failure. Circ Res 117:576–584

Perin EC, Sanz-Ruiz R, Sanchez PL, Lasso J, Perez-Cano R, Alonso-Farto JC, Perez-David E, Fernandez-Santos ME, Serruys PW, Duckers HJ et al (2014) Adipose-derived regenerative cells in patients with ischemic cardiomyopathy: the PRECISE trial. Am Heart J 168:88.e82–95.e82

Perin EC, Silva GV, Assad JA, Vela D, Buja LM, Sousa AL, Litovsky S, Lin J, Vaughn WK, Coulter S et al (2008) Comparison of intracoronary and transendocardial delivery of allogeneic mesenchymal cells in a canine model of acute myocardial infarction. J Mol Cell Cardiol 44:486–495

Perin EC, Willerson JT, Pepine CJ, Henry TD, Ellis SG, Zhao DX, Silva GV, Lai D, Thomas JD, Kronenberg MW et al (2012) Effect of transendocardial delivery of autologous bone marrow mononuclear cells on functional capacity, left ventricular function, and perfusion in chronic heart failure: the FOCUS-CCTRN trial. JAMA 307:1717–1726

Planat-Benard V, Menard C, Andre M, Puceat M, Perez A, Garcia-Verdugo JM, Penicaud L, Casteilla L (2004) Spontaneous cardiomyocyte differentiation from adipose tissue stroma cells. Circ Res 94:223–229

Poglajen G, Sever M, Cukjati M, Cernelc P, Knezevic I, Zemljic G, Haddad F, Wu JC, Vrtovec B (2014) Effects of transendocardial CD34+ cell transplantation in patients with ischemic cardiomyopathy. Circ Cardiovasc Interv 7:552–559

Reinecke H, Poppa V, Murry CE (2002) Skeletal muscle stem cells do not transdifferentiate into cardiomyocytes after cardiac grafting. J Mol Cell Cardiol 34:241–249

Reinecke H, Zhang M, Bartosek T, Murry CE (1999) Survival, integration, and differentiation of cardiomyocyte grafts: a study in normal and injured rat hearts. Circulation 100:193–202

Rodrigues M, Griffith LG, Wells A (2010) Growth factor regulation of proliferation and survival of multipotential stromal cells. Stem Cell Res Ther 1:1–12

Rosenstrauch D, Poglajen G, Zidar N, Gregoric ID (2005) Stem Cell Therapy for Ischemic Heart Failure. Tex Heart Inst J 32:339–347

Schachinger V, Assmus B, Britten MB, Honold J, Lehmann R, Teupe C, Abolmaali ND, Vogl TJ, Hofmann WK, Martin H et al (2004) Transplantation of progenitor cells and regeneration enhancement in acute myocardial infarction: final one-year results of the TOPCARE-AMI Trial. J Am Coll Cardiol 44:1690–1699

Schächinger V, Erbs S, Elsässer A, Haberbosch W, Hambrecht R, Hölschermann H, Yu J, Corti R, Mathey DG, Hamm CW et al (2006) Intracoronary bone marrow–derived progenitor cells in acute myocardial infarction. N Engl J Med 355:1210–1221

Schachinger V, Erbs S, Elsasser A, Haberbosch W, Hambrecht R, Holschermann H, Yu J, Corti R, Mathey DG, Hamm CW et al (2006) Improved clinical outcome after intracoronary administration of bone-marrow-derived progenitor cells in acute myocardial infarction: final 1-year results of the REPAIR-AMI trial. Eur Heart J 27:2775–2783

Schuleri KH, Feigenbaum GS, Centola M, Weiss ES, Zimmet JM, Turney J, Kellner J, Zviman MM, Hatzistergos KE, Detrick B et al (2009) Autologous mesenchymal stem cells produce reverse remodelling in chronic ischaemic cardiomyopathy. Eur Heart J 30:2722–2732

Segers VFM, Lee RT (2008) Stem-cell therapy for cardiac disease. Nature 451:937–942

Serruys PW, Morice MC, Kappetein AP, Colombo A, Holmes DR, Mack MJ, Stahle E, Feldman TE, van den Brand M, Bass EJ et al (2009) Percutaneous coronary intervention versus coronary-artery bypass grafting for severe coronary artery disease. N Engl J Med 360:961–972

Sethe S, Scutt A, Stolzing A (2006) Aging of mesenchymal stem cells. Ageing Res Rev 5:91–116

Sherman W, Martens TP, Viles-Gonzalez JF, Siminiak T (2006) Catheter-based delivery of cells to the heart. Nat Clin Pract Cardiovasc Med 3(Suppl 1):S57–S64

Silva GV, Fernandes MR, Cardoso CO, Sanz RR, Oliveira EM, Jimenez-Quevedo P, Lopez J, Angeli FS, Zheng Y, Willerson JT et al (2011) A dosing study of bone marrow mononuclear cells for transendocardial injection in a pig model of chronic ischemic heart disease. Tex Heart Inst J 38:219–224

Siminiak T, Fiszer D, Jerzykowska O, Grygielska B, Rozwadowska N, Kałmucki P, Kurpisz M (2005) Percutaneous trans-coronary-venous transplantation of autologous skeletal myoblasts in the treatment of post-infarction myocardial contractility impairment: the POZNAN trial. Eur Heart J 26:1188–1195

Stamm C, Westphal B, Kleine HD, Petzsch M, Kittner C, Klinge H, Schumichen C, Nienaber CA, Freund M, Steinhoff G (2003) Autologous bone-marrow stem-cell transplantation for myocardial regeneration. Lancet (Lond) 361:45–46

Strauer BE, Brehm M, Zeus T, Gattermann N, Hernandez A, Sorg RV, Kogler G, Wernet P (2001) Intracoronary, human autologous stem cell transplantation for myocardial regeneration following myocardial infarction. Dtsch Med Wochenschr 126:932–938

Strauer BE, Brehm M, Zeus T, Kostering M, Hernandez A, Sorg RV, Kogler G, Wernet P (2002) Repair of infarcted myocardium by autologous intracoronary mononuclear bone marrow cell transplantation in humans. Circulation 106:1913–1918

Suzuki E, Fujita D, Takahashi M, Oba S, Nishimatsu H (2015) Adipose tissue-derived stem cells as a therapeutic tool for cardiovascular disease. World J Cardiol 7:454–465

Takehara N, Tsutsumi Y, Tateishi K, Ogata T, Tanaka H, Ueyama T, Takahashi T, Takamatsu T, Fukushima M, Komeda M et al (2008) Controlled delivery of basic fibroblast growth factor promotes human cardiosphere-derived cell engraftment to enhance cardiac repair for chronic myocardial infarction. J Am Coll Cardiol 52:1858–1865

Tambara K, Sakakibara Y, Sakaguchi G, Lu F, Premaratne GU, Lin X, Nishimura K, Komeda M (2003) Transplanted skeletal myoblasts can fully replace the infarcted myocardium when they survive in the host in large numbers. Circulation 108(Suppl 1):Ii259–Ii263

Tang XL, Rokosh G, Sanganalmath SK, Tokita Y, Keith MC, Shirk G, Stowers H, Hunt GN, Wu W, Dawn B et al (2015) Effects of intracoronary infusion of escalating doses of cardiac stem cells in rats with acute myocardial infarction. Circ Heart Fail 8:757–765

Tang XL, Rokosh G, Sanganalmath SK, Yuan F, Sato H, Mu J, Dai S, Li C, Chen N, Peng Y et al (2010) Intracoronary administration of cardiac progenitor cells alleviates left ventricular dysfunction in rats with a 30-day-old infarction. Circulation 121:293–305

Tao B, Cui M, Wang C, Ma S, Wu F, Yi F, Qin X, Liu J, Wang H, Wang Z et al (2015) Percutaneous intramyocardial delivery of mesenchymal stem cells induces superior improvement in regional left ventricular function compared with bone marrow mononuclear cells in porcine myocardial infarcted heart. Theranostics 5:196–205

Tateishi K, Ashihara E, Honsho S, Takehara N, Nomura T, Takahashi T, Ueyama T, Yamagishi M, Yaku H, Matsubara H et al (2007a) Human cardiac stem cells exhibit mesenchymal features and are maintained through Akt/GSK-3beta signaling. Biochem Biophys Res Commun 352:635–641

Tateishi K, Ashihara E, Takehara N, Nomura T, Honsho S, Nakagami T, Morikawa S, Takahashi T, Ueyama T, Matsubara H et al (2007b) Clonally amplified cardiac stem cells are regulated by Sca-1 signaling for efficient cardiovascular regeneration. J Cell Sci 120:1791–1800

Thompson CA, Nasseri BA, Makower J, Houser S, McGarry M, Lamson T, Pomerantseva I, Chang JY, Gold HK, Vacanti JP et al (2003) Percutaneous transvenous cellular cardiomyoplasty. A novel nonsurgical approach for myocardial cell transplantation. J Am Coll Cardiol 41:1964–1971

Thompson CA, Reddy VK, Srinivasan A, Houser S, Hayase M, Davila A, Pomerantsev E, Vacanti JP, Gold HK (2005) Left ventricular functional recovery with percutaneous, transvascular direct myocardial delivery of bone marrow-derived cells. J Heart Lung Transplant 24:1385–1392

Tran KV, Gealekman O, Frontini A, Zingaretti MC, Morroni M, Giordano A, Smorlesi A, Perugini J, De Matteis R, Sbarbati A et al (2012) The vascular endothelium of the adipose tissue gives rise to both white and brown fat cells. Cell Metab 15:222–229

Tse HF, Kwong YL, Chan JK, Lo G, Ho CL, Lau CP (2003) Angiogenesis in ischaemic myocardium by intramyocardial autologous bone marrow mononuclear cell implantation. Lancet (Lond) 361:47–49

van Ramshorst J, Bax JJ, Beeres SL, Dibbets-Schneider P, Roes SD, Stokkel MP, de Roos A, Fibbe WE, Zwaginga JJ, Boersma E et al (2009) Intramyocardial bone marrow cell injection for chronic myocardial ischemia: a randomized controlled trial. JAMA 301:1997–2004

Vrtovec B, Poglajen G, Lezaic L, Sever M, Domanovic D, Cernelc P, Socan A, Schrepfer S, Torre-Amione G, Haddad F et al (2013a) Effects of intracoronary CD34+ stem cell transplantation in nonischemic dilated cardiomyopathy patients: 5-year follow-up. Circ Res 112:165–173

Vrtovec B, Poglajen G, Lezaic L, Sever M, Socan A, Domanovic D, Cernelc P, Torre-Amione G, Haddad F, Wu JC (2013b) Comparison of transendocardial and intracoronary CD34+ cell transplantation in patients with nonischemic dilated cardiomyopathy. Circulation 128:S42–S49

Wang Y, Zhang Z, Chi Y, Zhang Q, Xu F, Yang Z, Meng L, Yang S, Yan S, Mao A et al (2013) Long-term cultured mesenchymal stem cells frequently develop genomic mutations but do not undergo malignant transformation. Cell Death Dis 4:e950

Wernig M, Meissner A, Foreman R, Brambrink T, Ku M, Hochedlinger K, Bernstein BE, Jaenisch R (2007) In vitro reprogramming of fibroblasts into a pluripotent ES-cell-like state. Nature 448:318–324

Wert Gd, Mummery C (2003) Human embryonic stem cells: research, ethics and policy. Hum Reprod 18:672–682

Williams AR, Hare JM (2011) Mesenchymal stem cells: biology, patho physiology, translational findings, and therapeutic implications for cardiac disease. Circ Res 109:923–940

Williams AR, Hatzistergos KE, Addicott B, McCall F, Carvalho D, Suncion V, Morales AR, Da Silva J, Sussman MA, Heldman AW et al (2013) Enhanced effect of combining human cardiac stem cells and bone marrow mesenchymal stem cells to reduce infarct size and to restore cardiac function after myocardial infarction. Circulation 127:213–223

Wollert KC, Drexler H (2010) Cell therapy for the treatment of coronary heart disease: a critical appraisal. Nat Rev Cardiol 7:204–215

Wollert KC, Meyer GP, Lotz J, Ringes-Lichtenberg S, Lippolt P, Breidenbach C, Fichtner S, Korte T, Hornig B, Messinger D et al (2004) Intracoronary autologous bone-marrow cell transfer after myocardial infarction: the BOOST randomised controlled clinical trial. Lancet (Lond) 364:141–148

Xu M, Uemura R, Dai Y, Wang Y, Pasha Z, Ashraf M (2007) In vitro and in vivo effects of bone marrow stem cells on cardiac structure and function. J Mol Cell Cardiol 42:441–448

Yasuda T, Weisel RD, Kiani C, Mickle DA, Maganti M, Li RK (2005) Quantitative analysis of survival of transplanted smooth muscle cells with real-time polymerase chain reaction. J Thorac Cardiovasc Surg 129:904–911

Ye L, Haider HK, Jiang S, Tan RS, Ge R, Law PK, Sim EKW (2007) Improved angiogenic response in pig heart following ischaemic injury using human skeletal myoblast simultaneously expressing VEGF165 and angiopoietin-1. Eur J Heart Fail 9:15–22

Yee K, Malliaras K, Kanazawa H, Tseliou E, Cheng K, Luthringer DJ, Ho CS, Takayama K, Minamino N, Dawkins JF et al (2014) Allogeneic cardiospheres delivered via percutaneous transendocardial injection increase viable myocardium, decrease scar size, and attenuate cardiac dilatation in porcine ischemic cardiomyopathy. PLoS One 9:e113805

Yeh ETH, Zhang S (2006) A novel approach to studying the transformation of human stem cells into cardiac cells in vivo. Can J Cardiol 22:66B–71B

Zhang LX, DeNicola M, Qin X, Du J, Ma J, Tina Zhao Y, Zhuang S, Liu PY, Wei L, Qin G et al (2014) Specific inhibition of HDAC4 in cardiac progenitor cells enhances myocardial repairs. Am J Physiol Cell Physiol 307:C358–C372

Zhang S, Zhao L, Shen L, Xu D, Huang B, Wang Q, Lin J, Zou Y, Ge J (2012) Comparison of various niches for endothelial progenitor cell therapy on ischemic myocardial repair: coexistence of host collateralization and Akt-mediated angiogenesis produces a superior microenvironment. Arterioscler Thromb Vasc Biol 32:910–923

Zhou Y, Singh AK, Hoyt RF Jr, Wang S, Yu Z, Hunt T, Kindzelski B, Corcoran PC, Mohiuddin MM, Horvath KA (2014) Regulatory T cells enhance mesenchymal stem cell survival and proliferation following autologous cotransplantation in ischemic myocardium. J Thorac Cardiovasc Surg 148:1131–1137; discussion 1117

Zhu WZ, Hauch KD, Xu C, Laflamme MA (2009) Human embryonic stem cells and cardiac repair. Transplant Rev (Orlando) 23:53–68

Zohlnhofer D, Ott I, Mehilli J, Schomig K, Michalk F, Ibrahim T, Meisetschlager G, von Wedel J, Bollwein H, Seyfarth M et al (2006) Stem cell mobilization by granulocyte colony-stimulating factor in patients with acute myocardial infarction: a randomized controlled trial. JAMA 295:1003–1010

Chapter 11
Myocardial Tissue Engineering: A 5 Year—Update

Marie-Noelle Giraud and Inês Borrego

11.1 Introduction

Since the end of the nineteenth century, cardiovascular diseases (CVDs) have become and remained the N°1 serial killer. Among CVDs, coronary artery disease (CAD) represents the main etiology. The acute coronary ischemia induces myocardial infarction (MI) that progressively develops toward a chronic phase and severe heart failure. During the last decades, the identification and early treatment of risk factors, aggressive medical and interventional strategies allowed to significantly impact cardiovascular mortality after acute MI. Despite tremendous progress in diagnosis, prevention, and treatments, the chronic phase and the associated alarming progression of heart failure emphases the need for new therapies. Currently, no curative treatment exists for patients that survived acute MI; consequently, these patients face an excess risk of further cardiovascular events. Current therapeutic modalities including medical (life-style, drugs, psychology, etc.) and interventional procedures (percutaneous or surgical coronary revascularization, ventricular assistance, and implantable cardiac defibrillators with or without resynchronization therapy) intend to lower cardiovascular mortality and delay the progression of chronic heart failure by reducing the left ventricular remodelling. These therapies efficiently improve the clinical outcome and the quality of life of patients suffering acute MI. Nevertheless, despite these significant technological advances, the morbidity and mortality due to the progression of heart failure is still growing (Gjesdal et al. 2011). Therefore, new therapeutic options that will stop the progression of the disease to heart failure and foster cardiac regeneration are eagerly awaited.

Following myocardial injury, the recently identified regenerative capacity of the myocardium appears rapidly overloaded and clearly insufficient to repair the

M.-N. Giraud (✉) • I. Borrego
Department of Medicine, Cardiology, University of Fribourg, Fribourg, Switzerland
e-mail: marie-noelle.giraud@unifr.ch

damaged muscle. Heart function is initially maintained due to a compensatory mechanism that involves hypertrophy of the myocardium. When this physiological remodelling is overwhelmed, a decompensating phase results in left ventricle dilatation, thinning of the wall, remodelling of the mitral annulus, and appearance of heart failure. The possibility to stop the progression of the physio-pathological remodelling and/or to stimulate in situ repair mechanism has gained increasing interest during the last two decades. Reparative cell-based therapy has then emerged to become a clinical reality (Tongers et al. 2011). Initially, pre-clinical investigations provided compelling evidence of the beneficial effects of stem cell transplantation, including heart function recovery, decrease in infarct size and ventricle dilation, and increase in vascular density. Clinical application had rapidly followed. Bone marrow-derived cells, such as mesenchymal stem cells (MSC), circulating progenitors, adipose tissue-derived stem cells, resident cardiac stem cells (CSCs), and skeletal muscle cells, have been injected into patients with acute and chronic MI. Clinical trials revealed beneficial outcome, but pointed out important drawbacks that impair treatment efficacy. Concerns have been raised on low ability of the injured myocardium to permit cell retention and survival. Indeed, the damaged myocardium represents a rather hostile environment, due to hypoxia as well as intense immunologic and inflammatory activities. Improved cell retention and optimal delivery approaches have then been actively investigated. An interesting solution relies on the concomitant delivery of cells and exogenous matrix into the myocardium or the epicardial implantation of an in vitro engineered tissue. These promising alternatives have been shown to improve cell retention (Hamdi et al. 2011; Karam et al. 2012). Furthermore, the engineered tissues allow the creation of an adequate microenvironment favorable for (1) an improved cell survival, (2) the development of the contractile function resulting from the development of an engineered construct that mimic the myocardial structure, (3) a potential electrical pacing and/or coupling with the host myocardium, and (4) the packing of cytokines and growth factors involved in the paracrine related cardiac regeneration mechanism (Hwang and Kloner 2010). The emerging field of cardiac reparative tissue-based therapy offers unprecedented opportunities. Early attempt dealing with injection of isolated cells within the myocardium or via intracoronary delivery has nowadays been substituted with engineered tissues-based therapy with the exception of the injection of reprogrammed cells (Steppich et al. 2016).

Constant progress in biomaterials processes has fostered researchers' focus to develop advanced cardiac biografts. Alternatively, engineered tissues were developed without matrix and composed of a stack of cell monolayers, named cell sheets obtained with thermoresponsive cell culture dishes (Bel et al. 2010; Miyagawa et al. 2010).

A large majority of the actual engineered tissue-based therapies aim at replacing, repairing, or regenerating the damaged myocardium. Nevertheless, the ultimate goal of creating an entire beating heart has recently moved forward: Guyette et al. decellularized human heart that was used as a functional matrix to create de novo beating organ following recellularization with induced pluripotent stem cells (iPSCs) (Guyette et al. 2016).

11.2 Cells

Different types of cells are being evaluated for their capacity of promoting cardiac repair and cardiac regeneration after myocardial infarction. In the past, cardiomyocytes (CM) of neonatal or fetal origins have been initially studied to validate the proof of concept of engineered tissue-based cardiac therapy. Nowadays, focuses have been redirected on clinically relevant cell types including CM derived from pluripotent stem cells (embryonic stem cells (ESCs), iPSCs) and adult stem cells (MSCs, skeletal myoblasts (SMs), adipose-derived stem cells, and CSCs).

11.2.1 Pluripotent Stem Cells Derived-Cardiomyocytes

The main advantage of using pluripotent stem cells, such as ESCs and iPSCs, relies on their high capacity for self-renewal and the possibility to direct their differentiation into functional CM. Their potential to replace the lost CM and directly contribute to the host myocardium contraction were demonstrated with human embryonic stem cell-derived cardiomyocytes (hESC-CMs) transplanted in a guinea pig model; transplanted cells developed electric coupling with intact native myocardium. Nevertheless, the electrical integration was shown to be heterogeneous in injured hearts (Shiba et al. 2012). Furthermore, Chong et al. provided an evidence of cellengraftment, remuscularization, and electromechanical synchronization two to seven weeks following injection of one billion hESC-CMs into the hearts of primates with MI injury (Chong et al. 2014).

Besides the electromechanical coupling of CM derived from pluripotent stem cells, the paracrine regulation for neovascularization in particular has also been explored. Indeed, the transplantation of non-CM derived cells, such as human iPSC-derived ECs (hiPSC-ECs) and smooth muscle cells (hiPSC-SMCs) into ischemic porcine myocardial tissue, contributed to improvements of the perfusion, wall stress, and cardiac performance (Gu et al. 2012; Xiong et al. 2011).

11.2.1.1 Pluripotent Stem Cell Derived-Cardiomyocytes Associated with Matrices

The proof of concept of feasibility and importance of pluripotent stem cells in cardiac regeneration has been followed by investigation associated cell and various matrices. When transplanted into rodent hearts with chronic MI, ESC-CM injected with chitosan hydrogel repopulated ischemic and necrotic regions of dysfunctional myocardium, improved contractile performance resulting in improved cardiac function and cell survival (Lu et al. 2009). In addition, fibrin scaffold loaded with human ESC-derived cardiac progenitors improved contractility and attenuated remodelling up to 4 months in a chronic immune-deficient rat model. Nevertheless, the authors reported no sustained cell engraftment (Bellamy et al. 2015). However, when fibrin patches were loaded with insulin growth factor (IGF)-encapsulated microspheres and three types of human iPSC-derived cells (CMs, endothelial cells, and smooth muscle cells), cell integration was improved in the porcine infarcted myocardium after 4 weeks in comparison to scaffold-free cell transplantation. Main outcomes were an improvement in ventricular function, a reduction in the infarct size and apoptosis, an increase in angiogenesis and a normalised myocardial metabolism. No reverse effects such as arrhythmias were recorded (Ye et al. 2014).

Investigations reporting the implantation of iPSC-derived cardiomyocytes (iPSC-CM) organized in cell sheets in a chronic MI rodent model confirmed an increase of the cardiac function, neovascularization, engraftment of the cells, and a short-term survival. Remodelling and fibrosis were also reduced (Chang et al. 2014; Higuchi et al. 2015; Kawamura et al. 2012; Masumoto et al. 2014; Matsuo et al. 2015). Interstingly, Masumoto et al. (2012) provide evidence of the paracrine effect of cell sheets composed of purified CM, EC and mural cell derived from mouse iPCs. Improvement of the systolic function and neovascularisation in a rat MI model were sustained after the loss of the implanted cells. This effect was mediated by the presence of VEGF secreted by iPCS derived CM.

Engraftment of the cell sheet has been further improved when gelatin hydrogel microspheres were inserted between each cell sheet composed of iPSC-CM, endothelial cells, and vascular mural cells. Functional capillary network was improved and therefore allowed to increase the number of cell sheets stacked (Matsuo et al. 2015).

Moreover, an upscaling study established the feasibility of implanting iPSC-CM cell sheets in a large animal model and confirmed cardiac function improvement (Kawamura et al. 2012).

Although teratocarcinoma formation is a major concerned when implanting pluripotent stem cells derived CM, improved protocol for differentiation as well as cell purification may reduce the risk (Masumoto and Yamashita, 2016). Interestingly, Kawamura et al. (Kawamura 2016) provided evidence that the growth of malignant tumors from induced pluripotent stem cell-derived cardiac tissue constructs was blocked by host immune response.

The proof of concept of the safety of transplanting ESC-CM was followed by the recent launch of a clinical trial phase I. ESC-CMs (Isl-1+ and SSEA-1+ cells) associated with a fibrin scaffold were delivered to the infarct area of a first patient suffering from severe heart failure. After 3 months, the patient showed MI symptoms improvement and a new-onset contractility was observed by echocardiography. There have been no complications such as arrhythmias, tumor formation, or immunosuppression-related adverse events (Menasché et al. 2015).

11.2.2 Adult Stem Cells and Progenitors

Adult stem cells, found in many tissues, including the heart, are undifferentiated cells that can renew themselves and are capable of differentiating into specialized cell types. These cells are limited in their differentiation potential and are located in niches.

11.2.2.1 Mesenchymal Stem Cells

Mesenchymal Stem Cells also named Mesenchymal Stromal Cells are currently the most frequently used adult stem cell in regenerative medicine. Their therapeutic potential relies on several properties. They can be easily isolated from bone marrow, cord blood, peripheral blood, and fat tissue and largely amplified. They can differentiate into various lineages including adipocytes, chondrocytes, and osteoblasts. However, their differentiation into CM has been largely controversial and stays marginal. Of particular interest, MSCs exhibit immune regulatory activities both in vitro and in vivo, which are mediated by complex mechanisms that inhibit the function of different immune cell subpopulations of the innate and adaptive immunity. Finally, paracrine regulation mediated via MSC-secreted factors may contribute to the myocardial repair and have been investigated; identified factors include vascular endothelial growth factor (VEGF), hepatocyte growth factor, stromal cell-derived factor-1a, interleukin-6, macrophage inhibitory factor, and monocyte chemoattractant protein-1.

MCSs have also been associated with different scaffolds. Seeded in polymeric scaffolds of alginate and chitosan, MCSs maintain their proliferation and paracrine activity (Ceccaldi et al. 2014). Silk fibroin/hyaluronic acid MSCs patches reduced apoptosis, significantly promoted neovascularization and stimulated the secretions of various paracrine factors such as VEGF (Chi et al. 2012). Modified MSCs with insulin-like growth factor-1 (IGF-1) gene, loaded into a fibrin patch, were transplanted into a porcine model of MI and showed beneficial outcomes (Li et al. 2015).

11.2.2.2 Skeletal Myoblasts

They originate from satellite cells, which reside beneath the muscle fiber basal lamina. These myogenic precursor cells present several advantages for cell-based therapy including the possibility to expand them in vitro due to their proliferative capacity and their tolerance for a prolonged ischemia. Nevertheless, direct injection of myoblast in MI has raised several concerns such as the poor engraftment of injected cells and the risk of an arrhythmia when injected cell is not functionally coupled with the host myocardium (Narita et al. 2013). These shortcomings were overcome when SM, organized in cell sheets were implanted at the surface of the injured heart. Using acute and chronic MI rodent models, enhanced cell survival and improved cardiac function have been confirmed. Numerous investigations provided evidence of reduced infarct size, decreased fibrosis, attenuated cardiomyocyte hypertrophy, and increased neovascular formation (Narita et al. 2013; Pätilä et al. 2015; Shudo et al. 2014; Tano et al. 2016). Sheet of elastin secreting SM improved the long-term cardiac function and reduced the left ventricle end-diastolic dimensions and remodelling (Uchinaka et al. 2012).

The safety of SM sheet implantation and the absence of arrhythmia have been established in a porcine MI model (Terajima et al. 2014). Furthermore, SM sheet improved both systolic and diastolic function of severely damaged canine heart, especially by controlling the collagen I/III balance (Shirasaka et al. 2016). Following pre-clinical and safety evaluation, investigations on myoblast implanted as a stack of cell sheets have reached clinical phase II (Imamura et al. 2016). LV ejection fraction and heart failure symptoms improved significantly in the treated group, during the 6-month follow-up ($p < 0.05$) (Imamura et al. 2016).

11.2.2.3 Adipose-Derived Stem Cells

Adipose tissue includes a heterogeneous mixture of MSCs, hematopoietic stem cells, and endothelial progenitor cells. The major clinical advantage of this type of cells is their availability. Investigations reporting the implantation of adipose-derived stem cells organized in cell sheets in acute and chronic MI, confirmed an increased survival and engraftment. In addition, the secretion of a wide array of angiogenic and anti-apoptotic factors have been demonstrated and induced beneficial effects on perfusion and function in myocardial infarction models (Ishida et al. 2015; Hamdi et al. 2011; Yeh et al. 2014). The cardioprotective factors including HGF and VEGF contributed to the attenuation of the infarct size, inflammation, and left ventricular remodelling (Imanishi et al. 2011).

The implantation of adipose-derived stem cells sheet in a porcine model of chronic heart failure improved the left ventricular ejection fraction (Ishida et al. 2015).

An overexpression of VEGF in adipose-derived stem cell sheets promoted cell survival under hypoxia in vitro. When evaluated in a rabbit MI model, the authors observed a reduction in infarct size, an improved cardiac function, the suppression of fibrosis and an enhanced blood vessel formation (Yeh et al. 2014).

Adiposite-derived mesenchymal stem cells embedded in platelet-rich fibrin scaffold demonstrated superior effects compared to a direct implantation, improved left ventricular performance, and reduced left ventricular remodelling (Chen et al. 2015; Sun et al. 2014).

11.2.2.4 Cardiac Stem Cells

Cardiosphere-derived cells (CDCs) or cardiac progenitor cells (CPCs) can be isolated from postnatal heart. Their ability to self-renew and differentiate in CM have driven investigations for cardiac repair (Hosoyama et al. 2015). Beside direct injection of the CSC the regenerative capacity of engineered tissues composed of CSC and CDC sheets has been also studied. Their implantation in rodent MI models confirmed a reduced accumulation of interstitial fibrosis and an improved cardiac function (Alshammary et al. 2013; Hosoyama et al. 2015). Therapeutic efficacy was increased using hypoxic preconditioning cell sheets that augmented the angiogenesis and reducing the fibrosis (Hosoyama et al. 2015).

When delivered with poly(L-lactic acid) (PLLA) scaffold, CSC and VEGF showed modest effects on angiogenesis and cardiomyogenesis in the acutely infarcted hearts. The authors reported a reduction in cardiac remodelling and enhanced global cardiac function (Chung et al. 2015).

11.3 Matrices: A Multifaceted Substrate for the Cells

The development of fully organized and potentially contractile tissues for myocardium replacement request the selection of biomaterials that allow cell guidance and differentiation, mimic cardiac mechanical properties, and t permit electromechanical integration. Elaborating cardiac biografts have led to numerous progresses in biomaterials. INvestigations focussed on the modulation of their mechanic-electrical properties, the optimization of the cell–matrix interactions and the promotion of the long-term survival and vascularization after the implanted biograft.

The first challenge is the fine-tuning of the mechanical properties. In general, the e-modulus of solid matrix-based biografts stays markedly higher compared with the myocardial one. The maximal passive young moduli of the heart ventricle vary from 40 to 200 kPa depending on the axial strain (Hu et al. 2003). A stiff biomaterial would prompt a girdling effect and consequently limit the LV dilation. But, the rigidity of the scaffold may also hinder optimal heart contractility. When assessing the potential girdling effect of the PCL fibrous scaffold, Guex et al, found that the epicardial implantation of the matrix alone failed to prevent the progressive heart function decrease in a rat model of MI. Only when the e-spun fibers of PCL were seeded with MSC, a stabilization in heart function was recorded 4 weeks post implantation (Guex et al. 2014a).

The cell response to biomaterials properties is of paramount for the development of functional tissue. The substrate stiffness and/or the topography influence the cell differentiation and proliferation (Curtis and Russell 2011; Guex et al. 2012, 2013a; Valles et al. 2015) as well as the regulation of collagen and fibronectin deposit (Flück et al. 2003). Proteins complex such as focal adhesion sites and integrins provide the connections for mechanochemical transduction to signalling cascades and downstream the cellular adaptation. For example, Marsano et al. (2012) demonstrated that modulating the elasticity of poly(glycerol sebacate) matrices positively correlated with the contractile function of engineered cardiac constructs.

In parallel, the creation of semi-conductive matrices is rapidly evolving for excitable tissues. Indeed, the cardiac interstitial matrices play an important role in the propagation of the signal within the interconnecting ventricular myocyte layers (Coghlan et al. 2006). Consequently, new electroactive scaffolds have been developed. As examples, gold nanowire impregnated alginate scaffolds permitted electrical conductance and promoted CM connectivity through an increase in connexin 43 expression (Dvir et al. 2012). Electroactive polymer or eGel composed of electroactive nanoparticles greatly promoted excitable biografts development (Wang et al. 2016).

Finally, functionalization of the matrices and in particular the chemical surface modification of the scaffold and the addition of various molecules are necessary to respond to the complexity of muscle regeneration. Different approaches have been investigated. For examples Guex et al used a plasma coating process to enrichment with oxygen an electrospun polycaprolactone (PCL) fibrous matrix (Guex et al. 2012). This process modified the physical properties of the matrix and resulted in higher hydrophilicity and stability. The functionalized matrix presented a suitable environment for cell growth. It was demonstrated that oxygen enrichment significantly favored adhesion, orientation, and differentiation of SMs as well as bone marrow MSCs (Guex et al. 2012. Guex et al. 2014a). MSC adhered and spread on the matrix, producing a homogeneous cell layer. In addition, oxygen surface enrichment allowed the culture of ESC derived-cardiomyocytes, the cells maintained their cardiomyogenic phenotype with striated α-actinin mature sarcomeric structures and spontaneously beating areas, indicating of a mature differentiation into cardiomyocytes (Guex et al. 2013b).

Furthermore, active molecules (such as VEGF, IGF-1, FGF-2, thymosin β4, HGF, and FSLT1) have been added to the scaffolds in order to modify the microenvironment of the implanted cells, to stimulate cells survival and to promote stem cells recruitment as well as angiogenesis. This approach often provided a local delivery of the molecule and foster cardiac repair (Hwang and Kloner 2010; Segers and Lee 2011; Giraud et al. 2012; Guex et al. 2014b). In particular, collagen patches were dehydrated by compression and loaded with recombinant active molecules such as FSTL1 or conditioned medium. When suture on the surface of infarcted myocardium in mice, the authors documented a remarkably cardiac function recovery with a treatment-induced complete fractional shortening level reestablishment after 3 months. Interestingly, following implantation, FSTL1 loaded collagen patch promoted CM proliferation within the artificial matrix (Wei et al. 2015).

11.3.1 Solid Matrix: The Example of Electrospun Fibers

Solid matrices can be porous or fibrous with versatile pore size, fiber/pore diameters, and topographies. Processing matrices with electrospinning had become a primordial technique to engineered substrates for cardiac tissue. Electrospinning produces submicron fibers using an electrostatically driven jet of a polymer solution or a melt. A polymer solution is forced through a syringe needle where a high voltage (5–60 kV) is applied. The e-spun fibers are collected on a ground support and form of non-woven mats. E-spun matrices mimic the size and fibrous pattern of biologic extracellular matrices found in tissues. A large number of biomaterials composed of single or blended polymers have been electrospun (Zhao et al. 2015), and assessed for cardiac tissue. The versatility of this technique relies on the possibility to control the fiber composition, diameters and their alignment. Furtheremore, the simplicity of its implementation has contributed to the success of this technique. In addition, more sophisticated fibers have been developed such as a core/shell, electro-conductive, or functionalised fibers.

The cells are generally seeded on the surface of the e-spun scaffold and form a monolayer. A major drawback of this approach is the limited number of cells that can infiltrate the scaffold; efforts have been undertaken to control and increase the size of the pore of the fibrous matrices and improve the seeding method in order to develop more effective scaffold. As an alternative, cell electrospinning allowing 3D deposition of both cells and fibers have recently provide successful engineered cardiac like tissue with neonatal cardiomyocytes (Ehler and Jayasinghe 2014).

Interestingly, electrospinning allows incorporation of bioactive agents (proteins, enzymes, silver, etc.). The addition of cytokines or growth factor into the fiber may contribute to improve the cell survival, stem cell homing, or the vascularization of the patch. For example, fibronectin-immobilized PCL e-spun nanofibers allowed survival of umbilical-cord-blood-derived MSC in a rat MI model, induced cardiac function partial recovery, and reduced infarct (Kang et al. 2014). Interestingly, the authors provided evidence that fibronectin immobilization at the surface of the scaffold induced changes in the expression levels of genes involved in the paracrine regulation of cardiac repair including stem cell homing, angiogenesis and protection against apoptosis, inflammation, and fibrosis.

11.3.2 Fibrin as a Noteworthy Choice for Hydrogel Type Matrices

Hydrogels present implantable and injectable forms. The injectable hydrogel presents a liquid phase that becomes solid under temperature/pH changes or mix of two components and can be directly applied in combination of cells on the surface or within the myocardium. Hydrogels allow the control of the physical and chemical microenvironment of implanted cells and would determine their functionality (Li and Guan 2011). The hydrogel would modulate the reparative pathways involved in cell survival and gene expression after myocardial injury (Dobaczewski et al. 2010).

Among them, gel type scaffold such as fibrin formed from thrombin and fibrinogen has recently gained increasing interest for cardiac application (Roura et al. 2016). Other formulation of fibrin such as micro-particles has enlarged the options for fibrin-based cell and active molecules delivery. Fibrin is a natural matrix that is initially involved in physiological repair mechanisms. Following tissue injury, fibrinogen and plasma fibronectin extravagated from the vascular network provide a fibrin-based provisional matrix that plays a scaffolding role in inflammation regulation.

For example, fibrin patch-based transplantation of hESC-derived vascular cells in an MI porcine model allowed significant engraftment of cells and increased neovascularization and improved LV contractile function (Xiong et al. 2011). Fibrin has recently been associated with a large selection of cells such as endothelial progenitors (Atluri et al. 2014), ESC derived cardiac progenitors (Bellamy et al. 2015), MSC derived from bone marrow (Li et al. 2015), umbilical cord blood (Roura et al. 2015), or adipocyte tissue (Chen et al. 2015; Sun et al. 2014). Although, fibrin or plasma rich fibrin-based cardiac constructs have been successfully investigated in small and large animal model for cardiac salvage (Barsotti et al. 2011), clinical trials have not yet been initiated. Nevertheless, to date, optimal condition for cardiac biografts implantation remains undefined. The rapidly biodegradable fibrin may improve cell survival; however, the timing for the paracrine regulation is unknown, questions such as the time of implantation and the required minimal duration of cell survival stay open.

On the other hand, controlled delivery of bioactive molecules within fibrin has gained increasing interest. Recently, Ye et al. demonstrated that fibrin combined with encapsulated IGF-1 enhanced engraftment of induced pluripotent stem cell-derived cardiovascular cell populations injected into the injured myocardium of swine (Ye et al. 2014).

11.4 Conclusion

Experimental investigations of engineered cardiac tissue have provided a real enthusiasm. However, only a few clinical trials with engineered tissue have been so far initiated. Nevertheless, the numerous challenges such as the optimal timing for graft delivery, the minimal duration for cell survival necessary to initiate paracrine regulation of cardiac repair, and the identification of the mechanism of action should be carefully investigated to avoid the skepticism raised in the past from clinical trials using cell injection.

Acknowledgement This work was supported by the Swiss National Foundation [SNF 310030-149986] and the University of Fribourg.

References

Alshammary S, Fukushima S, Miyagawa S et al (2013) Impact of cardiac stem cell sheet transplantation on myocardial infarction. Surg Today 43:970–976

Atluri P, Miller JS, Emery RJ, Hung G, Trubelja A, Cohen JE, Lloyd K1, Han J1, Gaffey AC, MacArthur JW, Chen CS, Woo YJ (2014) Tissue-engineered, hydrogel-based endothelial progenitor cell therapy robustly revascularizes ischemic myocardium and preserves ventricular function. J Thorac Cardiovasc Surg 148(3):1090–1097; discussion 1097–1098. doi:10.1016/j.jtcvs.2014.06.038. Epub 2014 Jun 28

Barsotti MC, Felice F, Balbarini A, Di Stefano R (2011) Fibrin as a scaffold for cardiac tissue engineering. Biotechnol Appl Biochem 58(5):301–310

Bel A, Planat-Bernard V, Saito A, Bonnevie L, Bellamy V, Sabbah L et al (2010) Composite cell sheets: a further step toward safe and effective myocardial regeneration by cardiac progenitors derived from embryonic stem cells. Circulation 122(11 Suppl 1)

Bellamy V, Vanneaux V, Bel A et al (2015) Long-term functional benefits of human embryonic stem cell-derived cardiac progenitors embedded into a fibrin scaffold. J Hear Lung Transplant 34:1198–1207

Ceccaldi C, Bushkalova R, Alfarano C et al (2014) Evaluation of polyelectrolyte complex-based scaffolds for mesenchymal stem cell therapy in cardiac ischemia treatment. Acta Biomater 10:901–911

Chang D, Wen Z, Wang Y et al (2014) Ultrastructural features of ischemic tissue following application of a bio-membrane based progenitor cardiomyocyte patch for myocardial infarction repair. PLoS One 9:1–8

Chen YL, Sun CK, Tsai TH et al (2015) Adipose-derived mesenchymal stem cells embedded in platelet-rich fibrin scaffolds promote angiogenesis, preserve heart function, and reduce left ventricular remodeling in rat acute myocardial infarction. Am J Transl Res 7:781–803

Chi NH, Yang MC, Chung TW et al (2012) Cardiac repair achieved by bone marrow mesenchymal stem cells/silk fibroin/hyaluronic acid patches in a rat of myocardial infarction model. Biomaterials 33:5541–5551

Chong JJH, Yang X, Don CW et al (2014) Human embryonic stem cell-derived cardiomyocytes regenerate non-human primate hearts james. Nature 510:273–277

Chung HJ, Kim JT, Kim HJ et al (2015) Epicardial delivery of VEGF and cardiac stem cells guided by 3-dimensional PLLA mat enhancing cardiac regeneration and angiogenesis in acute myocardial infarction. J Control Release 205:218–230

Coghlan HC, Coghlan AR, Buckberg GD, Cox JL (2006) "The electrical spiral of the heart": its role in the helical continuum. The hypothesis of the anisotropic conducting matrix. Eur J Cardio Thoracic Surg 29(Suppl 1):S178–S187

Curtis MW, Russell B (2011) Micromechanical regulation in cardiac myocytes and fibroblasts: implications for tissue remodeling. Pflugers Arch 462(1):105–117

Dobaczewski M, Gonzalez-Quesada C, Frangogiannis NG (2010) The extracellular matrix as a modulator of the inflammatory and reparative response following myocardial infarction. J Mol Cell Cardiol 48(3):504–511

Dvir T, Timko BP, Brigham MD, Naik SR, Sandeep S, Levy O et al (2012) Nanowired three dimensional cardiac patches. Nat Nanotechnol 6(11):720–725

Ehler E, Jayasinghe SN (2014) Cell electrospinning cardiac patches for tissue engineering the heart. Analyst 139(18):4449–4452

Flück M, Giraud MN, Tunç V, Chiquet M (2003) Tensile stress-dependent collagen XII and fibronectin production by fibroblasts requires separate pathways. Biochim Biophys Acta 1593(2–3):239–248

Giraud MN, Guex AG, Tevaearai HT (2012) Cell therapies for heart function recovery: focus on myocardial tissue engineering and nanotechnologies. Cardiol Res Pract 2012:971614

Gjesdal O, Bluemke DA, Lima JA (2011) Cardiac remodeling at the population level--risk factors, screening, and outcomes. Nat Rev Cardiol 8(12):673–685

Gu M, Nguyen PK, Lee AS et al (2012) Microfluidic single cell analysis show porcine induced pluripotent stem cell-derived endothelial cells improve myocardial function by paracrine activation mingxia. Circ Res 111:882–893

Guex AG, Kocher FM, Fortunato G, Körner E, Hegemann D, Carrel TP et al (2012) Fine-tuning of substrate architecture and surface chemistry promotes muscle tissue development. Acta Biomater 8(4):1481–1489

Guex AG, Birrer DL, Fortunato G, Tevaearai HT, Giraud M-N (2013a) Anisotropically oriented electrospun matrices with an imprinted periodic micropattern: a new scaffold for engineered muscle constructs. Biomed Mater 8(2):021001

Guex AG, Romano F, Marcu IC, Tevaearai HT, Ullrich ND, Giraud M-N (2013b) Culture of Cardiogenic Stem Cells on PCL-Scaffolds: Towards the Creation of Beating Tissue Constructs. IASTED Conference Proceedings, Vol. 791

Guex AG, Frobert A, Valentin J, Fortunato G, Hegemann D, Cook S et al (2014a) Plasma-functionalized electrospun matrix for biograft development and cardiac function stabilization. Acta Biomater 10(7):2996–3006

Guex AG, Hegemann D, Giraud MN, Tevaearai HT, Popa AM, Rossi RM et al (2014b) Covalent immobilisation of VEGF on plasma-coated electrospun scaffolds for tissue engineering applications. Colloids Surf B Biointerfaces 123:724–733

Guyette JP, Charest J, Mills RW, Jank B, Moser PT, Gilpin SE et al (2016) Bioengineering human myocardium on native extracellular matrix. Circ Res 118(1):56–72

Hamdi H, Planat-Benard V, Bel A et al (2011) Epicardial adipose stem cell sheets results in greater post-infarction survival than intramyocardial injections. Cardiovasc Res 91:483–491

Higuchi T, Miyagawa S, Pearson JT et al (2015) Functional and electrical integration of induced pluripotent stem cell-derived cardiomyocytes in a myocardial infarction rat heart. Cell Transplant 24:2479–2489

Hosoyama T, Samura M, Kudo T et al (2015) Cardiosphere-derived cell sheet primed with hypoxia improves left ventricular function of chronically infarcted heart. Am J Transl Res 7:2738–2751

Hu Z, Metaxas D, Axel L (2003) In vivo strain and stress estimation of the heart left and right ventricles from MRI images. Med Image Anal 7(4):435–444

Hwang H, Kloner RA (2010) Improving regenerating potential of the heart after myocardial infarction: factor-based approach. Life Sci 86(13–14):461–472

Imamura T, Kinugawa K, Sakata Y et al (2016) Improved clinical course of autologous skeletal myoblast sheet (TCD-51073) transplantation when compared to a propensity score-matched cardiac resynchronization therapy population. J Artif Organs 19:80–86

Imanishi Y, Miyagawa S, Maeda N et al (2011) Induced adipocyte cell-sheet ameliorates cardiac dysfunction in a mouse myocardial infarction model: a novel drug delivery system for heart failure. Circulation 124:10–17

Ishida O, Hagino I, Nagaya N et al (2015) Adipose-derived stem cell sheet transplantation therapy in a porcine model of chronic heart failure. Transl Res 165:631–639

Kang B-J, Kim H, Lee SK, Kim J, Shen Y, Jung S et al (2014) Umbilical cord blood-derived mesenchymal stem cells seeded onto fibronectin-immobilized PCL nanofiber improve the cardiac function. Acta Biomater 10:3007–3017

Karam JP, Muscari C, Montero-Menei CN (2012) Combining adult stem cells and polymeric devices for tissue engineering in infarcted myocardium. Biomaterials 33(23):5683–5695

Kawamura M, Miyagawa S, Miki K et al (2012) Feasibility, safety, and therapeutic efficacy of human induced pluripotent stem cell-derived cardiomyocyte sheets in a porcine ischemic cardiomyopathy model. Circulation 126:29–37

Kawamura A, Miyagawa S, Fukushima S et al (2016) Teratocarcinomas arising from allogeneic induced pluripotent stem cell-derived cardiac tissue constructs provoked host immune rejection in mice. Sci Rep 6:1–13

Li Z, Guan J (2011) Hydrogels for cardiac tissue engineering. Polym (Basel) 3(2):740–761

Li J, Zhu K, Yang S et al (2015) Fibrin patch-based insulin-like growth factor-1 gene-modified stem cell transplantation repairs ischemic myocardium. Exp Biol Med 240:585–592

Lu W-N, Lü S-H, Wang H-B et al (2009) Functional improvement of infarcted heart by co-injection of embryonic stem cells with temperature-responsive chitosan hydrogel. Tissue Eng Part A 15:1437–1447

Marsano A, Maidhof R, Wan LQ, Wang Y (2012) NIH Public Access 26(5):1382–1390

Masumoto H, Yamashita JK (2016) Exploiting human iPS cell-derived cardiovascular cell populations toward cardiac regenerative therapy. Stem Cell Transl Invest 3:e1226

Masumoto H, Matsuo T, Yamamizu K, Uosaki H, Narazaki G, Katayama S, Marui A, Shimizu T, Ikeda T, Okano T et al (2012) Pluripotent stem cell-engineered cell sheets re-assembled with defined cardiovascular populations ameliorate reduction in infarct heart function through cardiomyocyte-mediated neovascularization. Stem Cells 30:1196–1205

Masumoto H, Ikuno T, Takeda M et al (2014) Human iPS cell-engineered cardiac tissue sheets with cardiomyocytes and vascular cells for cardiac regeneration. Sci Rep 4:1–7

Matsuo T, Masumoto H, Tajima S et al (2015) Efficient long-term survival of cell grafts after myocardial infarction with thick viable cardiac tissue entirely from pluripotent stem cells. Sci Rep 5:1–14

Menasché P, Vanneaux V, Hagege A et al (2015) Human embryonic stem cell-derived cardiac progenitors for severe heart failure treatment: first clinical case report. Eur Heart J 36:2011–2017

Miyagawa S, Saito A, Sakaguchi T, Yoshikawa Y, Yamauchi T, Imanishi Y et al (2010) Impaired myocardium regeneration with skeletal cell sheets--a preclinical trial for tissue-engineered regeneration therapy. Transplantation 90(4):364–372

Narita T, Shintani Y, Ikebe C et al (2013) The use of scaffold-free cell sheet technique to refine mesenchymal stromal cell-based therapy for heart failure. Mol Ther 21:860–867

Pätilä T, Miyagawa S, Imanishi Y et al (2015) Comparison of arrhythmogenicity and proinflammatory activity induced by intramyocardial or epicardial myoblast sheet delivery in a rat model of ischemic heart failure. PLoS One 10:1–15

Roura S, Soler-Botija C, Bagó JR, Llucià-Valldeperas A, Férnandez MA, Gálvez-Montón C, Prat-Vidal C, Perea-Gil I, Blanco J, Bayes-Genis A (2015) Postinfarction functional recovery driven by a three-dimensional engineered fibrin patch composed of human umbilical cord blood-derived mesenchymal stem cells. Stem Cells Transl Med 4(8):956–966

Roura S, Gálvez-Montón C, Bayes-Genis A (2016) Fibrin, the preferred scaffold for cell transplantation after myocardial infarction? An old molecule with a new life. J Tissue Eng Regen Med. doi:10.1002/term.2129 [Epub ahead of print]

Segers VF, Lee RT (2011) Biomaterials to enhance stem cell function in the heart. Circ Res 109(8):910–922

Shiba Y, Fernandes S, Zhu W et al (2012) hESC-derived cardiomyocytes electrically couple and suppress arrhythmias in injured hearts. Nature 489:322–325

Shirasaka T, Miyagawa S, Fukushima S et al (2016) Skeletal myoblast cell sheet implantation ameliorates both systolic and diastolic cardiac performance in canine dilated cardiomyopathy model. Transplantation 100:295–302

Shudo Y, Miyagawa S, Ohkura H et al (2014) Addition of mesenchymal stem cells enhances the therapeutic effects of skeletal myoblast cell-sheet transplantation in a rat ischemic cardiomyopathy model. Tissue Eng Part A 20:728–739

Steppich B1, Hadamitzky M, Ibrahim T, Groha P, Schunkert H, Laugwitz KL, Kastrati A, Ott I (2016) Regenerate vital myocardium by vigorous activation of bone marrow stem cells (REVIVAL-2) study investigators. Stem cell mobilisation by granulocyte-colony stimulating factor in patients with acute myocardial infarction. Long-term results of the REVIVAL-2 trial. Thromb Haemost 115(4):864–868

Sun CK, Zhen YY, Leu S et al (2014) Direct implantation versus platelet-rich fibrin-embedded adipose-derived mesenchymal stem cells in treating rat acute myocardial infarction. Int J Cardiol 173:410–423

Tano N, Kaneko M, Ichihara Y et al (2016) Allogeneic mesenchymal stromal cells transplanted onto the heart surface achieve therapeutic myocardial repair despite immunologic. J Am Heart Assoc 5:1–14

Terajima Y, Shimizu T, Tsuruyama S, Sekine H (2014) Autologous skeletal myoblast sheet therapy for porcine myocardial infarction without increasing risk of arrhythmia. Cell Med 6:99–109

Tongers J, Losordo DW, Landmesser U (2011) Stem and progenitor cell-based therapy in ischaemic heart disease: Promise, uncertainties, and challenges. Eur Heart J 32(10):1197–1206

Uchinaka A, Kawaguchi N, Hamada Y et al (2012) Transplantation of elastin-secreting myoblast sheets improves cardiac function in infarcted rat heart. Mol Cell Biochem 368:203–214

Valles G, Bensiamar F, Crespo L, Arruebo M, Vilaboa N, Saldaña L (2015) Topographical cues regulate the crosstalk between MSCs and macrophages. Biomaterials 37:124–133

Wang Q, Wang Q, Teng W (2016) Injectable, degradable, electroactive nanocomposite hydrogels containing conductive polymer nanoparticles for biomedical applications. Int J Nanomed 11:131–144

Wei K, Serpooshan V, Hurtado C, Diez-Cunado M, Zhao M, Maruyama S et al (2015) Epicardial FSTL1 reconstitution regenerates the adult mammalian heart. Nature 525(7570):479–485

Xiong Q, Hill KL, Li Q et al (2011) A fibrin patch-based enhanced delivery of human embryonic stem cell-derived vascular cell transplantation in a porcine model of postinfarction left ventricular remodeling. Stem Cells 29:367–375

Ye L, Chang Y-H, Xiong Q et al (2014) Cardiac repair in a porcine model of acute myocardial infarction with human induced pluripotent stem cell-derived cardiovascular cell populations. Cell Stem Cell 15:750–761

Yeh TS, Dean Fang YH, Lu CH et al (2014) Baculovirus-transduced, VEGF-expressing adipose-derived stem cell sheet for the treatment of myocardium infarction. Biomaterials 35:174–184

Zhao G, Zhang X, Lu TJ, Xu F (2015) Recent advances in electrospun nanofibrous scaffolds for cardiac tissue engineering. Adv Funct Mater 25(36):5726–5738

Index

A

Acute respiratory distress syndrome (ARDS)
 acute respiratory failure, 88
 allogeneic adipose-derived MSCs, 90
 alveoli inflammation after fluid accumulation, 88, 89
 antimicrobial peptide secretion, 89
 biomarkers, 88
 ClinicalTrials.gov, 90
 complications, 123
 diagnosis, 123
 diagnostic criteria, 88
 early phase clinical trials, 89
 factors influencing, 123
 histopathological manifestations, 123
 inflammatory mediators, 89
 pathogenesis, 88, 123
 principles, 130
 public health problem, 88
 pulmonary and systemic inflammation, 89
 stages/phases, 123
 treatment, 129
Adipose derived stem cells (ADSCs)
 advantages, 202
 cardioprotective factors, 202
 overexpression of, 202
 platelet-rich fibrin scaffold, 202
 stem cell therapy, 170
Adult stem cells, 201–202
 CDCs//CPCs, 202
 MSCs, 201
 SM (*see* Skeletal myoblasts (SM))
 in tissues, 200
Airway restoration
 ad hoc large animal model, 71
 autologous MSCs, 72
 bronchial stump closure, 72
 goat model, 71, 72
 MCS endoscopical transplantation, 72
 water submersion test, 72
Airway smooth muscle (ASM), 95
Alcoholic liver disease (ALD)
 ethanol, pathways, 18
 high alcohol consumption, 18
 reactive oxygen species (ROS), 18
Angiogenic cells
 adipose tissue, 167
 blood vessel formation, 167
 endothelial cells (ECs), 167
 heart injury repair, 167
 mononuclear cells, 167
Animal trials
 HSCs, 45–46
 MSCs, 46–47
ARDS. *See* Acute respiratory distress syndrome (ARDS)
Arrhythmia, 148–149
Arrhythmogenicity, 148
ASM. *See* Airway smooth muscle (ASM)
Asthma
 airway during symptoms, 95
 ASM, 95
 bronchial hyper-responsiveness, 95
 MSCs, 96–97
 pulmonary airways and bronchial hyper-responsiveness, 94

B

Bone marrow derived MSCs (BM-MSCs), 78
Bone marrow derived stem cells (BMSCs), 170

BPD. *See* Bronchopulmonary dysplasia (BPD)
BPF. *See* Bronchopleural fistula (BPF)
Brain expressed X-linked 2 (Bex2) gene
   colony-formation ability, 7
   during embryonic development, 6
   endodermal organs and kidneys, 6
   hematopoietic stem/progenitor cell fraction, 8
   hepatectomy/chemical-induced injury, 8
   mesenchymal cells, 7
   stem/progenitor cells, analysis of, 6
   TCEAL genes, 6
Bronchopleural fistula (BPF)
   incidence, pleuropulmonary surgery, 70
   mortality rate, 72
   post-resectional, 70, 71
Bronchopulmonary dysplasia (BPD)
   bronchioles and alveoli, 101, 102
   cell-based therapy, 101
   chronic respiratory disease, 101
   diffuse pulmonary inflammation, 101
   lung maturation, 101
   MSCs, 102–105
   neonatal RDS, 101
   pathogenesis, 101

C
Cardiac progenitor cells (CPCs), 202
Cardiac repair, 159–161
   artificial heart tissue
      cell–cell connections, 161
      continuous electrical stimulation, 161
      disadvantages, 160
      micro-channels, 160
      myocardial infarction, 159
      neonatal cardiomyocytes, 159
      regional myocardial defects, 160
      sequential implanted cell sheets, 160
   AUGMENT HF trial, 159
   description, 158
   EHTs (*see* Engineered heart tissue (EHTs))
   hiPSCs, 161
   immature cells, 158
   proper cell source detection, 158
Cardiosphere derived cells (CDCs)
   cardiac functions, 182
   pig model, 184
   engineered cardiac tissues, 202
   postnatal heart, 202
Cardiovascular disease
   atherosclerotic fragments, obstruction of, 165
   cause, 165

   coronary artery bypass graft surgery, 166
   coronary artery disease, 165
   lifestyle changes, 165–166
   medical therapy, 166
   stent procedures, 166
CDCs. *See* Cardiosphere derived cells (CDCs)
Cell sheet technique, 155
Cell sheet transplantation, 147
Cell therapy, 25–28
   ESCs, 25
   hAECs, 30
   hepatocytes, 24–25
   iPSCs (*see* Induced pluripotent stem cells (iPSCs))
   macrophages, 25
   MSCs (*see* Mesenchymal stem cells (MSCs))
Cell transplantation, 172–173
   allogeneic and autologous MSCs, 175
   autologous AC133+ bone marrow cells, 175
   cell dose, 175
   direct intra-myocardial injection (*see* Direct intra-myocardial injection)
   myoblast autologous grafting, 175
   transplanted pig models, 176
   trans-vascular infusion (*see* Trans-vascular infusion)
Chronic liver disease, 18–30
   ALD (*see* Alcoholic liver disease (ALD))
   annual cost, 16
   cirrhosis, 19
   description, 19
   ECM molecular composition, 18
   HBV antiviral drug, 23
   HBV and HCV, 17
   hepatic fibrogenesis, 19
   hepatic macrophages (*see* Hepatic macrophages)
   hepatocyte replication and ECM proteins, 16
   hepatocytes, 16
   HSCs (*see* Hepatic stellate cells (HSCs))
   ILCs (*see* Innate lymphoid cells (ILCs))
   LPCs (*see* Liver progenitor cells (LPCs))
   NAFLD (*see* Non-alcoholic fatty liver disease (NAFLD))
   OLT, 24
   Penicillamine, 24
   progressive hepatocyte injury, 18
   sustained virological response (SVR), 23
   treatment (*see* Cell therapy)
   weight reduction and healthy life style, 24

Index

Chronic obstructive pulmonary disease (COPD)
  anti-inflammatory drugs, 91
  causative factor, 124
  cigarette smoking, 91
  extracellular matrix protein production, 124
  inflammation, 91
  neutrophils, 124
  normal lung and lung with, comparison, 91
  persistent airflow limitation, 91
  treatment, 129–130
  Wnt and Notch signaling pathway, 124
  normal lung and lung with, comparison, 91
Coronary vein injection, 172–173
CPCs. *See* Cardiac progenitor cells (CPCs)

D

Decellularized heart tissue, 147
  cardiac cells, 154
  Langendorff-perfusion with Triton X-100 and SDS, 154
  reseeding process, 154
Direct intra-myocardial injection
  animal and human clinical studies, 172
  coronary vein, 172–173
  trans-endocardial, 172
  trans-epicardial, 172

E

ECFCs. *See* Endothelial colony forming cells (ECFCs)
EHT. *See* Engineered heart tissue (EHT)
Electrospun fibers
  bioactive agents, 205
  core/shell, electro-conductive/functionalised fibers, 204
  engineered substrates, cardiac tissue, 204
  e-spun fibers, 204
  fibronectin immobilization, 205
  neonatal cardiomyocytes, 204
Embryonic stem cells (ESCs), 25, 170
Endothelial colony forming cells (ECFCs), 183
Endothelial progenitor cells (EPCs), 125
Engineered heart tissue (EHT)
  fibrin-based, 156
  hydrogel-based, 159
  in *in vitro* culture, 156
  native cardiomyocytes, 155
  native heart muscle tissue, 157
  from hPSC, 161
  sarcomeric organization, 157
ESCs. *See* Embryonic stem cells (ESCs)

F

Fibroblast growth factor (FGF)-10, 125
Fluorescence-activated cell sorting (FACS), 143, 144
Foetal stem cells, 48

G

G-CSF. *See* Granulocyte-colony stimulating factor (G-CSF)
Granulocyte-colony stimulating factor (G-CSF), 43

H

hAECs. *See* Human amnion epithelial cells (hAECs)
Haematopoietic stem cells (HSCs)
  animal trials, 45–46
  human trials, 46
hBMMSCs. *See* Human bone marrow MSCs (hBMMSCs)
Heart regeneration, 140–143, 145–149
  bioengineering, human myocardium
    decellularization, heart, 147
    three-dimensional bioprinting, 148
  cell transplantation, adverse effects
    arrhythmia, 148–149
    immunological rejection, 148
  purification strategy, PSC-derived cardiomyocytes, 143–145
  regenerative cardiomyocytes
    biomaterials, 145–147
    cell sheet, 147
    direct cell injection, 145
  regenerative cardiomyocytes, stem cell therapies
    cardiac differentiation protocol, 140
    cardiomyocytes phenotype, 142, 143
    mass culture system, 140–142
Hepatic growth factor (HGF), 46
Hepatic macrophages
  liver injury, 21
  monocyte-derived, 21
  pro-fibrogenic Ly6C^{hi} CD11b$^+$F480$^+$ population, 21
  proinflammatory cytokines and chemokines, 20

Hepatic stellate cells (HSCs)
　and inhibition, 20
　endothelin-1, 20
　fibrogenic cells, 19
　hepatocytes and sinusoidal endothelial
　　cells, 17
　initiation stage, 19
　Ito/fat storing cells, 17
　liver regeneration, 17
　mitogenic factors, 20
　perpetuation stage, 19
　proinflammatory cytokines
　　and chemokines, 19
Hepatic stem/progenitor cells, 6–11
　adult hepatocytes, 4
　BeX2 gene (see Brain expressed X-linked
　　2 (Bex2) gene)
　cholangiocytic markers, 4
　conventional culture system, 5
　extracellular matrix and integrin, 4
　hematopoietic cells, 4
　HPSCs (see Human pluripotent stem
　　cells (iPSs))
　MEFs, 5
　MEK-ERK pathway, 6
　oncostatin M (OSM), 4
　pancreatic acinar cell organization, 4
　transcription factors, 4
Hepatocytes
　cell therapy, 24–25
Hepatocyte transplantation therapy, 52
HGF. See Hepatic growth factor (HGF)
HSCs. See Haematopoietic stem cells (HSCs)
　(see Hepatic stellate cells (HSCs))
hUC-MSCs. See Human umbilical cord
　blood-derived MSCs (hUC-MSCs)
Human amnion epithelial cells (hAECs), 30
Human bone marrow MSCs (hBMMSCs), 28
Human iPS. See Human pluripotent stem
　cells (iPSs)
Human pluripotent stem cells (iPSs)
　anti-human Fn14 blocking antibody, 10
　bile ductal disease model, 11
　$CD13^+CD133^+$ cells, 9
　CD221, CD340 and CD266, 9
　cholangiocytic differentiation, 9, 10
　erbB family, 10
　paracrine factors, 10
　patient-derived, 11
　pluripotent, 8
　positive markers, 9
　primary germ layers, 8
　rodent animals, 8
　sequential stimulation, 9

somatic stem cells, 8
spheroids, 9
Human trials
　HSCs, 46
　MSCs, 47
Human umbilical cord blood-derived MSCs
　(hUC-MSCs)
　stem cell therapy, 171
Hydrogel technique, 154–155
Hydrogel type matrices
　cardiac biografts implantation, 205
　fibrin, 205
　implantable and injectable forms, 205

I

Idiopathic pulmonary fibrosis (IPF)
　activated myofibroblasts, 98
　biopsy, 122
　definition, 97, 121
　inhalation, exogenous noxious
　　compounds, 121
　lung volume, 121
　lungs airways and air sacs, 97, 98
　MSCs, 99–101
　nonpharmacologic therapies, 99
　pathogenesis, 121, 122
　subpleural and paraseptal
　　parenchyma, 98
　treatment, 128–129
ILCs. See Innate lymphoid cells (ILCs)
Induced pluripotent stem cells (iPSCs),
　　139–140, 149
　autologous adult cells, 25
　autologous transplantation, 148
　chronic/genetic disease, 26
　clinical trials, 26
　derived hepatocytes, 26
　dimethylnitrosamine (DMN), 26
　human
　　differentiation control, 140
　　iPSCs Bank, 139–140
　　regenerative cardiomyocytes, 149
　　reprogramming vectors and somatic
　　　cell source, 139
　stem cell therapy, 171
　undifferentiated, 141
Innate lymphoid cells (ILCs)
　cytotoxic and non-cytotoxic
　　helper cells, 22
　IL-33 signaling, 22
　innate and adaptive immune systems, 22
　NK and LTi cells, 22
　role in, 22

Intracoronary infusion, 173
Intravenous infusion, 173
IPF. *See* Idiopathic pulmonary fibrosis (IPF)
iPSCs. *See* Induced pluripotent stem cells (iPSCs)
Ischemic heart disease
    cell transplantation on clinical trials, 178–181
    clinical trials, 184–186
    cultured hUC-MSCs, 177
    MCAE, 176
    myoblast transplantation, 177
    preclinical trials, 176–184

L

Liver
    cholesterol homeostasis and lipoprotein metabolism, 15
    coagulation factors, 15
    damaged cells regeneration, 51
    functions, 51
    hepatocytes, 16
    non-parenchymal cells, 16
    partial hepatectomy, 15
Liver cirrhosis, 42, 43, 53–54
    bile production, 42
    complications
        fat-soluble vitamin deficiencies, 42
        hepatocellular carcinoma, 43
        portal hypertension, 42
    definition, 51
    epidemiology, 41
    FGF2, 41
    hepatocyte transplantation therapy, 52
    intra-hepatic and extra-hepatic biliary tree, 42
    Kupffer cells and HeSCs, 52
    liver regeneration, 52, 53
    metalloproteinase, 52
    MSCs (*see* Mesenchymal stem cells (MSCs))
    OLT, 52
    pathophysiology, 42
    stem cells (*see* Stem cells)
    treatment, 43
    types, 51
Liver development
    cell surface molecules, 3 (*see also* Hepatic stem/progenitor cells)
    hepatoblasts, 3
    parenchymal and non-parenchymal cells, 3
    progenitor-marker gene, 3

Liver progenitor cells (LPCs), 17
    HCC development, 22
    liver regeneration, 21
    mature hepatocyte and bile duct cells, 22
    role, 22
Liver sinusoidal endothelial cells (LSECs)
    endothelial cells, 16
    receptor mediated endocytosis, 16
LPCs. *See* Liver progenitor cells (LPCs)
LSECs. *See* Liver sinusoidal endothelial cells (LSECs)
LTi. *See* Lymphoid tissue inducer (LTi) cells
Lung, 122–124
    developmental stages, 121
    disorders, 121–122
        ARDS, 122, 123
        COPD, 124
        IPF (*see* Idiopathic pulmonary fibrosis (IPF))
Lung diseases, 88, 91–92, 94, 97, 101–102, 105–106
    ARDS (*see* Acute respiratory distress syndrome (ARDS))
    asthma (*see* Asthma)
    BM-derived MSCs, 84
    BPD (*see* Bronchopulmonary dysplasia (BPD))
    COPD (*see* Chronic obstructive pulmonary disease (COPD))
    development proliferative myofibroblast lesions, 85
    dysregulated cell adhesion, 84
    epithelial branching morphogenesis, 85
    IPF (*see* Idiopathic pulmonary fibrosis (IPF))
    PAH (*see* Pulmonary arterial hypertension (PAH))
    PDGF-BB signalling, 83, 84
    WIF-1 and SFRPs, 84
    Wnt proteins, 84
Lung regeneration, 125, 128–130
    progenitor cells
        bronchiolar smooth muscles, 125
        EPCs, 125
        FGF-10, 125
    treatment
        ARDS, 129
        COPD, 129–130
        IPF, 128–129
Lung resection, 70
    BPF (*see* Bronchopleural fistula (BPF))
    impaired bronchial stump healing, 70
Lymphoid tissue inducer (LTi) cells, 22

M

Macrophages
  cell therapy, 25
Major cardiac adverse events (MCAE), 176
MCAE. *See* Major cardiac adverse events (MCAE)
MEFs. *See* Mouse embryonic fibroblasts (MEFs)
Mesenchymal and Tissue Stem Cell International Committee, 81
Mesenchymal stem cell therapy
  airway restoration, 71–72
  cell tracking, 72–73
  impaired bronchial stump healing, 70
  post-resectional BPF, 70
Mesenchymal stem cells (MSCs), 73, 74
  adult human lung allografts, 81
  adult multipotent, 26
  adult tissues, 81
  ameliorates inflammation, 56
  in animal models, 27
  animal trials, 46–47
  apoptotic and antiapoptotic genes, 93
  array of tissues, 80
  BM-MSCs, 27, 78
  cell–cell contact mechanisms, 57
  cell-specific markers/antigens, 79
  cell surface markers, 26, 81
  cell types, 26
  characterization, 54
  clinical trials of, 28, 29, 59, 60
  ClinicalTrials.gov, 94
  conditioned media, 82
  consensus criteria, 81
  C-reactive protein levels, 92
  definition, 53
  endoscopical transplantation, 72
  engraftment level, host organs, 94
  episomal or microsomal particles release, 71
  exosomes, role of, 27
  extracellular vesicles (EVs), 82
  fibrogenesis, 58
  future therapeutic studies, 79
  harvest, 71
  hBMMSCs, 28
  heme oxygenase-1 enzyme, 93
  HGF-expressing, 82
  histological localization, 80
  Hoechst33342^{dim}CD45^{neg} cells, 82
  homing process, 27
  human tissues, 26
  human trials, 47
  IGF-1 gene, 201
  immune system, 27
  immune-enhancing/immunosuppressive properties, 78
  immunomodulation and paracrine functions, 83
  immunomodulatory therapy, 81
  immunosuppressive and anti-inflammatory properties, 54
  in vitro and *in vivo*, 55–56, 79
  inflammatory mediators, 27, 92
  intrapulmonary administration, 92
  ISCT, 26
  isolation and characterization, 54
  leukocyte migration, 54
  lineage tracing studies, 79, 81
  lung inflammation, 71 (*see also* Lung diseases)
  lung injury and repair, 77
  lung regeneration, 83
  mesodermal origin, 81
  microparticles/exosomes, 57
  mitochondria, tunneling nanotubes, 58
  multipotent, 81
  myofibroblasts, 77, 82
  oxidative stress, 93
  pathologic micro-environmental stimuli, 78
  pharmacological/genetic manipulations, 79
  phase clinical trials, 92
  potential mitogens, 57
  preclinical studies, 27, 58, 59, 83
  proliferative capacity, 81
  pulmonary administration, 92 (*see also* Regeneration of MSCs)
  regenerative medicine, 54
  repair series, 79, 80
  scaffolds, 201
  SPARC, 58
  stromal bone marrow cells, 53
  tissues, 79, 82, 83
  trophic factors, 57
  VEGF signalling pathway, 93
  ventricular myocytes and cardiomyocytes, 81
MI. *See* Myocardial infarction (MI)
Mononuclear stem cells, 47
Mouse embryonic fibroblasts (MEFs), 5
MSCs. *See* Mesenchymal stem cells (MSCs)
Murine bleomycin model, 99
Myocardial infarction (MI), 197
Myocardial tissue engineering, 154–157, 199–202
  acute coronary ischemia, 197
  in adequate microenvironment, 198
  adult (*see* Adult stem cells)
  advanced cardiac biografts, 198
  bone marrow-derived cells, 198 (*see also* Cardiac repair)
  cardiac reparative tissue-based therapy, 198

cell sheet technique, 155
cell source, 157–158
decellularized heart tissue (*see* Decellularized heart tissue)
heart function, 198
heart transplantation, 153
hydrogel technique, 154–155
hydrogel type matrices, 205–206
medical and interventional procedures, 197
pluripotent stem cells (*see* Pluripotent stem cells)
prefabricated matrix approach, 155
requirements
   cell maturation, 157
   hypoxia resistance, 156
   non-cardiomyocytes, 156
   strain, 157
   vascularization, 155–156
solid matrices, 204, 205

N
NAFLD. *See* Non-alcoholic fatty liver disease (NAFLD)
Neonatal rat cardiomyocytes, 156
Non-alcoholic fatty liver disease (NAFLD)
   liver fibrosis, 18
   steatosis, presence of, 18
   two-hit hypothesis, 18
Non-cardiomyocytes, 156

O
OLT. *See* Orthotopic liver transplantation (OLT)
Oncostatin M (OSM)
   and extracellular matrices, 4
   in hepatic cells, 4
   interleukin 6 family cytokine, 4
   maturation factor, 4
Orthotopic liver transplantation (OLT), 24, 52
OSM. *See* Oncostatin M (OSM)

P
PAH. *See* Pulmonary arterial hypertension (PAH)
PDGF-BB. *See* Platelet-derived growth factor subunit B (PDGF-BB) signalling
Pericytes. *See* Hepatic stellate cells (HSCs)
Platelet-derived growth factor subunit B (PDGF-BB) signalling
   lung diseases through MSCs, 83
   and PDGFRE, 84
   vascular neo-intimal thickening, 84

Pluripotent stem cells
   cardiomyocytes, 199–200
   engraftment, remuscularization and electromechanical synchronization, 199
   ESCs and induced iPSCs, 199
   non-CM derived cells, 199
Prefabricated matrix approach, 155
Progenitor cells, 125
   lung regeneration
      EPCs, 125
      FGF-10, 125
Pulmonary arterial hypertension (PAH)
   classifications, 105
   heterogeneous disease, 105
   histological and functional abnormalities, 105
   MSCs, 106, 107
   pulmonary arteries, 105
   vasodilators, 105

R
RAGE. *See* Receptor for advanced glycosylation end products (RAGE)
RDS. *See* Respiratory distress syndrome (RDS)
Receptor for advanced glycosylation end products (RAGE), 88–89
Regeneration of MSCs, 85–86
   immunomodulatory properties, 86–88
   tissue engineering (*see* Tissue engineering)
   tissue repair, 86
Regenerative cardiomyocytes, 140–143, 145–147
   stem cell therapies
      cardiac differentiation protocol, 140
      cardiomyocytes phenotype, 142–143
      mass culture system, 140–142
   transplantation strategies
      biomaterials, 145–147
      cell sheet, 147
      direct cell injection, 145
Resident cardiac stem cells, 171
Resident hepatic macrophages (Kupffer cells), 16
Respiratory distress syndrome (RDS), 78

S
Saphenous vein derived pericytes (SVPs), 182
SCNT. *See* Somatic cell nucleus transfer (SCNT)
Secreted frizzled related proteins (SFRPs), 84

Secreted protein acidic and rich in cysteine (SPARC), 58
SFRPs. *See* Secreted frizzled related proteins (SFRPs)
Single-photon emission computed tomography (SPECT), 73
Skeletal muscle-derived stem cells (SMSCs), 168
Skeletal myoblasts (SM), 202
    acute and chronic MI rodent models, 201
    adipose-derived (*see* Adipose-derived stem cells)
    LV ejection fraction and heart failure symptoms, 202
    muscle fiber basal lamina, 201
    sheet implantation, 201
SM. *See* Skeletal myoblasts (SM)
SMSCs. *See* Skeletal muscle-derived stem cells (SMSCs)
Solid matrices. *See* Electrospun fibers
Somatic cell nucleus transfer (SCNT), 158
SPARC. *See* Secreted protein acidic and rich in cysteine (SPARC)
SREP1c. *See* Sterol regulatory element-binding protein (SREP1c)
Stem cell therapy
    ADSCs, 170
    BMSCs, 170
    cardiac cell necrosis, 175 (*see also* Cardiovascular disease) (*see also* Cell transplantation)
    direct differentiation mechanism, 174
    ESCs, 170
    hUC-MSCs, 171
    indirection mechanism, 174
    initial evaluation, 169
    iPSCs, 171 (*see also* Ischemic heart disease)
    resident cardiac stem cells, 171
    sources of, 169
Stem cells, 45–47
    allogenic transplantation, 44
    angiogenic cells, 167
    autologous transplantation, 44
    CCl_4 and BMSCs, 45
    cell-based therapies, 78
    c-kit stem cell antigen marker, 78
    classification, 167
    collection and reinfusion, 43–44
    definition, 167
    endogenous role, 78
    FAH deficiency, 45
    foetal, 48
    G-CSF, 43
    host and donor cells, 45
    HSCs (*see* Haematopoietic stem cells (HSCs))
    from ileum, 44
    mesoderm- and non-mesoderm-derived tissues, 78
    mononuclear, 47
    MSCs (*see* Mesenchymal stem cells (MSCs))
    myocardial cells, 168–169
    population of undifferentiated cells, 78
    techniques, 44
    tissue regeneration, 79
    undifferentiated bone marrow, 45
Sterol regulatory element-binding protein (SREP1c), 18
Super paramagnetic iron oxide (SPIO) nanoparticles, 73
SVPs. *See* Saphenous vein derived pericytes (SVPs)

T

TCEAL. *See* Transcription elongation factor A (SII)-like (TCEAL) genes
Three-dimensional bioprinting, 148
TIMPs. *See* Tissue inhibitors of MMPs (TIMPs)
Tissue engineering
    bone marrow/dipose tissue aspirates, 86
    de-cellularized/synthetic scaffolds, 85
    homoeostasis, 85
    in vivo transplantation, 85
Tissue inhibitors of MMPs (TIMPs), 57
TNF-like weak inducer of apoptosis (TWEAK), 17, 21–22
TNF-related apoptosis-inducing ligand (TRAIL), 21
TRAIL. *See* TNF-related apoptosis-inducing ligand (TRAIL)
Transcription elongation factor A (SII)-like (TCEAL) genes, 6
Trans-endocardial injection, 172
Trans-vascular infusion
    intracoronary, 173
    intravenous, 173
TWEAK. *See* TNF-like weak inducer of apoptosis (TWEAK)

V

von Willebrand factor, 88

W

Wnt inhibitory factor (WIF-1), 84

Printed by Printforce, the Netherlands